AF560385

Kinetic Methods of Analysis in Analytical Chemistry

Kinetic Methods of Analysis in Analytical Chemistry

Dr. Hirdayesh Kumar Vatsa

Kinetic Methods of Analysis in Analytical Chemistry

ISBN 978-93-5111-621-9

Published in 2015 in India by

RANDOM PUBLICATIONS

4376-A/4B, Gali Murari Lal, Ansari Road
New Delhi-110 002
Phone : +9111-43580356, 011-23289044, 011-43142548
e-mail: sales@randompublications.com,
info@randompublications.com, randomexports@gmail.com

Reprinted 2022

Type Setting by : Friends Media, Delhi-110089
Digitally Printed at: Replika Press Pvt. Ltd.

Preface

Analytical chemistry is often described as the area of chemistry responsible for characterizing the composition of matter, both qualitatively (what is present) and quantitatively (how much is present). This description is misleading.

The basic types of reactions used for determinative purpose encompass the traditional four in equilibrium-based measurements: precipitation (ion exchange), acid-base (proton exchange), redox (electron exchange) and complexation (ligand exchange). These four basic types, or cases that can be reduced to them, are also found in kinetic-based measurements with some distinguishable trends.

The influence of concentration on the position of a chemical equilibrium is described in quantitative terms by means of an equilibrium-constant expression. Such expressions are important because they permit the chemist to predict the direction and completeness of a chemical reaction. However, the size of one equilibrium constant tells us nothing about the rate (the kinetic) of the reaction. A large equilibrium constant does not imply that a reaction is fast. In fact, we sometimes encounter reactions that have highly favorable equilibrium constants but are of slight analytical use because their rates are low.

A reaction in which no reaction intermediates have been detected, or need to be postulated in order to describe the chemical reaction on a molecular scale. Until evidence to the contrary is discovered, an elementary reaction is assumed to occur in a single step and to pass through a single transition state.

A mode of measurement in a kinetic method of analysis, in which the period of time, required to bring about the same predetermined change in he concentration of a reactant or product, is measured. The use of the term "variable time method" is not recommended.

This book provides a clear and concise understanding of the principles, applications and limitations of the various techniques involved in analytical chemistry.

I would like to thank my team for standing beside me throughout my career and writing this book. My special thanks go to "Random Publications" who have published the book.

– Dr. Hirdayesh Kumar Vatsa

Contents

1

Analytical Chemistry

INTRODUCTION

We begin this section with a deceptively simple question. What is analytical chemistry? Like all fields of chemistry, analytical chemistry is too broad and active a discipline for us to easily or completely define in an introductory textbook. Instead, we will try to say a little about what analytical chemistry is, as well as a little about what analytical chemistry is not.

Analytical chemistry is often described as the area of chemistry responsible for characterizing the composition of matter, both qualitatively (what is present) and quantitatively (how much is present). This description is misleading. After all, almost all chemists routinely make qualitative or quantitative measurements. The argument has been made that analytical chemistry is not a separate branch of chemistry, but simply the application of chemical knowledge.1 In fact, you probably have performed quantitative and qualitative analyses in other chemistry courses. For example, many introductory courses in chemistry include qualitative schemes for identifying inorganic ions and quantitative analyses involving titrations.

Unfortunately, this description ignores the unique perspective that analytical chemists bring to the study of chemistry. The craft of analytical chemistry is not in performing a routine analysis on a routine sample (which is more appropriately called chemical analysis), but in improving established methods, extending existing methods to new types of samples, and developing new methods for measuring chemical phenomena.

Here's one example of this distinction between analytical chemistry and chemical analysis. Mining engineers evaluate the economic feasibility of extracting an ore by comparing the cost of removing the ore with the value of its contents. To estimate its value they analyse a sample of the ore. The challenge of developing and validating the method providing this information is the analytical chemist's responsibility. Once developed, the routine, daily application of the method becomes the job of the chemical analyst. Another distinction between analytical chemistry and chemical analysis is that analytical chemists work to improve established methods. For example, several factors complicate the quantitative analysis of Ni^{2+} in ores, including the presence of a complex heterogeneous mixture of silicates and oxides, the low concentration of Ni_2+ in ores, and the presence of other metals that may interfere in the analysis. Figure is a schematic outline of one standard method in use during the late nineteenth century.3 After dissolving a sample of the ore in a mixture of H_2SO_4 and HNO3, trace metals that interfere with the analysis, such as Pb^{2+}, Cu^{2+} and Fe^{3+}, are removed by precipitation. Any cobalt and nickel in the sample are reduced to Co and Ni, isolated by filtration and weighed (point A). After dissolving the mixed solid, Co is isolated and weighed (point B). The amount of nickel in the ore sample is determined from the difference in the masses at points A and B.

$$\%Ni = \frac{\text{mass point A} - \text{mass pointB}}{\text{mass sample}} \times 100$$

The combination of determining the mass of Ni^{2+} by difference, coupled with the need for many reactions and filtrations makes this procedure both time-consuming and difficult to perform accurately. The development, in 1905, of dimethylgloxime (DMG), a reagent that selectively precipitates Ni^{2+} and Pd^{2+}, led to an improved analytical method for determining Ni^{2+} in ores. As shown in Figure, the mass of Ni^{2+} is measured directly, requiring fewer manipulations and less time. By the 1970s, the standard method for the analysis of Ni_2^+ in ores progressed from precipitating $Ni(DMG)^2$ to flame atomic absorption spectrophotometry, 5 resulting in an even more rapid analysis. Current interest is directed towards using inductively coupled plasmas for determining trace metals in ores.

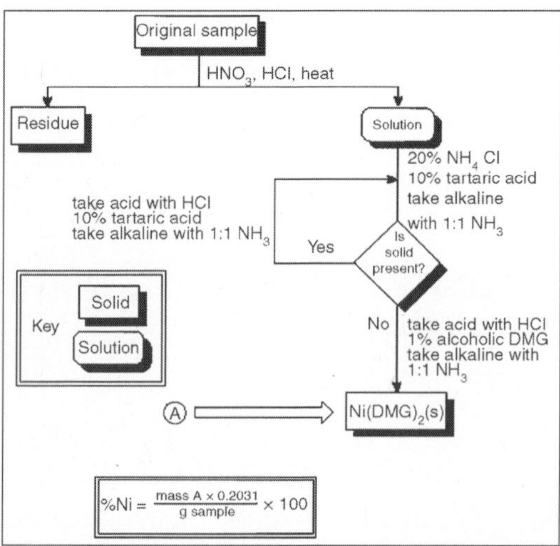

Fig. Analytical Scheme for the Gravimetric Analysis of Ni in Ores (DMG = Dimethylgloxime). The Factor of 0.2031 in the Equation for per centNi Accounts for the Difference in the Formula Weights of Ni(DMG)2 and Ni.

In summary, a more appropriate description of analytical chemistry is "... the science of inventing and applying the concepts, principles, and... strategies for measuring the characteristics of chemical systems and species."Analytical chemists typically operate at the extreme edges of analysis, extending and improving the ability of all chemists to make meaningful measurements on smaller samples, on more complex samples, on shorter time scales, and on species present at lower concentrations.

Throughout its history, analytical chemistry has provided many of the tools and methods necessary for research in the other four traditional areas of chemistry, as well as fostering multidisciplinary research in, to name a few, medicinal chemistry, clinical chemistry, toxicology, forensic chemistry, material science, geochemistry, and environmental chemistry.

You will come across numerous examples of qualitative and quantitative methods in this text, most of which are routine examples of chemical analysis. It is important to remember, however, that nonroutine problems prompted analytical chemists to develop these methods. Whenever possible, we will try to place these methods in their appropriate historical context. In addition, examples of current research problems in analytical chemistry are scattered throughout the text.

The next time you are in the library, look through a recent issue of an analytically oriented journal, such as *Analytical Chemistry.* Focus on the titles and abstracts of the research articles. Although you will not recognize all the terms and methods, you will begin to answer for yourself the question "What is analytical chemistry".

HISTORY

Analytical chemistry has been important since the early days of chemistry, providing methods for determining which elements and chemicals are present in the object in question. During this period significant analytical contributions to chemistry include the development of systematic elemental analysis by Justus von Liebig and systematized organic analysis based on the specific reactions of functional groups.

The first instrumental analysis was flame emissive spectrometry developed by Robert Bunsen and Gustav Kirchhoff who discovered rubidium (Rb) and caesium (Cs) in 1860.Most of the major developments in analytical chemistry take place after 1900. During this period instrumental analysis becomes progressively dominant in the field. In particular many of the basic spectroscopic and spectrometric techniques were discovered in the early 20th century and refined in the late 20th century.

The separation sciences follow a similar time line of development and also become increasingly transformed into high performance instruments. In the 1970s many of these techniques began to be used together to achieve a complete characterization of samples.

Starting in approximately the 1970s into the present day analytical chemistry has progressively become more inclusive of biological questions (bioanalytical chemistry), whereas it had previously been largely focused on inorganic or small organic molecules. Lasers have been increasingly used in chemistry as probes and even to start and influence a wide variety of reactions. The late 20th century also saw an expansion of the application of analytical chemistry from somewhat academic chemical questions to forensic, environmental, industrial and medical questions, such as in histology.

Modern analytical chemistry is dominated by instrumental analysis. Many analytical chemists focus on a single type of instrument. Academics tend to either focus on new applications and discoveries or on new methods of analysis. The discovery of a chemical present in blood that increases the risk of cancer would be a discovery that an analytical chemist might be involved in. An effort to develop a new method might involve the use of a tunable laser to increase the specificity and sensitivity of a spectrometric method. Many methods, once developed, are kept purposely static so that data can be compared over long periods of time. This is particularly true in industrial quality assurance (QA), forensic and environmental applications. Analytical chemistry plays an increasingly important role in the pharmaceutical industry where, aside from QA, it is used in discovery of new drug candidates and in clinical applications where understanding the interactions between the drug and the patient are critical.

CLASSICAL METHODS

Although modern analytical chemistry is dominated by sophisticated instrumentation, the roots of analytical chemistry and some of the principles used in modern instruments are from traditional techniques many of which are still used today. These techniques also tend to form the backbone of most undergraduate analytical chemistry educational labs.

Qualitative Analysis

A qualitative analysis determines the presence or absence of a particular compound, but not the mass or concentration. By definition, qualitative analyses do not measure quantity.

Chemical Tests

There are numerous qualitative chemical tests, for example, the acid test for gold and the Kastle-Meyer test for the presence of blood.

Flame Test

Inorganic qualitative analysis generally refers to a systematic scheme to confirm the presence of certain, usually aqueous, ions or elements by performing a series of reactions that eliminate ranges of possibilities and then confirms suspected ions with a confirming test. Sometimes small carbon containing ions are included in such schemes. With modern instrumentation these tests are rarely used but can be useful for educational purposes and in field work or other situations where access to state-of-the-art instruments are not available or expedient.

QUANTITATIVE ANALYSIS

Gravimetric Analysis

Gravimetric analysis involves determining the amount of material present by weighing the sample before and/or after some transformation. A common example used in undergraduate education is the determination of the amount of water in a hydrate by heating the sample to remove the water such that the difference in weight is due to the loss of water.

Volumetric Analysis

Titration involves the addition of a reactant to a solution being analysed until some equivalence point is reached. Often the amount of material in the solution being analysed may be determined. Most familiar to those who have taken chemistry during secondary education is the acid-base titration involving a colour changing indicator. There are many other types of titrations, for example potentiometric titrations. These titrations may use different types of indicators to reach some equivalence point.

INSTRUMENTAL METHODS

Spectroscopy

Spectroscopy measures the interaction of the molecules with electromagnetic radiation. Spectroscopy consists of many different applications such as atomic absorption spectroscopy, atomic emission spectroscopy, ultraviolet-visible spectroscopy, x-ray fluorescence spectroscopy, infrared spectroscopy, Raman spectroscopy, dual polarisation interferometry, nuclear magnetic resonance spectroscopy, photoemission spectroscopy, Mössbauer spectroscopy and so on.

Mass Spectrometry

Mass spectrometry measures mass-to-charge ratio of molecules using electric and magnetic fields. There are several ionization methods: electron impact, chemical ionization, electrospray, fast atom bombardment, matrix assisted laser desorption ionization, and others. Also, mass spectrometry is categorized by approaches of mass analysers: magnetic-sector, quadrupole mass analyser, quadrupole ion trap, time-of-flight, Fourier transform ion cyclotron resonance, and so on.

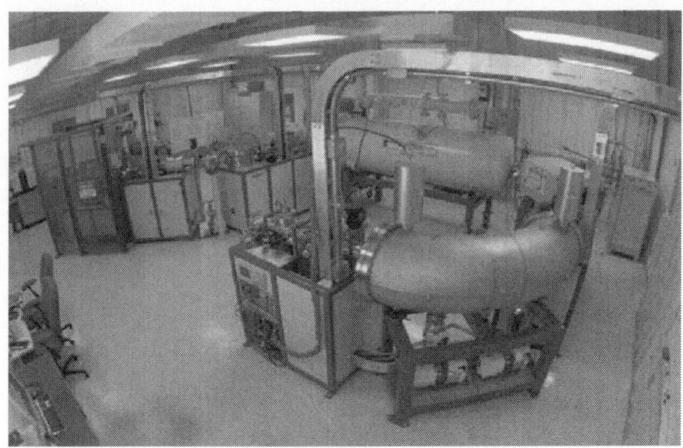

Fig. An accelerator mass spectrometer used for radiocarbon dating and other analysis.

Electrochemical Analysis

Electroanalytical methods measure the potential (volts) and/or current (amps) in an electrochemical cell containing the analyte. These methods can be categorized according to which aspects of the cell are controlled and which are measured. The three main categories arepotentiometry (the difference in electrode potentials is measured), coulometry (the cell's current is measured over time), and voltammetry (the cell's current is measured while actively altering the cell's potential).

Thermal Analysis

Calorimetry and thermogravimetric analysis measure the interaction of a material and heat.

Separation

Fig. Separation of black ink on a thin layer chromatography plate.

Separation processes are used to decrease the complexity of material mixtures. Chromatography, electrophoresis and Field Flow Fractionation are representative of this field.

Hybrid Techniques

Combinations of the above techniques produce a "hybrid" or "hyphenated" technique. Several examples are in popular use today and new hybrid techniques are under development. For example, gas chromatography-mass spectrometry, gas chromatography-infrared spectroscopy, liquid chromatography-mass spectrometry, liquid chromatography-NMR spectroscopy. liquid chromagraphy-infrared spectroscopy and capillary electrophoresis-mass spectrometry.

Hyphenated separation techniques refers to a combination of two (or more) techniques to detect and separate chemicals from solutions. Most often the other technique is some form of chromatography. Hyphenated techniques are widely used in chemistry and biochemistry. A slash is sometimes used instead of hyphen, especially if the name of one of the methods contains a hyphen itself.

Microscopy

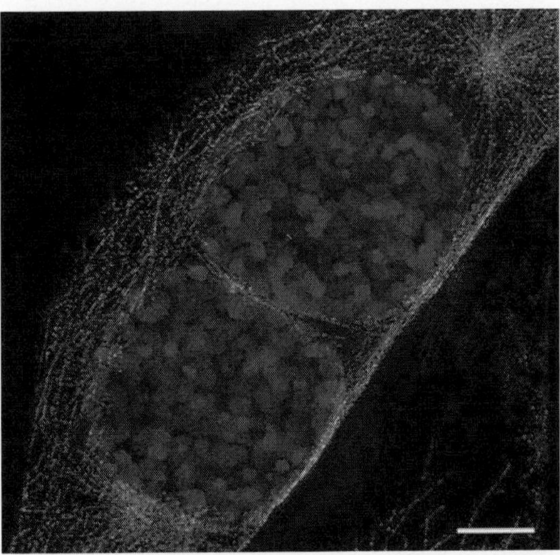

Fig. Fluorescence microscopeimage of two mouse cell nuclei inprophase (scale bar is 5 μm).

The visualization of single molecules, single cells, biological tissues and nanomaterials is an important and attractive approach in analytical science. Also, hybridization with other traditional analytical tools is revolutionizing analytical science. Microscopy can be categorized into three different fields: optical microscopy, electron microscopy, and scanning probe microscopy. Recently, this field is rapidly progressing because of the rapid development of the computer and camera industries.

Lab-on-a-Chip

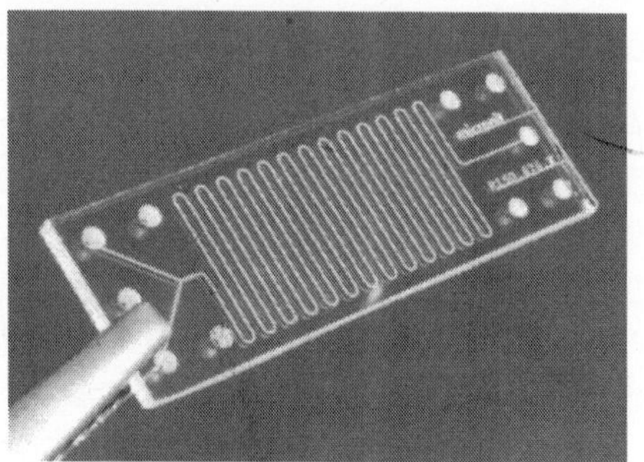

Fig. A glass microreactor

Devices that integrate (multiple) laboratory functions on a single chip of only millimetres to a few square centimetres in size and that are capable of handling extremely small fluid volumes down to less than picolitres.

STANDARDS

Standard Curve

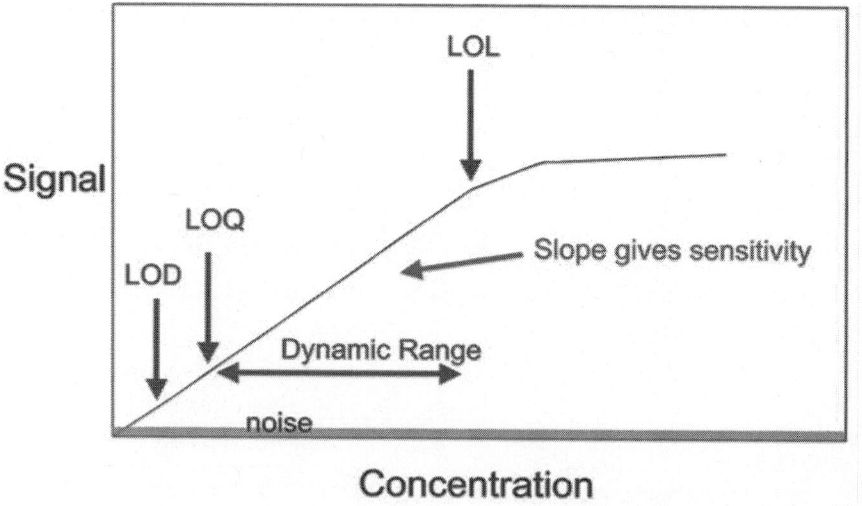

Fig. A calibration curve plot showinglimit of detection (LOD), limit of quantification (LOQ), dynamic range, and limit of linearity (LOL).

A general method for analysis of concentration involves the creation of a calibration curve. This allows for determination of the amount of a chemical in a material by comparing the results of unknown sample to those of a series of known standards. If the concentration of element or compound in a sample is too high for the detection range of the technique, it can simply be diluted in a pure solvent. If the amount in the sample is below an instrument's range of measurement, the method of addition can be used. In this method a known quantity of the element or compound under study is added, and the difference between the concentration added, and the concentration observed is the amount actually in the sample.

Internal Standards

Sometimes an internal standard is added at a known concentration directly to an analytical sample to aid in quantitation. The amount of analyte present is then determined relative to the internal standard as a calibrant. An ideal internal standard is isotopically-enriched analyte which gives rise to the method of isotope dilution.

Standard Addition

The method of standard addition is used in instrumental analysis to determine concentration of a substance (analyte) in an unknown sample by comparison to a set of samples of known concentration, similar to using a calibration curve. Standard addition can be applied to most analytical techniques and is used instead of a calibration curve to solve the matrix effect problem.

Signals and Noise

One of the most important components of analytical chemistry is maximizing the desired signal while minimizing the associated noise. The analytical figure of merit is known as the signal-to-noise ratio (S/N or SNR). Noise can arise from environmental factors as well as from fundamental physical processes.

Thermal Noise

Thermal noise results from the motion of charge carriers (usually electrons) in an electrical circuit generated by their thermal motion. Thermal noise is white noise meaning that the power spectral density is constant throughout the frequency spectrum.

The root mean square value of the thermal noise in a resistor is given by

$$v_{RMS} = \sqrt{4k_B TR\Delta f},$$

where k_B is Boltzmann's constant, T is the temperature, R is the resistance, and Δf is the bandwidth of the frequency f.

Shot Noise

Shot noise is a type of electronic noise that occurs when the finite number of particles (such as electrons in an electronic circuit or photons in an optical device) is small enough to give rise to statistical fluctuations in a signal.

Shot noise is a Poisson process and the charge carriers that make up the current follow a Poisson distribution. The root mean square current fluctuation is given by

$$i_{RMS} = \sqrt{2eI\Delta f}$$

where e is the elementary charge and I is the average current. Shot noise is white noise.

Flicker Noise

Flicker noise is electronic noise with a $1/f$ frequency spectrum; as f increases, the noise decreases. Flicker noise arises from a variety of sources, such as impurities in a conductive channel, generation and recombination noise in a transistor due to base current, and so on. This noise can be avoided by modulation of the signal at a higher frequency, for example through the use of a lock-in amplifier.

Environmental Noise

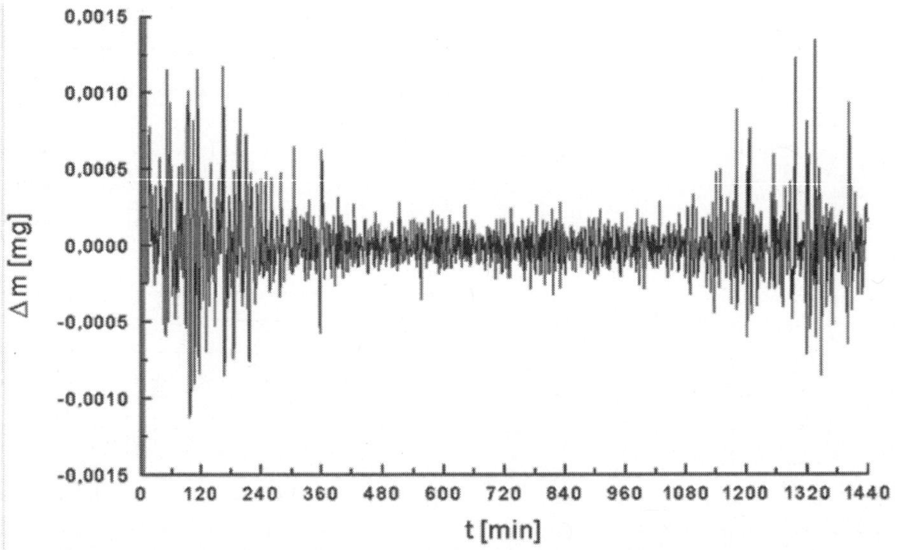

Fig. Noise in a thermogravimetric analysis; lower noise in the middle of the plot results from less human activity (and environmental noise) at night.

Environmental noise arises from the surroundings of the analytical instrument. Sources of electromagnetic noise are power lines, radio and television stations, wireless devices, Compact fluorescent lamps and electric motors.

Many of these noise sources are narrow bandwidth and therefore can be avoided. Temperature and vibration isolation may be required for some instruments.

Noise Reduction

Noise reduction can be accomplished either in computer hardware or software. Examples of hardware noise reduction are the use of shielded cable, analog filtering, and signal modulation. Examples of software noise reduction are digital filtering, ensemble average, boxcar average, and correlation methods.

COMMON ANALYTICAL PROBLEMS

In Section 1A we indicated that analytical chemistry is more than a collection of qualitative and quantitative methods of analysis. Nevertheless, many problems on which analytical chemists work ultimately involve either a qualitative or quantitative measurement. Other problems may involve characterizing a sample's chemical or physical properties. Finally, many analytical chemists engage in fundamental studies of analytical methods. In this section we briefly discuss each of these four areas of analysis.

Many problems in analytical chemistry begin with the need to identify what is present in a sample. This is the scope of a qualitative analysis, examples of which include identifying the products of a chemical reaction, screening an athlete's urine for the presence of a performance-enhancing drug, or determining the spatial distribution of Pb on the surface of an airborne particulate.

Much of the early work in analytical chemistry involved the development of simple chemical tests to identify the presence of inorganic ions and organic functional groups. The classical laboratory courses in inorganic and organic qualitative analysis,9 still taught at some schools, are based on this work.

Currently, most qualitative analyses use methods such as infrared spectroscopy, nuclear magnetic resonance, and mass spectrometry. These qualitative applications of identifying organic and inorganic compounds are covered adequately elsewhere in the undergraduate curriculum and, so, will receive no further consideration in this text.

Perhaps the most common type of problem encountered in the analytical lab is a quantitative analysis. Examples of typical quantitative analyses include the elemental analysis of a newly synthesized compound, measuring the concentration of glucose in blood, or determining the difference between the bulk and surface concentrations of Cr in steel. Much of the analytical work in clinical, pharmaceutical, environmental, and industrial labs involves developing new methods for determining the concentration of targeted species in complex samples. Most of the examples in this text come from the area of quantitative analysis. Another important area of analytical chemistry, which receives some attention in this text, is the development of new methods for characterizing physical and chemical properties. Determinations of chemical structure, equilibrium constants, particle size, and surface structure are examples of a characterization analysis.

The purpose of a qualitative, quantitative, and characterization analysis is to solve a problem associated with a sample. A fundamental analysis, on the other hand, is directed towards improving the experimental methods used in the other areas of analytical chemistry. Extending and improving the theory on which a method is based, studying a method's limitations, and designing new and modifying old methods are examples of fundamental studies in analytical chemistry.

ANALYTICAL PROBLEMS AND PROCEDURES

ANALYTICAL PROBLEMS

The most important aspect of an analysis is to ensure that it will provide useful and reliable data on the qualitative and/or quantitative composition of a material or structural information about the individual compounds present.

The analytical chemist must often communicate with other scientists and nonscientists to establish the amount and quality of the information required, the time-scale for the work to be completed and any budgetary constraints.

The most appropriate analytical technique and method can then be selected from those available or new ones devised and validated by the analysis of substances of known composition and/or structure. It is essential for the analytical chemist to have an appreciation of the objectives of the analysis and an understanding of the capabilities of the various analytical techniques at his/her disposal without which the most appropriate and cost-effective method cannot be selected or developed.

ANALYTICAL PROCEDURES

The stages or steps in an overall analytical procedure can be summarised as follows.

- *Definition of the problem:* Analytical information and level of accuracy required. Costs, timing, availability of laboratory instruments and facilities.
- *Choice of technique and method:* Selection of the best technique for the required analysis, such as chromatography, infrared spectrometry, titrimetry, thermogravimetry. Selection of the method (*i.e.* the detailed stepwise instructions using the selected technique).
- *Sampling:* Selection of a small sample of the material to be analysed. Where this is heterogeneous, special procedures need to be used to ensure that a genuinely representative sample is obtained.
- *Sample pre-treatment or conditioning:* Conversion of the sample into a form suitable for detecting or measuring the level of the analyte(s) by the selected technique and method. This may involve dissolving it, converting the analyte(s) into a specific chemical form or separating the analyte(s) from other components of the sample (the sample matrix) that could interfere with detection or quantitative measurements.
- *Qualitative analysis:* Tests on the sample under specified and controlled conditions. Tests on reference materials for comparison. Interpretation of the tests.

- *Quantitative analysis:* Preparation of standards containing known amounts of the analyte(s) or of pure reagents to be reacted with the analyte(s). Calibration of instruments to determine the responses to the standards under controlled conditions. Measurement of the instrumental response for each sample under the same conditions as for the standards. All measurements may be replicated to improve the reliability of the data, but this has cost and time implications. Calculation of results and statistical evaluation.
- *Preparation of report or certificate of analysis:* This should include a summary of the analytical procedure, the results and their statistical assessment, and details of any problems encountered at any stage during the analysis.
- *Review of the original problem:* The results need to be discussed with regard to their significance and their relevance in solving the original problem. Sometimes repeat analyses or new analyses may be undertaken.

THE ANALYTICAL PERSPECTIVE

Having noted that each field of chemistry brings a unique perspective to the study of chemistry, we now ask a second deceptively simple question. What is the "analytical perspective"? Many analytical chemists describe this perspective as an analytical approach to solving problems.7 Although there are probably as many descriptions of the analytical approach as there are analytical chemists, it is convenient for our purposes to treat it as a five-step process:

- Identify and define the problem.
- Design the experimental procedure.
- Conduct an experiment, and gather data.
- Analyse the experimental data.
- Propose a solution to the problem.

Figure shows an outline of the analytical approach along with some important considerations at each step. Three general features of this approach deserve attention. First, steps 1 and 5 provide opportunities for analytical chemists to collaborate with individuals outside the realm of analytical chemistry.

In fact, many problems on which analytical chemists work originate in other fields. Second, the analytical approach is not linear, but incorporates a "feedback loop" consisting of steps 2, 3, and 4, in which the outcome of one step may cause a reevaluation of the other two steps. Finally, the solution to one problem often suggests a new problem.

Analytical chemistry begins with a problem, examples of which include evaluating the amount of dust and soil ingested by children as an indicator of environmental exposure to particulate based pollutants, resolving contradictory evidence regarding the toxicity of perfluoro polymers during combustion, or developing rapid and sensitive detectors for chemical warfare agents.

At this point the analytical approach involves a collaboration between the analytical chemist and the individuals responsible for the problem. Together they decide what information is needed.

It is also necessary for the analytical chemist to understand how the problem relates to broader research goals.

The type of information needed and the problem's context are essential to designing an appropriate experimental procedure.

Designing an experimental procedure involves selecting an appropriate method of analysis based on established criteria, such as accuracy, precision, sensitivity, and detection limit; the urgency with which results are needed; the cost of a single analysis; the number of samples to be analyzed; and the amount of sample available for analysis. Finding an appropriate balance between these parameters is frequently complicated by their interdependence.

For example, improving the precision of an analysis may require a larger sample. Consideration is also given to collecting, storing, and preparing samples, and to whether chemical or physical interferences will affect the analysis. Finally, a good experimental procedure may still yield useless information if there is no method for validating the results.

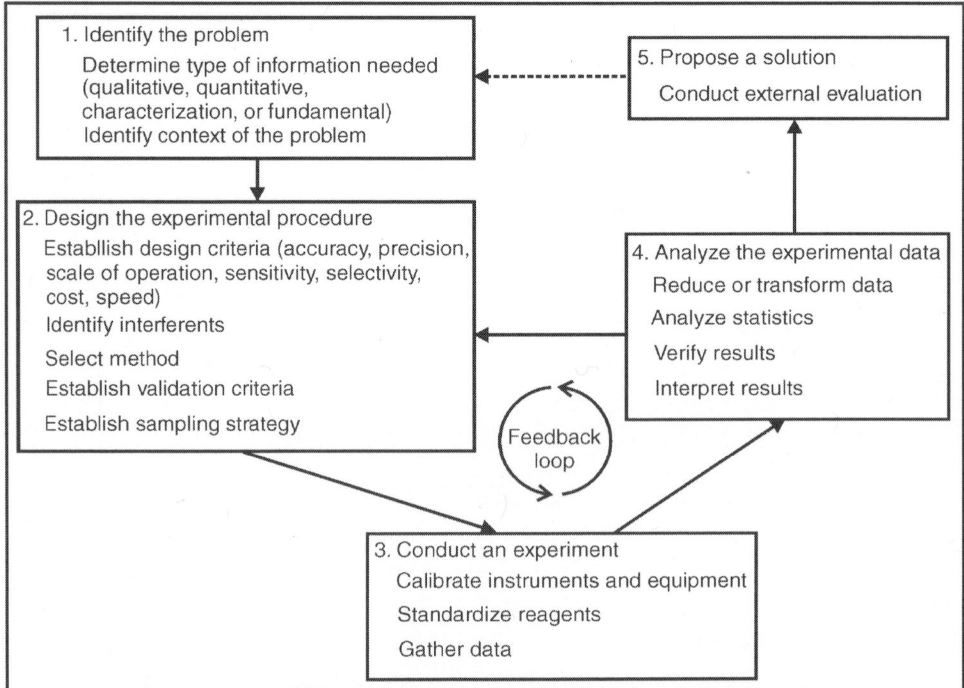

Fig. Flow Diagram for the Analytical Approach to Solving Problems.

The most visible part of the analytical approach occurs in the laboratory. As part of the validation process, appropriate chemical or physical standards are used to calibrate any equipment being used and any solutions whose concentrations must be known. The selected samples are then analyzed and the raw data recorded.

The raw data collected during the experiment are then analyzed. Frequently the data must be reduced or transformed to a more readily analyzable form. A statistical treatment of the data is used to evaluate the accuracy and precision of the analysis and to validate the procedure.

These results are compared with the criteria established during the design of the experiment, and then the design is reconsidered, additional experimental trials are run, or a solution to the problem is proposed. When a solution is proposed, the results are subject to an external evaluation that may result in a new problem and the beginning of a new analytical cycle.

Before continuing, take some time to read the article, locating the discussions pertaining to each of the five steps outlined in Figure. In addition, consider the following questions:

- What is the analytical problem?
- What type of information is needed to solve the problem?
- How will the solution to this problem be used?
- What criteria were considered in designing the experimental procedure?
- Were there any potential interferences that had to be eliminated? If so, how were they treated?
- Is there a plan for validating the experimental method?
- How were the samples collected?
- Is there evidence that steps 2, 3, and 4 of the analytical approach are repeated more than once?
- Was there a successful conclusion to the problem?

According to our model, the analytical approach begins with a problem. The motivation for this research was to develop a method for monitoring the transport of solid aerosol particulates following their release from a high-temperature combustion source. Because these particulates contain significant concentrations of toxic heavy metals and carcinogenic organic compounds, they represent a significant environmental hazard.

An aerosol is a suspension of either a solid or a liquid in a gas. Fog, for example, is a suspension of small liquid water droplets in air, and smoke is a suspension of small solid particulates in combustion gases. In both cases the liquid or solid particulates must be small enough to remain suspended in the gas for an extended time.

Solid aerosol particulates, which are the focus of this problem, usually have micrometer or submicrometer diameters. Over time, solid particulates settle out from the gas, falling to the Earth's surface as dry deposition.

Existing methods for monitoring the transport of gases were inadequate for studying aerosols. To solve the problem, qualitative and quantitative information were needed to determine the sources of pollutants and their net contribution to the total dry deposition at a given location. Eventually the methods developed in this study could be used to evaluate models that estimate the contributions of point sources of pollution to the level of pollution at designated locations.

Following the movement of airborne pollutants requires a natural or artificial tracer (a species specific to the source of the airborne pollutants) that can be experimentally measured at sites distant from the source. Limitations placed on the tracer, therefore, governed the design of the experimental procedure. These limitations included cost, the need to detect small quantities of the tracer, and the absence of the tracer from other natural sources.

In addition, aerosols are emitted from high-temperature combustion sources that produce an abundance of very reactive species. The tracer, therefore, had to be both thermally and chemically stable. On the basis of these criteria, rare earth isotopes, such as those of Nd, were selected as tracers. The choice of tracer, in turn, dictated the analytical method (thermal ionization mass spectrometry, or TIMS) for measuring the isotopic abundances of Nd in samples. Unfortunately, mass spectrometry is not a selective technique. A mass spectrum provides information about the abundance of ions with a given mass. It cannot distinguish, however, between different ions with the same mass. Consequently, the choice of TIMS required developing a procedure for separating the tracer from the aerosol particulates.

Validating the final experimental protocol was accomplished by running a model study in which 148Nd was released into the atmosphere from a 100-MW coal utility boiler. Samples were collected at 13 locations, all of which were 20 km from the source. Experimental results were compared with predictions determined by the rate at which the tracer was released and the known dispersion of the emissions. Finally, the development of this procedure did not occur in a single, linear pass through the analytical approach. As research progressed, problems were encountered and modifications made, representing a cycle through steps 2, 3, and 4 of the analytical approach.

Others have pointed out, with justification, that the analytical approach outlined here is not unique to analytical chemistry, but is common to any aspect of science involving analysis. Here, again, it helps to distinguish between a chemical analysis and analytical chemistry. For other analytically oriented scientists, such as physical chemists and physical organic chemists, the primary emphasis is on the problem, with the results of an analysis supporting larger research goals involving fundamental studies of chemical or physical processes. The essence of analytical chemistry, however, is in the second, third, and fourth steps of the analytical approach. Besides supporting broader research goals by developing and validating analytical methods, these methods also define the type and quality of information available to other research scientists. In some cases, the success of an analytical method may even suggest new research problems.

QUALITY IN ANALYTICAL LABORATORIES

QUALITY CONTROL

Analytical data must be of demonstrably high quality to ensure confidence in the results. Quality control (QC) comprises a system of planned activities in an analytical laboratory whereby analytical methods are monitored at every stage to verify compliance with validated procedures and to take steps to eliminate the causes of unsatisfactory performance. Results are considered to be of sufficiently high quality if

- They meet the specific requirements of the requested analytical work within the context of a defined problem.
- There is confidence in their validity.
- The work is cost effective.

To implement a QC system, a complete understanding of the chemistry and operations of the analytical method and the likely sources and magnitudes of errors at each stage is essential. The use of reference materials during method validation ensures that results are traceable to certified sources. QC processes should include:

- Checks on the accuracy and precision of the data using statistical tests.
- Detailed records of calibration, raw data, results and instrument performance.
- Observations on the nature and behaviour of the sample and unsatisfactory aspects of the methodology.
- Control charts to determine system control for instrumentation and repeat analyses.
- Provision of full documentation and traceability of results to recognized reference materials through recorded identification.
- Maintenance and calibration of instrumentation to manufacturers' specifications.
- Management and control of laboratory chemicals and other materials including checks on quality.
- Adequate training of laboratory personnel to ensure understanding and competence.
- External verification of results wherever possible.
- Accreditation of the laboratory by an independent organization.

QUALITY ASSURANCE

The overall management of an analytical laboratory should include the provision of evidence and assurances that appropriate QC procedures for laboratory activities are being correctly implemented. Quality assurance (QA) is a managerial responsibility that is designed to ensure that this is the case and to generate confidence in the analytical results. Part of QA is to build confidence through the laboratory participating in interlaboratory studies where several laboratories analyze one or more identical homogeneous materials under specified conditions.

Proficiency testing is a particular type of study to assess the performance of a laboratory or analyst relative to others, whilst method performance studies and certification studies are undertaken to check a particular analytical method or reference material respectively. The results of such studies and their statistical assessment enable the performances of individual participating laboratories to be demonstrated and any deficiencies in methodology and the training of personnel to be addressed.

ACCREDITATION SYSTEM

Because of differences in the interpretation of the term quality, which can be defined as fitness for purpose, QC and QA systems adopted by analyical laboratories in different industries and fields of activity can vary widely. For this reason, defined quality standards have been introduced by a number of organizations throughout the world. Laboratories can design and implement their own quality systems and apply to be inspected and accredited by the organization for the standard most appropriate to their activity. A number of organizations that offer accreditation suitable for analytical laboratories and their corresponding quality standards are given in *Table*.

Accreditation organizations and their quality standards

Name of accreditation organization	Quality standard
Organization for Economic Co-operation and Development (OECD)	Good Laboratory Practice (GLP)
The International Organization for Standardization (ISO)	ISO 9000 series of quality standards ISO Guide 25 general requirements for competence of calibration and testing laboratories
European Committee for Standardization (CEN)	EN 29000 series EN 45000 series
British Standards Institution (BSI)	BS 5750 quality standard BS 7500 series
National Measurement Accreditation Service (NAMAS)	NAMAS

2

Kinetics and Equilibrium Chemistry

INTRODUCTION

The connection between the equilibrium constant for a reaction and the rate constants of the elementary steps by which it occurs is an important one. We begin by stating the relationship. We will then demonstrate, for a specific case, that it is true. The equilibrium constant for a net reaction is the ratio of the product of the forward rate constants for all steps in the mechanism to the product of the reverse rate constants for all steps in the mechanism:

$$K_{eq} = k_1k_2k_{3..}/k_{-1}\,k_{-2}\,k_{-3..}$$

We now show for a specific net reaction and a specific (proposed) mechanism that this statement is true. Consider again the reaction of NO_2 and F_2 to produce NO_2F, reaction:

$$2NO_2 + F_2 \text{ Û } 2NO_2F$$

We have written the reaction with the double arrow because we are now interested in the forward and reverse rates at equilibrium. At equilibrium, the overall forward rate and overall reverse rate must be equal:

$$\text{rate}_f = \text{rate}_r$$

This must be true regardless of the detailed pathway by which the reaction occurs. All available experimental evidence indicates that the mechanism is the two-step process:

$$NO_2 + F_2 \text{ Û } NO_2F + F\ [k_1, k_{-1}]$$
$$NO_2 + F \rightarrow NO_2F\ [k_2, k_{-2}]$$

When the overall reaction is in equilibrium, so must each elementary step be. Consequently, we have used double arrows in the elementary steps as well. The equality of the forward and reverse rates of the individual steps is expressed mathematically as follows.

$$k_1[NO_2]_e[F_2]_e = k_{-1}[NO_2F]_e[F]_e$$
$$k_2[NO_2]_e\,[F]_e = k_{-2}[NO_2F]_e$$

Eliminating the concentration of the intermediate, F.

$$[NO_2F]_e^{\,2}/[NO_2]_e[F_2]_e = k_1k_2/k_{-2}\,k_{-2} = K_{eq}$$

The statement made at the outset is thus seen to be true for this specific case: the equilibrium constant is the ratio of the product of forward rate constants to the product of reverse rate constants. For the general overall reaction occuring by an n-step mechanism,

$$K_{eq} = k_1 k_{2}.. k_n / k_{-1} k_{-2}.. k_{-n}$$

Objection is frequently made to the method of derivation that we have just used, because the mechanism for is not known with certainty; it is and will remain a hypothesis, however well based in experiment.

If it is found that the two step mechanism is indeed NOT correct, then the relationship is invalid. This is certainly true. However, even if the mechanism above is incorrect, Reaction must occur by some mechanism. At equilibrium each step in the mechanism must be balanced in rate, enabling us to eliminate concentrations of any and all intermediates just as we did above, to arrive at a modified version. The validity of the general relationship is independent of the details of mechanism.

Thus the intuitively-expected connection between k and K does indeed exist. In fact, there are many examples in which the equilibrium constant for a reaction has been obtained from kinetics studies. This process can never be reversed however; it is not possible to obtain k by performing equilibrium studies. Before leaving this matter, we make one more very important point. There is a tendency to believe that reactions with large equilibrium constants are fast, whereas those with small equilibrium constants are slow. However, neither of these beliefs is true.

A large value for K_{eq} means that the product of forward rate constants is much larger than the product of reverse rate constants. It does not follow, however, that the forward rate constants are LARGE. It means only that they are larger than the reverse rate constants.

$k_1 k_{2}.. k_n >> k_{-1} k_{-2}.. k_{-n}$ does not mean that
$k_1, k_2..., k_n$ are large in the absolute sense.

A large K_{eq} can result from the ratio of a slow forward rate to an even slower reverse rate.

CATALYSIS

A number of chemical reactions that ordinarily occur slowly can be induced to occur more rapidly by the addition of a suitable substance called a catalyst. We will define a catalyst as a substance that speeds up a reaction without itself being consumed or chemically changed in the overall reaction process. This definition is in practice rather restrictive, because eventually, all real catalysts become deactivated by "poisoning" or via irreversible structural changes; thus they cannot function indefinitely.

For our purposes, here, however, this definition will serve. The reactant molecule that is affected by the catalyst is called the substrate. The process by which a catalyst effects its action is called catalysis. Catalysis is one of the most intensely studied areas in science because it is of tremendous biological and industrial importance. We begin our exploration of catalysts by discussing how they work.

The Mode of Catalyst Function

A catalyst functions either by entering into a slow step of the uncatalyzed reaction mechanism, or by creating an entirely new mechanism for the reaction. In either case, it is thought that the catalyzed pathway has a lower activation energy than the uncatalyzed path. We can illustrate catalytic action generically in terms of the following reaction scheme, where *A, B, D*, and *F* are reactants and products in the overall reaction, and *C* is a catalyst for the reaction.

$$A + B \rightarrow D + F \text{ [overall]}$$

Uncatalyzed mechanism:

$$A + B \rightarrow AB \text{ [slow]}$$
$$AB \rightarrow D + F$$

Catalyzed mechanism:

$$A + C \rightarrow CA \text{ [fast]}$$
$$CA + B \rightarrow CAB$$

[less slow than first step of uncatalyzed mechanism]

$$CAB \rightarrow \mathrm{D} + CF \text{ [fast]}$$
$$CF \rightarrow C + F \text{ [fast]}$$

The effect of the catalyst on the activation energy of the slow step of the reaction is shown schematically. The catalyst affects only E_a; it does not affect the energy of either the reactants or products of the reaction. The intermediate, *CAB*, in which both reactants are bound to the catalyst, is of particular interest. A possible (and fairly common) structural motif for this intermediate.

Notice that, although *A* and *B* are bound to different sites on the catalyst, *C*, they are in proximity and can effectively interact. Binding of *A* and *B* to *C* causes shifts in electron density that may facilitate bond breaking within *A* and/or *B* and bond formation between *A* and *B* or fragments of them.

Thus the catalyst provides an organizing centre for *A* and *B*, facilitating their interaction. In this manner the catalyst can overcome the orientation factor. As soon as the molecule of catalyst is regenerated in the last step of the 4-step mechanism, it may bind another molecule of *A* and proceed once again through the sequence, converting *A* and *B* to D and *F*. This sequence is repeated a large number of times, so that generally only a small amount of catalyst is required to convert a substantial quantity of reactant to product.

Chemists call the process by which the catalyst cycles through the reaction over and over again a catalytic loop. The unbound form of the catalyst, *C*, is placed at the 12 oclock position of the loop, and the various forms in which the catalyst is found are placed more or less evenly around the remainder of the loop. Arrows show the direction of reaction around the loop, with reactants brought in from outside the loop, and products ejected out of the loop. This is a very effective visual presentation of catalyst action. The number of times that a molecule of catalyst cycles through the loop per unit time is called the turnover number of the catalyst. An effective measure of turnover number is moles product produced per mole catalyst present per time. Let's look at a specific example of the effect of a catalyst on a simple reaction, from which we can draw some general conclusions.

Example: The reaction of ethanol (ethyl alcohol) with bromide ion to produce ethyl bromide and hydroxide ion.

$$C_2H_5OH + Br^- \rightarrow C_2H_5Br + OH^-$$

The reaction is quite slow in neutral or basic solution, but occurs quite rapidly in acidic solution because it is catalyzed by the hydronium ion, H_3O^+. Discuss the mechanisms for the uncatalyzed and catalyzed processes.

Solution: The uncatalyzed reaction is thought to occur in a single elementary step, in which bromide ion attacks the hydroxyl carbon atom of ethanol while the hydroxide group simultaneosusly departs:

$$Br^- + C_2H_5OH \rightarrow \text{activated complex} \rightarrow C_2H_5Br + OH^-$$

$$\text{Rate} = k_{obs}[C_2H_5OH][Br^-]$$

This mechanism is proposed based on the experimental rate law, which is overall second order. The observed rate law for the catalyzed process is

$$\text{Rate} = k_{obs}[Br^-][C_2H_5OH][H_3O^+]$$

The following three-step mechanism is consistent with this rate law if k_{obs} is identified with k_2K_1:

$$C_2H_5OH + H_3O^+ \rightarrow C_2H_5OH_2^+ + H_2O \text{ [fast, } K_1]$$

$$C_2H_5OH_2^+ + Br^- \rightarrow C_2H_5Br + H_2O \text{ [slow } k_2]$$

$$2H_2O \rightarrow H_3O^+ + OH^-$$

The reaction coordinate diagrams for the uncatalyzed and catalyzed pathways. Several general statements about catalysis are evident from this example.

- Because the catalyst does not appear in the overall equation for the reaction, it does not affect the equilibrium; it affects only the kinetics. The catalyst speeds up not only the forward reaction, but also the reverse, by the same factor. The position of equilibrium remains the same.
- the activation barrier is lower in the catalyzed mechanism. Even though the catalyzed path involves more steps, the overall reaction occurs more rapidly.
- The concentration of the catalyst appears in the rate law, even though the catalyst does not appear as a reactant or product in the overall equation.
- Although a catalyst and an intermediate share the property of not appearing in the overall equation, they are different types of things. First, the catalyst appears in the rate law; an intermediate cannot. Second, the catalyst appears first as a reactant, then is generated

later as a product. An intermediate appears first as a product, then later as a reactant.

Examples of Catalysis

We will now briefly discuss several examples of catalysis that are actually used on a mammoth scale in the chemical industry. Each of these processes, in its own way, has a major impact on the quality of our lives.

The Polymerization of Ethylene

Ethylene is a very simple molecule with formula C_2H_4. It is a gas, obtained as a byproduct during the catalytic cracking of petroleum. Under certain conditions, ethylene molecules can be made to join together end-to-end to form very long chain-like molecules of polyethylene, so called because it consists of many (poly) ethylenes. The process is represented.

$$2n\ CH_2 = CH_2 \xrightarrow{TiCl_3} (CH_2CH_2CH_2CH_2)_n^-$$

Even at high temperature and pressure of ethylene, this process occurs negligibly slowly. In the presence of a small amount of a modified form of $TiCl_3$ (called tickle-3 in the plastics industry), however, it occurs rapidly at only moderately high temperature and pressure. The process is referred to as Ziegler-Natta catalysis after its two coinventors, who jointly received the Nobel Prize in Chemistry in 1963. Since the discovery of this process in the 1950's, the entire plastics industry has developed and grown to huge proportions.

To this day, the mechanism of Ziegler-Natta catalysis is not fully understood. Mechanistic studies are difficult for several reasons, one of which is that the process is heterogeneous; that is, the catalyst ($TiCl_3$) and substrate (ethylene) are in different phases.

The catalytic process takes place on the surface of crystals of $TiCl_3$, which rapidly become covered with and blocked from view by the resulting polyethylene. Much effort is ongoing in the US chemical industry to develop more efficient and easily handled Ziegler-Natta catalysts.

The Production of Sulfuric Acid

Year after year, sulfuric acid ranks first on the list of the top ten chemical substances produced in the United States: billions of pounds are produced annually. Sulfuric acid is synthesized by the so-called Contact Process, which involves the four sequential steps below:

$$S + O_2 \rightarrow SO_2 \text{ [fast]}$$

$$SO_2 + O_2 \rightarrow SO_3$$

[slow, because it occurs by a termolecular elementary process catalyzed by V_2O_5]

$$SO_3 + H_2SO_4 \rightarrow H_2S_2O_7 \text{ [fast]}$$

$$H_2S_2O_7 + H_2O \rightarrow 2H_2SO_4 \text{ [fast]}$$

The reaction of SO_3 with water is very exothermic and causes extensive spattering and production of a fine mist of highly acidic water.

For this reason, direct reaction of SO_3 with water in the third step is impractical. Instead, SO_3 is bubbled into pure sulfuric acid, with which it reacts smoothly to give fuming sulfuric acid, $H_2S_2O_7$. This can then be treated with the stoichiometrically correct amount of water to give sulfuric acid.

The contact process has been so perfected that sulfuric acid is very inexpensive to produce. Consequently, it is used in any industrial process requiring acid. It finds it major uses in the production of phosphate fertilizers; in paper manufacture; in the petroleum industry; in steel production; and in the production of detergents. The importance of these products in our lives is obvious.

Catalytic Converters

For some years now, catalysts have been placed within the exhaust systems of automobiles to reduce the amount of poisonous or otherwise harmful emissions. These noble metal catalysts (based on platinum and palladium) carry out a dual function. First, they facilitate oxidation of carbon monoxide, resulting from incomplete hydrocarbon combustion, to carbon dioxide:

$$CO(g) + 1/2\ O_2(g) \xrightarrow{Pt} CO_2(g)$$

Second, they catalyze decomposition of nitric oxide, produced during engine operation and oxidized rapidly to toxic NO_2 by atmospheric oxygen, to N_2 and O_2. Unfortunately, catalytic converters also facilitate oxidation of SO_2 to SO_3, which is the precursor of acid rain. Low-sulfur petroleum distillates are therefore essential.

The Haber Process. Vast quantities of ammonia are synthesized each year for use as fertilizer. Currently the most efficient process for ammonia synthesis is the Haber-Bosch Process, developed during the ten-year period preceding 1913, in which nitrogen and hydrogen react directly at high temperature and pressure and in the presence of an activated iron catalyst to form ammonia.

$$N_2(g) + 3H_2(g) \xrightarrow{Fe} 2NH_3(g)$$

A catalyst and high temperature are necessary to cause the reaction to go at a reasonable rate. Unfortunately, high temperature makes the exothermic reaction less favoured, so very high pressure is used to favour products. Even with conditions optimized, reaction is incomplete, and unreacted hydrogen gas is recycled for maximum efficiency of ammonia production.

Nitrogen fertilizers produced from ammonia are largely responsible for the incredible growth in agricultural production over the decades since the First World War, when the process was first put on line in Germany. Ironically, the original motivation for development of the process was the requirement of explosives for the war effort.

The processes discussed above have at least three features in common. First, catalysis is heterogeneous; the catalyst is in all cases a solid, interacting with the substrate in the gas phase. Second, the mechanisms for these processes are incompletely understood. Thus catalysts are used successfully on a huge scale, even though we do not understand how they work. Third, in all cases the catalyst involves a transition metal, from the D block of the periodic table.

Transition metals are often versatile catalysts because they are flexible in coordination number (that is, the number of atoms, ions, or molecules to which they may bind in a Lewis acid-base interaction); in stereochemistry (that is, in the shapes that their adducts assume); and in oxidation state (that is, in the charge that they carry). Nature has chosen transition metals to serve as the centerpieces for many of its catalysts, the enzymes, most probably for these same reasons.

REVERSIBLE REACTIONS AND CHEMICAL EQUILIBRIA

In 1798, the chemist Claude Berthollet accompanied a French military expedition to Egypt. While visiting the Natron Lakes, a series of salt water lakes carved from limestone, Berthollet made an observation that contributed to an important discovery. Upon analyzing water from the Natron Lakes, Berthollet found large quantities of common salt, NaCl, and soda ash, Na_2CO_3, a result he found surprising. Why would Berthollet find this result surprising and how did it contribute to an important discovery? Answering these questions provides an example of chemical reasoning and introduces the topic of this chapter.

Berthollet "knew" that a reaction between Na_2CO_3 and $CaCl_2$ goes to completion, forming NaCl and a precipitate of $CaCO_3$ as products.

$$Na_2CO_3 + CaCl_2 \rightarrow 2NaCl + CaCO_3$$

Understanding this, Berthollet expected that large quantities of NaCl and Na2CO3 could not coexist in the presence of $CaCO_3$. Since the reaction goes to completion, adding a large quantity of $CaCl_2$ to a solution of Na_2CO_3 should produce NaCl and $CaCO_3$, leaving behind no unreacted Na_2CO_3. In fact, this result is what he observed in the laboratory. The evidence from Natron Lakes, where the coexistence of NaCl and Na_2CO_3 suggests that the reaction has not gone to completion, ran counter to Berthollet's expectations. Berthollet's important insight was recognizing that the chemistry occurring in the Natron Lakes is the reverse of what occurs in the laboratory.

$$CaCO_3 + 2NaCl \rightarrow Na_2CO_3 + CaCl_2$$

Using this insight Berthollet reasoned that the reaction is reversible, and that the relative amounts of "reactants" and "products" determine the direction in which the reaction occurs, and the final composition of the reaction mixture.

We recognize a reaction's ability to move in both directions by using a double arrow when writing the reaction.

$$Na_2CO_3 + CaCl_2 \rightleftharpoons 2NaCl + CaCO_3$$

Berthollet's reasoning that reactions are reversible was an important step in understanding chemical reactivity. When we mix together solutions of Na_2CO_3 and $CaCl_2$, they react to produce NaCl and $CaCO_3$. If we monitor the mass of dissolved Ca^{2+} remaining and the mass of $CaCO_3$ produced as a function of time, the result will look something like the graph in Figure.

At the start of the reaction the mass of dissolved Ca^{2+} decreases and the mass of $CaCO_3$ increases. Eventually, however, the reaction reaches a point after which no further changes occur in the amounts of these species. Such a condition is called a state of equilibrium.

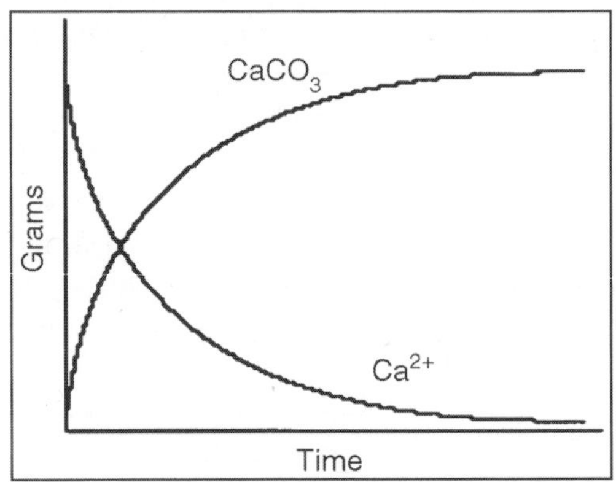

Fig. Change in Mass of Undissolved Ca^{2+} and Solid $CaCO_3$ Over Time During the Precipitation of $CaCO_3$.

Although a system at equilibrium appears static on a macroscopic level, it is important to remember that the forward and reverse reactions still occur. A reaction at equilibrium exists in a "steady state," in which the rate at which any species forms equals the rate at which it is consumed.

THERMODYNAMICS AND EQUILIBRIUM CHEMISTRY

Thermodynamics is the study of thermal, electrical, chemical, and mechanical forms of energy. The study of thermodynamics crosses many disciplines, including physics, engineering, and chemistry. Of the various branches of thermodynamics, the most important to chemistry is the study of the changes in energy occurring during a chemical reaction. Consider, for example, the general equilibrium reaction shown in equation, involving the solutes A, B, C, and D, with stoichiometric coefficients *a, b, c,* and *d.*

$$aA + bB \rightleftharpoons cC + dD$$

By convention, species to the left of the arrows are called reactants, and those on the right side of the arrows are called products. As Berthollet discovered, writing a reaction in this fashion does not guarantee that the reaction of A and B to produce C and D is favorable.

Depending on initial conditions, the reaction may move to the left, to the right, or be in a state of equilibrium. Understanding the factors that determine the final position of a reaction is one of the goals of chemical thermodynamics.

Chemical systems spontaneously react in a fashion that lowers their overall free energy. At a constant temperature and pressure, typical of many bench-top chemical reactions, the free energy of a chemical reaction is given by the Gibb's free energy function

$$\Delta G = \Delta H - T\Delta S$$

where T is the temperature in kelvins, and ΔG, ΔH, and ΔS are the differences in the Gibb's free energy, the enthalpy, and the entropy between the products and reactants. Enthalpy is a measure of the net flow of energy, as heat, during a chemical reaction. Reactions in which heat is produced have a negative ΔH and are called exothermic. Endothermic reactions absorb heat from their surroundings and have a positive ΔH.

Entropy is a measure of randomness, or disorder. The entropy of an individual species is always positive and tends to be larger for gases than for solids and for more complex rather than simpler molecules. Reactions that result in a large number of simple, gaseous products usually have a positive ΔS. The sign of ΔG can be used to predict the direction in which a reaction moves to reach its equilibrium position.

A reaction is always thermodynamically favored when enthalpy decreases and entropy increases. Substituting the inequalities $\Delta H < 0$ and $\Delta S > 0$ into equation shows that ΔG is negative when a reaction is thermodynamically favored. When ΔG is positive, the reaction is unfavorable as written (although the reverse reaction is favorable). Systems at equilibrium have a ΔG of zero.

As a system moves from a nonequilibrium to an equilibrium position, ΔG must change from its initial value to zero. At the same time, the species involved in the reaction undergo a change in their concentrations. The Gibb's free energy, therefore, must be a function of the concentrations of reactants and products. As shown in equation, the Gibb's free energy can be divided into two terms.

$$\Delta G = \Delta G° + RT \ln Q$$

The first term, $\Delta G°$, is the change in Gibb's free energy under standard-state conditions; defined as a temperature of 298 K, all gases with partial pressures of 1 atm, all solids and liquids pure, and all solutes present with 1 M concentrations. The second term, which includes the reaction quotient, Q, accounts for nonstandard-state pressures or concentrations. For reaction 6.1 the reaction quotient is

$$Q = \frac{[C]^c[D]^d}{[A]^a[B]^b}$$

where the terms in brackets are the molar concentrations of the solutes. Note that the reaction quotient is defined such that the concentrations of products are placed in the numerator, and the concentrations of reactants are placed in the denominator.

In addition, each concentration term is raised to a power equal to its stoichiometric coefficient in the balanced chemical reaction. Partial pressures are substituted for concentrations when the reactant or product is a gas. The concentrations of pure solids and pure liquids do not change during a chemical reaction and are excluded from the reaction quotient.

At equilibrium the Gibb's free energy is zero, and equation simplifies to

$$\Delta G^\circ = -RT \ln K$$

where *K* is an equilibrium constant that defines the reaction's equilibrium position. The equilibrium constant is just the numerical value obtained when substituting the concentrations of reactants and products at equilibrium into equation; thus,

$$K = \frac{[C]_{ed}^{c}[D]_{eq}^{d}}{[A]_{eq}^{a}[B]_{eq}^{b}}$$

where the subscript "eq" indicates a concentration at equilibrium. Although the subscript "eq" is usually omitted, it is important to remember that the value of *K* is determined by the concentrations of solutes at equilibrium.

As written, equation is a limiting law that applies only to infinitely dilute solutions, in which the chemical behaviour of any species in the system is unaffected by all other species. Corrections to equation are possible and are discussed in more detail at the end of the chapter.

MANIPULATING EQUILIBRIUM CONSTANTS

We will use two useful relationships when working with equilibrium constants. First, if we reverse a reaction's direction, the equilibrium constant for the new reaction is simply the inverse of that for the original reaction. For example, the equilibrium constant for the reaction

$$A + 2B \rightleftharpoons AB_2 \quad K_1 = \frac{[AB_2]}{[A][B^2]}$$

is the inverse of that for the reaction

$$AB_2 \rightleftharpoons A + 2B \quad K_2 = \frac{1}{K_1} = \frac{[A][B]^2}{[AB_2]}$$

Second, if we add together two reactions to obtain a new reaction, the equilibrium constant for the new reaction is the product of the equilibrium constants for the original reactions.

$$A + C \rightleftharpoons AC \quad K_1 = \frac{[AC]}{[A][C]}$$

$$AC + C \rightleftharpoons AC_2 \quad K_2 = \frac{[AC_2]}{[AC][C]}$$

$$A + 2C \rightleftharpoons AC_2 \quad K_3 = K_1K_2 = \frac{[AC]}{[A][C]} \times \frac{[AC_2]}{[AC][C]} = \frac{[AC_2]}{[A][C]^2}$$

EQUILIBRIUM CONSTANTS FOR CHEMICAL REACTIONS

Several types of reactions are commonly used in analytical procedures, either in preparing samples for analysis or during the analysis itself. The most important of these are precipitation reactions, acid–base reactions, complexation reactions, and oxidation–reduction reactions. In this section we review these reactions and their equilibrium constant expressions.

Precipitation Reactions

A precipitation reaction occurs when two or more soluble species combine to form an insoluble product that we call a precipitate. The most common precipitation reaction is a metathesis reaction, in which two soluble ionic compounds exchange parts. When a solution of lead nitrate is added to a solution of potassium chloride, for example, a precipitate of lead chloride forms. We usually write the balanced reaction as a net ionic equation, in which only the precipitate and those ions involved in the reaction are included. Thus, the precipitation of $PbCl_2$ is written as

$$Pb^{2+}(aq) + 2Cl^-(aq) \rightleftharpoons PbCl_2(s)$$

In the equilibrium treatment of precipitation, however, the reverse reaction describing the dissolution of the precipitate is more frequently encountered.

$$PbCl_2(s) \rightleftharpoons Pb^{2+}(aq) + 2Cl^-(aq)$$

The equilibrium constant for this reaction is called the solubility product, K_{sp}, and is given as

$$K_{sp} = [Pb^{2+}][Cl^-]^2 = 1.7 \times 10^{-5}$$

Note that the precipitate, which is a solid, does not appear in the K_{sp} expression. It is important to remember, however, that equation is valid only if $PbCl_2(s)$ is present and in equilibrium with the dissolved Pb^{2+} and Cl^-.

Acid Base Reactions

A useful definition of acids and bases is that independently introduced by Johannes Brønsted (1879–1947) and Thomas Lowry (1874–1936) in 1923. In the Brønsted-Lowry definition, acids are proton donors, and bases are proton acceptors. Note that these definitions are interrelated. Defining a base as a proton acceptor means an acid must be available to provide the proton. For example, in reaction acetic acid, CH_3COOH, donates a proton to ammonia, NH3, which serves as the base.

$$CH_3COOH(aq) \rightleftharpoons CH_3COO^-(aq) + NH_4 + (aq)$$

Strong and Weak Acids The reaction of an acid with its solvent (typically water) is called an acid dissociation reaction. Acids are divided into two categories based on the ease with which they can donate protons to the solvent. Strong acids, such as HCl, almost completely transfer their protons to the solvent molecules.

$$HCl(aq) + H_2O(\ell) \rightarrow H_3O^+Cl^-(aq)$$

In this reaction H_2O serves as the base. The hydronium ion, H_3O^+, is the conjugate acid of H_2O, and the chloride ion is the conjugate base of HCl. It is the hydronium ion that is the acidic species in solution, and its concentration determines the acidity of the resulting solution. We have chosen to use a single arrow in place of the double arrows (t) to indicate that we treat HCl as if it were completely dissociated in aqueous solutions. A solution of 0.10 M HCl is effectively 0.10 M in H_3O^+ and 0.10 M in Cl^-. In aqueous solutions, the common strong acids are hydrochloric acid (HCl), hydroiodic acid (HI), hydrobromic acid (HBr), nitric acid (HNO_3), perchloric acid ($HClO_4$), and the first proton of sulfuric acid (H_2SO_4). Weak acids, of which aqueous acetic acid is one example, cannot completely donate their acidic protons to the solvent. Instead, most of the acid remains undissociated, with only a small fraction present as the conjugate base.

$$CH_3COOH(aq) + H_2O(\ell) \rightleftharpoons H_3O^+(aq) + CH_3COO^-(aq)$$

The equilibrium constant for this reaction is called an acid dissociation constant,

K_a, and is written as

$$K_a = \frac{[H_3O^+][CH_3COO^-]}{[CH_3COOH]} = 1.75 \times 10^{-5}$$

Note that the concentration of H_2O is omitted from the *K*a expression because its value is so large that it is unaffected by the dissociation reaction. The magnitude of *K*a provides information about the relative strength of a weak acid, with a smaller *K*a corresponding to a weaker acid. The ammonium ion, for example, with a *K*a of 5.70×10^{-10}, is a weaker acid than acetic acid.

Monoprotic weak acids, such as acetic acid, have only a single acidic proton and a single acid dissociation constant. Some acids, such as phosphoric acid, can donate more than one proton and are called polyprotic weak acids. Polyprotic acids are described by a series of acid dissociation steps, each characterized by it own acid dissociation constant. Phosphoric acid, for example, has three acid dissociation reactions and acid dissociation constants.

$$H_3PO_4(aq) + H_2O(\ell) \rightleftharpoons H_3O^+(aq) + H_2PO_4^-(aq)$$

$$K_{a1} = \frac{[H_2PO_4^-][H_3O^+]}{[H_3PO_4]} = 7.11 \times 10^{-3}$$

$$H_2PO_4^-(aq) + H_2O(\ell) \rightleftharpoons H_3O^+(aq) + HPO_4^-(aq)$$

$$K_{a2} = \frac{[HPO_4^{2-}][H_3O^+]}{[H_2PO_4^-]} = 6.32 \times 10^{-8}$$

$$HPO_4^{2-}(aq) + H_2O(\ell) \rightleftharpoons H_3O^+(aq) + PO_4^{3-}(aq)$$

$$K_{a3} = \frac{[PO_4^{3-}][H_3O^+]}{[HPO_4^{2-}]} = 4.5 \times 10^{-13}$$

The decrease in the acid dissociation constant from *K*a1 to Ka_3 tells us that each successive proton is harder to remove. Consequently, H_3PO_4 is a stronger acid than $H_2PO_4^-$, and $H_2PO_4^-$ is a stronger acid than HPO_4^{2-}.

Strong and Weak Bases Just as the acidity of an aqueous solution is a measure of the concentration of the hydronium ion, H_3O^+, the basicity of an aqueous solution is a measure of the concentration of the hydroxide ion, OH^-. The most common example of a strong base is an alkali metal hydroxide, such as sodium hydroxide, which completely dissociates to produce the hydroxide ion.

$$NaOH(aq) \rightarrow Na^+(aq) + OH^-(aq)$$

Weak bases only partially accept protons from the solvent and are characterized by a base dissociation constant, *K*b. For example, the base dissociation reaction and base dissociation constant for the acetate ion are

$$CH_3COO^-(aq) + H_2O(\ell) \rightleftharpoons OH^-(aq) + CH_3COOH(aq)$$

$$K_b = \frac{[CH_3COOH][OH^-]}{[CH_3COO^-]} = 5.71 \times 10^{-10}$$

Polyprotic bases, like polyprotic acids, also have more than one base dissociation reaction and base dissociation constant.

Amphiprotic Species Some species can behave as either an acid or a base. For example, the following two reactions show the chemical reactivity of the bicarbonate ion, HCO^{3-}, in water.

$$HCO_3^-(aq) + H_2O(\ell) \rightleftharpoons H_3O^+(aq) + CO_3^{2-}(aq)$$

$$HCO_3^-(aq) + H_2O(\ell) \rightleftharpoons HO^-(aq) + H_2CO_3(aq)$$

A species that can serve as both a proton donor and a proton acceptor is called amphiprotic. Whether an amphiprotic species behaves as an acid or as a base depends on the equilibrium constants for the two competing reactions. For bicarbonate, the acid dissociation constant for reaction 6.8

$$K_{a2} = 4.69 \times 10^{-11}$$

is smaller than the base dissociation constant for reaction.

$$K_{b2} = 2.25 \times 10^{-8}$$

Since bicarbonate is a stronger base than it is an acid, we expect that aqueous solutions of HCO^{3-} will be basic.

Dissociation of Water Water is an amphiprotic solvent in that it can serve as an acid or a base. An interesting feature of an amphiprotic solvent is that it is capable of reacting with itself as an acid and a base.

$$H_2O(\ell) + H_2O(\ell) \rightleftharpoons H_3O^+(aq) + OH^-(aq)$$

The equilibrium constant for this reaction is called water's dissociation constant, *K*w,

$$K_w = [H_3O^+][OH^-]$$

which has a value of 1.0000×10^{-14} at a temperature of 24°C. The value of *K*w varies substantially with temperature. For example, at 20°C, *K*w is 6.809×10^{-15}, but at 30°C *K*w is 1.469×10^{-14}. At the standard state temperature of 25°C, *K*w is 1.008×10^{-14}, which is sufficiently close to 1.00×10^{-14} that the latter value can be used with negligible error.

The pH Scale An important consequence of equation is that the concentrations of H_3O^+ and OH^- are related. If we know $[H_3O^+]$ for a solution, then $[OH^-]$ can be calculated using equation. Equation also allows us to develop a pH scale that indicates the acidity of a solution. When the concentrations of H_3O^+ and OH^- are equal, a solution is neither acidic nor basic; that is, the solution is neutral. Letting

$$[H_3O^+] = [OH^-]$$

and substituting into equation leaves us with

$$K_w = [H_3O^+]^2 = 1.00 \times 10^{-14}$$

Solving for $[H_3O^+]$ gives

$$[H_3O^+] = \sqrt{1.00 \times 10^{-14}} = 1.00 \times 10^{-7}$$

A neutral solution has a hydronium ion concentration of 1.00×10^{-7} M and a pH of 7.00. For a solution to be acidic, the concentration of H_3O^+ must be greater than that for OH^-, or

$$[H_3O^+] > 1.00 \times 10^{-7} \text{ M}$$

The pH of an acidic solution, therefore, must be less than 7.00. A basic solution, on the other hand, will have a pH greater than 7.00. Figure shows the pH scale along with pH values for some representative solutions.

Tabulating Values for *Ka* and *Kb* A useful observation about acids and bases is that the strength of a base is inversely proportional to the strength of its conjugate acid. Consider, for example, the dissociation reactions of acetic acid and acetate.

$$CH_3COOH(aq) + H_2O(\ell) \rightleftharpoons H_3O^+(aq) + CH_3COO^-(aq)$$

$$CH_3COO^-(aq) + H_2O(\ell) \rightleftharpoons CH_3COOH(aq) + OH^-(aq)$$

Adding together these two reactions gives

$$2H_2O(\ell) \rightleftharpoons H_3O^+(aq) + OH^-(aq)$$

The equilibrium constant for equation is *K*w. Since equation is obtained by adding together reactions, *K*w may also be expressed as the product of *K*a for CH_3COOH and *K*b for CH_3COO^-. Thus, for a weak acid, HA, and its conjugate weak base, A^s,

$$K_w = K_a \times K_b$$

This relationship between K_a and K_b simplifies the tabulation of acid and base dissociation constants. The corresponding values of K_b for their conjugate weak bases are determined using equation.

Complexation Reactions

A more general definition of acids and bases was proposed by G. N. Lewis (1875–1946) in 1923. The Bronsted–Lowry definition of acids and bases focuses on an acid's proton-donating ability and a base's proton-accepting ability. Lewis theory, on the other hand, uses the breaking and forming of covalent bonds to describe acid–base characteristics. In this treatment, an acid is an electron pair acceptor, and a base is an electron pair donor. Although Lewis theory can be applied to the treatment of acid–base reactions, it is more useful for treating complexation reactions between metal ions and ligands.

The following reaction between the metal ion Cd^{2+} and the ligand NH3 is typical of a complexation reaction.

$$Cd^{2+}(aq) + 4(:NH_3)(aq) \rightleftharpoons Cd(:NH_3)_4^{2+}(aq)$$

The product of this reaction is called a metal–ligand complex. In writing the equation for this reaction, we have shown ammonia as:NH3 to emphasize the pair of electrons it donates to Cd^{2+}. In subsequent reactions we will omit this notation. The formation of a metal–ligand complex is described by a formation constant, K_f. The complexation reaction between Cd^{2+} and NH_3, for example, has the following equilibrium constant

$$K_f = \frac{[Cd(NH_3)_4^{2+}]}{[Cd^{2+}][NH_3]^4} = 5.5 \times 10^7$$

The reverse of reaction is called a dissociation reaction and is characterized by a dissociation constant, *Kd*, which is the reciprocal of K_f.

Many complexation reactions occur in a stepwise fashion. For example, the reaction between Cd^{2+} and NH_3 involves four successive reactions

$$Cd^{2+}(aq) + NH_3(aq) \rightleftharpoons Cd(NH_3)^{2+}(aq)$$

$$Cd(NH_3)^{2+}(aq) + NH_3(aq) \rightleftharpoons Cd(NH_3)_2{}^{2+}(aq)$$

$$Cd(NH_3)^{2+}(aq) + NH_3(aq) \rightleftharpoons Cd(NH_3)_3{}^{2+}(aq)$$

$$Cd(NH_3)^{2+}(aq) + NH_3(aq) \rightleftharpoons Cd(NH_3)_4{}^{2+}(aq)$$

This creates a problem since it no longer is clear what reaction is described by a formation constant. To avoid ambiguity, formation constants are divided into two categories.

Stepwise formation constants, which are designated as K_i for the *i*th step, describe the successive addition of a ligand to the metal–ligand complex formed in the previous step. Thus, the equilibrium constants for reactions are, respectively, K_1, K_2, K_3, and K_4. Overall, or cumulative formation constants, which are designated as S_i, describe the addition of *i* ligands to the free metal ion. The equilibrium constant expression given in equation, therefore, is correctly identified as β_4, where

$$\beta_4 = K_1 \times K_2 \times K_3 \times K_4$$

In general

$$\beta_i = K_1 \times K_2 \times .... \times K_i$$

Equilibrium constants for complexation reactions involving solids are defined by combining appropriate K_{sp} and K_f expressions. For example, the solubility of AgCl increases in the presence of excess chloride as the result of the following complexation reaction

$$AgCl(s) + Cl^-(aq) \rightleftharpoons AgC_2^-(aq)$$

This reaction can be separated into three reactions for which equlibrium constants are known—the solubility of AgCl, described by its K_{sp}

$$AgCl(s) \rightleftharpoons Ag^+(aq) + C^-(aq)$$

and the stepwise formation of $AgCl_2{}^-$, described by K_1 and K_2

$$Ag^+(aq) + Cl^-(aq) \rightleftharpoons AgCl(aq)$$

$$AgCl(aq) + Cl^-(aq) \rightleftharpoons AgCl_2{}^-(aq)$$

The equilibrium constant for reaction, therefore, is equal to $K_{sp} \times K_1 \times K_2$.

OxidationÐReduction Reactions

In a complexation reaction, a Lewis base donates a pair of electrons to a Lewis acid. In an oxidation–reduction reaction, also known as a redox reaction, electrons are not shared, but are transferred from one reactant to another. As a result of this electron transfer, some of the elements involved in the reaction undergo a change in oxidation state. Those species experiencing an increase in their oxidation state are oxidized, while those experiencing a decrease in their oxidation state are reduced. For example, in the following redox reaction between Fe3+ and oxalic acid, $H_2C_2O_4$, iron is reduced since its oxidation state changes from +3 to +2.

$$2Fe^{3}(aq) + H_2C_2O_4(aq) + 2H_2O(\ell) \rightleftharpoons 2Fe^{2+}(aq) + 2CO_2(g) + 2H_3O^+(aq)$$

Oxalic acid, on the other hand, is oxidized since the oxidation state for carbon increases from +3 in $H_2C_2O_4$ to +4 in CO_2.

Redox reactions, such as that shown in equation, can be divided into separate half-reactions that individually describe the oxidation and the reduction processes.

$$H_2C_2O_4(aq) + 2H_2O(\ell) \rightarrow 2CO_2(g) + 2H_3O^+(aq) + 2e^-$$

$$Fe^{3+}(aq) + e^- \rightarrow Fe^{2+}(aq)$$

It is important to remember, however, that oxidation and reduction reactions always occur in pairs. This relationship is formalized by the convention of calling the species being oxidized a reducing agent, because it provides the electrons for the reduction half-reaction.

Conversely, the species being reduced is called an oxidizing agent. Thus, in reaction, Fe^{3+} is the oxidizing agent and $H_2C_2O_4$ is the reducing agent.

The products of a redox reaction also have redox properties. For example, the Fe^{2+} in reaction can be oxidized to Fe^{3+}, while CO_2 can be reduced to $H_2C_2O_4$. Borrowing some terminology from acid–base chemistry, we call Fe2+ the conjugate reducing agent of the oxidizing agent Fe^{3+} and CO_2 the conjugate oxidizing agent of the reducing agent $H_2C_2O_4$.

Unlike the reactions that we have already considered, the equilibrium position of a redox reaction is rarely expressed by an equilibrium constant. Since redox reactions involve the transfer of electrons from a reducing agent to an oxidizing agent, it is convenient to consider the thermodynamics of the reaction in terms of the electron.

The free energy, ΔG, associated with moving a charge, Q, under a potential, E, is given by

$$\Delta G = EQ$$

Charge is proportional to the number of electrons that must be moved. For a reaction in which one mole of reactant is oxidized or reduced, the charge, in coulombs, is

$$Q = nF$$

where n is the number of moles of electrons per mole of reactant, and F is Faraday's constant. The change in free energy (in joules per mole; J/mol) for a redox reaction, therefore, is

$$\Delta G = -nFE$$

where ΔG has units of joules per mole. The appearance of a minus sign in equation is due to a difference in the conventions for assigning the favored direction for reactions. In thermodynamics, reactions are favored when ΔG is negative, and redox reactions are favored when E is positive. The relationship between electrochemical potential and the concentrations of reactants and products can be determined by substituting equation

$$-nFE = -nFE° + RT \ln Q$$

where $E°$ is the electrochemical potential under standard-state conditions. Dividing through by $-nF$ leads to the well-known Nernst equation.

$$E = E° - \frac{RT}{nF} \ln Q$$

Substituting appropriate values for R and F, assuming a temperature of 25 °C (298 K), and switching from ln to log* gives the potential in volts as

$$E = E° - \frac{0.05916}{n} \log Q$$

The standard-state electrochemical potential, $E°$, provides an alternative way of expressing the equilibrium constant for a redox reaction. Since a reaction at equilibrium has a ΔG of zero, the electrochemical potential, E, also must be zero. Substituting into equation and rearranging shows that

$$E° = \frac{RT}{nF} \log K$$

Standard-state potentials are generally not tabulated for chemical reactions, but are calculated using the standard-state potentials for the oxidation, $E°$ox, and reduction half-reactions, $E°$red. By convention, standard-state potentials are only listed for reduction half-reactions, and $E°$ for a reaction is calculated as

$$E°_{reac} = E°_{red} - E°_{ox}$$

where both $E°$red and $E°$ox are standard-state reduction potentials.

Since the potential for a single half-reaction cannot be measured, a reference halfreaction is arbitrarily assigned a standard-state potential of zero. All other reduction potentials are reported relative to this reference. The standard half-reaction is

$$2H_3O^+(aq) + 2e^- \rightleftharpoons 2H_2O(\ell) + H_2(g)$$

The more positive the standard-state reduction potential, the more favorable the reduction reaction will be under standard-state conditions. Thus, under standard-state conditions, the reduction of Cu^{2+} to Cu ($E°$ = + 0.3419) is more favorable than the reduction of Zn^{2+} to Zn ($E°$ = –0.7618).

LE CHATELIERS PRINCIPLE

The equilibrium position for any reaction is defined by a fixed equilibrium constant, not by a fixed combination of concentrations for the reactants and products. This is easily appreciated by examining the equilibrium constant expression for the dissociation of acetic acid.

$$K_a = \frac{[H_3O^+][CH_3COO^-]}{[CH_3COOH]} = 1.75 \times 10^{-5}$$

As a single equation with three variables, equation does not have a unique solution for the concentrations of CH3COOH, CH3COO–, and H_3O^+. At constant temperature, different solutions of acetic acid may have different values for $[H_3O^+]$, $[CH_3COO^-]$ and $[CH_3COOH]$, but will always have the same value of K_a.

If a solution of acetic acid at equilibrium is disturbed by adding sodium acetate, the $[CH_3COO^-]$ increases, suggesting an apparent increase in the value of *K*a. Since *K*a must remain constant, however, the concentration of all three species in equation must change in a fashion that restores *K*a to its original value. In this case, equilibrium is reestablished by the partial reaction of CH_3COO^- and H_3O^+ to produce additional CH_3COOH.

The observation that a system at equilibrium responds to a stress by reequilibrating in a manner that diminishes the stress, is formalized as Le Châtelier's principle. One of the most common stresses that we can apply to a reaction at equilibrium is to change the concentration of a reactant or product.

We already have seen, in the case of sodium acetate and acetic acid, that adding a product to a reaction mixture at equilibrium converts a portion of the products to reactants. In this instance, we disturb the equilibrium by adding a product, and the stress is diminished by partially reacting the excess product. Adding acetic acid has the opposite effect, partially converting the excess acetic acid to acetate.

In our first example, the stress to the equilibrium was applied directly. It is also possible to apply a concentration stress indirectly. Consider, for example, the following solubility equilibrium involving AgCl

$$AgCl(s) \rightleftharpoons Ag^+(aq) + Cl^-(aq)$$

The effect on the solubility of AgCl of adding $AgNO_3$ is obvious, but what is the effect of adding a ligand that forms a stable, soluble complex with Ag+? Ammonia, for example, reacts with Ag+ as follows

$$Ag^+(aq) + 2NH_3(aq) \rightleftharpoons Ag(NH_3)_2{}^+(aq)$$

Adding ammonia decreases the concentration of Ag^+ as the $Ag(NH3)^{2+}$ complex forms. In turn, decreasing the concentration of Ag^+ increases the solubility of AgCl as reaction reestablishes its equilibrium position. Adding together reactions clarifies the effect of ammonia on the solubility of AgCl, by showing that ammonia is a reactant.

$$AgCl(s) + 2NH_3(aq) \rightleftharpoons Ag(NH_3)_2{}^+(aq) + Cl^-(aq)$$

Increasing or decreasing the partial pressure of a gas is the same as increasing or decreasing its concentration.

The effect on a reaction's equilibrium position can be analyzed as described in the preceding example for aqueous solutes. Since the concentration of a gas depends on its partial pressure, and not on the total pressure of the system, adding or removing an inert gas has no effect on the equilibrium position of a gas-phase reaction. Most reactions involve reactants and products that are dispersed in a solvent. If the amount of solvent is changed, either by diluting or concentrating the solution, the concentrations of all reactants and products either decrease or increase. The effect of these changes in concentration is not as intuitively obvious as when the concentration of a single reactant or product is changed. As an example, let's consider how dilution affects the equilibrium position for the formation of the aqueous silver-amine complex. The equilibrium constant for this reaction is

$$\beta_2 = \frac{[Ag(NH_3)_2^+]_{eq}}{[Ag^+]_{eq}[NH_3]_{eq}^2}$$

where the subscript "equation" is included for clarification. If a portion of this solution is diluted with an equal volume of water, each of the concentration terms in equation is cut in half. Thus, the reaction quotient becomes

$$Q = \frac{(0.5)[Ag(NH_3)_2^+]_{eq}}{(0.5)[Ag^+]_{eq}(0.5)^2[NH_3]_{eq}^2}$$

Since Q is greater than Q^2, equilibrium must be reestablished by shifting the reaction to the left, decreasing the concentration of $Ag(NH_3)^{2+}$.

Furthermore, this new equilibrium position lies towards the side of the equilibrium reaction with the greatest number of solutes (one Ag^+ ion and two molecules of NH_3 versus the single metal–ligand complex). If the solution of $Ag(NH_3)^{2+}$ is concentrated, by evaporating some of the solvent, equilibrium is reestablished in the opposite direction. This is a general conclusion that can be applied to any reaction, whether gas-phase, liquid-phase, or solid-phase.

Increasing volume always favors the direction producing the greatest number of particles, and decreasing volume always favors the direction producing the fewest particles. If the number of particles is the same on both sides of the equilibrium, then the equilibrium position is unaffected by a change in volume.

PRINCIPLES OF CHEMICAL EQUILIBRIUM

Equilibrium. In this chapter the dynamic equilibrium established in chemical reactions is explored, and its similarity to phase equilibrium is emphasized. The chapter culminates in a relationship between two signposts of spontaneity: the standard free energy of reaction and the equilibrium constant.

SPECIFIC CONCEPTS

Dynamic equilibrium is a balance of opposing processes occurring at equal rates.

- Chemical reactions achieve dynamic equilibrium similar to phase equilibrium and characterized by an equilibrium constant, K_{eq}, analogous to vapour pressure.

  Equilibrium as a balance of opposing rates. This state of balance gives the appearance of stasis at the macroscopic level, despite the continuous and dynamic activity at the microscopic level. Phase equilibrium is dynamic equilibrium. We summarize here the characteristics of dynamic equilibrium:
- It occurs only in closed systems.
- At least 2 states of a system must be simultaneously present (coexist).
- It is established only over a finite time interval, required to establish the equality of opposing rates.
- Values of macroscopic observables remain constant with time, but at the molecular level there is constant transit among the states present
- When subjected to a perturbation from outside, it responds to minimize the effect of the perturbation and returns to equilibrium, again over a finite time interval.
- It represents a compromise between the tendencies to minimize potential energy and maximize disorder.

In this, we extend the dynamic equilibrium concept to chemical reactions. We will see that, like a pure substance in a closed container, a chemical reaction in a closed system reaches a state of dynamic equilibrium, a state in which conversion of reactants to products is balanced by conversion of products to reactants. At the macroscopic scale, there is no change in the amount of a particular reactant or product with time. In contrast, at the molecular level there is a continuous reshuffling of atoms.

CHEMICAL EQUILIBRIUM

We introduce chemical equilibrium with a simple reaction, the dimerization of nitrogen dioxide to dinitrogen tetroxide, in equation:

$$2\ NO_2 \rightarrow N_2O_4$$

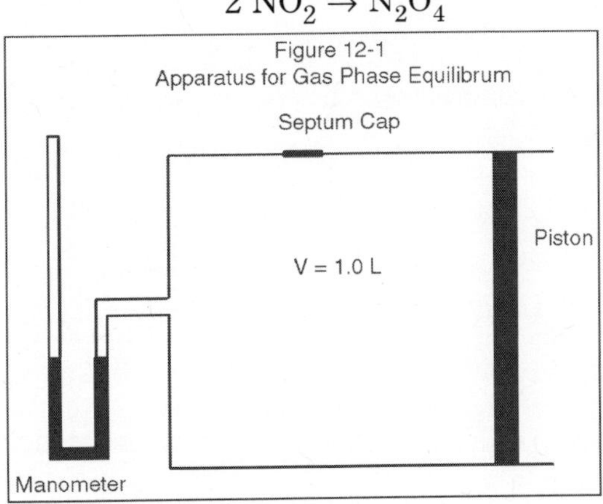

Figure 12-1
Apparatus for Gas Phase Equilibrum

As we have done several times before, we will study the reaction by carrying out several thought experiments designed to reveal the Behaviour of the reaction under various conditions. Throughout the following discussion, bear in mind that the experiments could be (and in fact have been) carried out in the Labouratory (although in a somewhat different mánner than described).

Above figure shows an initially evacuated container of volume 1.0 L, with an attached manometre. The manometre allows the total gas pressure in the container to be measured. The temperature of the container is 298 K. The container is fitted with a piston that can be freely moved or locked in place. The following experiments are sequentially performed.

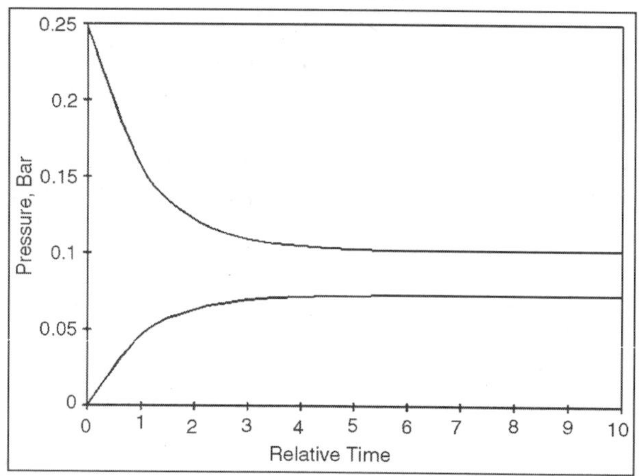

Expt 1. With the piston locked in place, 0.0460 g (0.0100 moles) of NO2 gas (red-brown in Colour, and toxic!) is injected into the container through a septum cap. The manometre quickly registers a pressure of 0.248 atm (188 mm of Hg). Then, over several minutes, the red-brown Colour fades and the pressure of the system falls, at first rapidly, then more and more slowly, until it levels off at a value of 0.175 atm.

Concept

The results of experiments on many different reactions over many years allow us to generalize as follows. For the generic reaction, in which the participating species are gases, the partial pressures at equilibrium obey the simple expressions, independent of the initial pressures of the gases.

$$a\,A + bB \rightarrow dD + fF$$

$$K_{eq} = P_D{}^d P_F{}^f / P_A{}^a P_B{}^b$$

In words, the equation says that if we divide the product of the pressures of the products, each raised to the power of its stoichiometric coefficient in the equation for the reaction, by a similar product of reactant pressures, the result is a constant, K_{eq}, that depends only on temperature. Every chemical reaction, simple or complex, obeys an equilibrium constant expression. This is a very simple result that makes possible calculations involving the direction and extent of reaction. If the value of K_{eq} for a reaction is large, the pressures of products will be large relative to those of reactants at equilibrium — the equilibrium lies to the right. If K_{eq} is small, pressures of products will be small relative to those of reactants at equilibrium — the equilibrium lies to the left.

Our conclusions are not restricted to gas phase reactions. An expression analogous as applied to reactions in solution, except that molarities must be used in place of pressures. In fact, when we recognize that gas pressure is proportional to n/V (molarity) through the ideal gas law, we realise the equivalence of expressions in terms of pressure and molar concentration. Equation below therefore applies to all reactions, in the gas phase, in solution, or in some combination:

$$K_{eq} = [D]^d[F]^f/[A]^a[B]^b$$

The brackets signify molarity. One caution is in order for gas phase reactions. The numerical value and the units of the equilibrium constant depend on whether pressures or molarities are used. If K_{eq} is specified in units of bar, pressures are used in calculations; if in moles/L, concentrations are used. If K_{eq} has no units, either pressure or molarity may be used.

LE CHATELIER'S PRINCIPLE

This principle applies as well to chemical equilibrium as to phase equilibrium. In fact, the principle is a qualitative manifestation of the quantitative characteristics of the equilibrium constant. Three stresses are particularly relevant to chemical equilibrium.

- Change in concentration (pressure) of a participant. For example, consider the reaction

$$N_2 + 3H_2 \rightarrow 2NH_3$$

Assume the reaction to be initially at equilibrium. Suppose that we add some NH_3 to the system. This imposes a stress on the system that will be relieved, according to Le Chatelier, by a shift that relieves the stress — *i.e.*, uses up NH_3. The reaction shifts left until enough NH_3 is used and N_2 and H_2 produced to again make $Q = K_{eq}$. When the new equilibrium is reached, the pressures of all three gases will be larger than they were before the stress. Note that addition of NH_3 to the system does not change the value of K_{eq}. The same value applies before and after the stress.

Molecular View of Chemical Equilibrium

Consider now the general reaction in equations.

$$A + B \rightarrow 2C$$

$$K_{eq} = P_C{}^2/P_AP_B$$

Why do the stoichiometric coefficients appear as exponents in the expression for K_{eq}? Chemical equilibrium is like phase equilibrium; it results from a balance of opposing rates. In this case, the rates involved are those of the reaction in the forward and reverse directions:

At equilibrium, $R_f = R_r$

We assume now, and explore more fully later, that molecules must collide in order to react. This is certainly sensible, because it is not possible for atom rearrangements to occur unless the reacting atom aggregates are in close proximity. The rate of a chemical reaction is thus expected to be proportional to the number of collisions between molecules per unit time: rate of rxn proportional to #collisions/time

For the forward and reverse processes, we write,

$$R_f = k_f\ (\#\text{collisions of A and B})$$
$$R_r = k_r'(\#\text{collisions of C and C})$$

The rate constants, k_f and k_r', are proportionality constants in the same way that k_E and k_C are. If there are N_A molecules of A and N_B of B in a container, a particular A molecule may collide with any of the N_B molecules of B. The total number of ways in which A and B may collide is thus $N_A \times N_B$.

Then,

$$R_f = k_f N_A N_B$$

N_C molecules of C can collide in a total of $N_C(N_C-1)/2$ ways. This arises because each molecule can collide with N_C-1 others. However, the simple product of $N_C$and N_C-1 double counts the collisions (*i.e.*, it includes collisions of molecule x with y and of y with x; however, there can be only one collision between two particular molecules). We divide by 2 to eliminate double counting.

Finally, because N_C is very large, $N_C - 1 = N_C$. It follows that:

$$R_r = k_r' N_C^2/2 = k_r N_C^2$$

At chemical equilibrium, then,

$$k_f N_A N_B = k_r N_C^2$$

Rearranging gives,

$$k_f/k_r = N_C^2/N_A N_B = K_{eq}$$

Since pressure is proportional to number of molecules, this equation is readily translated to pressure terms, giving the equilibrium constant expression in equation. Notice that, as was true for phase equilibrium, the equilibrium constant is a ratio of rate constants for forward and reverse processes. Chemical equilibrium, like phase equilibrium, is dynamic rather than static. Although it appears at the macroscopic level that nothing is happening since concentrations remain constant, both forward and reverse reactions continue at equal rates at the molecular level.

FREE ENERGY AND EQUILIBRIUM

To this point we have encountered two quantities that indicate the extent to which a chemical reaction will proceed: its spontaneity. These are K_{eq} and the standard free energy change of reaction, $DG_R{}^o$. We now explore the relationship between these quantities. The sign of $DG_R{}^o$ indicates the spontaneity or non-spontaneity of a process in which reactants in their standard states are converted to products in their standard states. The magnitude of $DG_R{}^o$ tells us how far the standard state is from the equilibrium state. A process for which $DG_R{}^o$ is negative is spontaneous under standard conditions, and will proceed left to right to reach equilibrium. At equilibrium, the reactants will be present at less than the standard state pressure (1 atm) or concentration (1 M), because they have been used up to some extent in achieving equilibrium. The products will be present at greater than the standard state pressure or concentration, because they have been formed in achieving equilibrium. In the equilibrium state, products will be present in relatively large amounts, reactants will be present in relatively small amounts, consistent with $K_{eq} > 1$.

Thus $DG_R{}^o < 0$ corresponds to $K_{eq} > 1$. Further, for a process with $DG_R{}^o = -50$ kJ, the standard state is much farther away from equilibrium than is the standard state for a process with $DG_R{}^o = -5$ kJ. The first reaction will have to proceed further to reach the equilibrium state than will the second reaction, and will thus have a larger K_{eq}. We can make very similar statements about processes for which $DG_R{}^o > 0$, except in reverse. The larger the magnitude of $DG_R{}^o$, the further the standard state is from the equilibrium position, and the further reaction will have to proceed to reach equilibrium from the standard state. For $DG_R{}^o > 0$, reaction must proceed from right to left towards equilibrium, decreasing the pressures or concentrations of products and increasing those of reactants. Thus $K_{eq} < 1$. The larger the magnitude of $DG_R{}^o$, the smaller K_{eq}. Finally, when $DG_R{}^o = 0$, the process is equally spontaneous in both forward and reverse directions; the standard state is the equilibrium state.

Supplement Applications Consider the Reaction Below

$$2NO_2(g) \rightarrow N_2O_4(g) \quad K_{eq} = 7.0 \text{ atm}^{-1}$$

If you mix 0.12 atm NO_2 and 0.43 atm N_2O_4 in a closed bulb, will any reaction occur? In which direction? (you must provide convincing support for your answers.)

What will the partial pressures of NO_2 and N_2O_4 be once equilibrium is attained? Consider the reaction below:

$$2NOBr(g) \rightarrow 2NO(g) + Br_2(g)$$

Over a time interval between t_0 and t_1, the reaction is at equilibrium at some temperature, with constant partial pressures of reactants and products as At time t_1, the volume of the reaction system is instantly halved. Show on the graphs the immediate effect of the volume decrease on the partial pressures of species, and the manner in which these pressures change as equilibrium is reestablished. Your graphs should be as quantitatively correct as possible. You are studying the equilibrium $P_4(g)$ $2P_2(g)$. In a previously evacuated vessel, you place a quantity of diatomic phosphorus gas.

What is the value of the reaction quotient Q? Which direction will reaction proceed to achieve equilibrium? You allow the system to come to equilibrium and then decide to tamper with it. In which direction would the equilibrium shift if you add $P_4(g)$? Increase the pressure of the system (at const T)? Increase the volume of the system (at const T)? Increase the T of the system (at const P)? What do you know about a system at equilibrium when you know that the value of the equilibrium constant is 0.34 at a particular temperature?

SOLVING EQUILIBRIUM PROBLEMS

Ladder diagrams are a useful tool for evaluating chemical reactivity, usually providing a reasonable approximation of a chemical system's composition at equilibrium. When we need a more exact quantitative description of the equilibrium condition, a ladder diagram may not be sufficient.

In this case we can find an algebraic solution. Perhaps you recall solving equilibrium problems in your earlier coursework in chemistry. We will learn how to set up and solve equilibrium problems. We will start with a simple problem and work towards more complex ones.

A SIMPLE PROBLEM: SOLUBILITY OF $PB(IO_3)_2$ IN WATER

When an insoluble compound such as $Pb(IO_3)_2$ is added to a solution a small portion of the solid dissolves. Equilibrium is achieved when the concentrations of Pb^{2+} and IO^{3-} are sufficient to satisfy the solubility product for $Pb(IO_3)_2$. At equilibrium the solution is saturated with $Pb(IO_3)_2$. How can we determine the concentrations of Pb^{2+} and IO^{3-}, and the solubility of $Pb(IO_3)_2$ in a saturated solution prepared by adding $Pb(IO_3)_2$ to distilled water? We begin by writing the equilibrium reaction

$$Pb(IO_3)_2(s) \rightleftharpoons Pb^{2+}(aq) + 2IO_3^-(aq)$$

and its equilibrium constant

$$K_{sp} = [Pb^{2+}][IO_3^-]^2 = 2.5 \times 10^{-13}$$

As equilibrium is established, two IO^{3-} ions are produced for each ion of Pb^{2+}. If we assume that the molar concentration of Pb^{2+} at equilibrium is *x* then the molar concentration of IO^{3-} is $2x$. To help keep track of these relationships, we can use the following table.

	Pbl2(s)	$\rightleftharpoons$	$Pb^{2+}(aq)$	+	$2IO_3^-(aq)$
Initial concentration	solid		0		0
Change in concentration	solid		+x		+2x
Equilibrium concentration	solid		0 + x = x		0 + 2x = 2x

Substituting the equilibrium concentrations into equation

$$(x)(2x)^2 = 2.5 \times 10^{-13}$$

and solving gives

$$4x^3 = 2.5 \times 10^{-13}$$

$$x = 3.97 \times 10^{-5}$$

The equilibrium concentrations of Pb^{2+} and IO_3^-, therefore, are

$$[Pb^{2+}] = x = 4.0 \times 10^{-5} \text{ M}$$

$$[I^-] = 2x = 7.9 \times 10^{-5} \text{ M}$$

Since one mole of $Pb(IO_3)_2$ contains one mole of Pb^{2+}, the solubility of $Pb(IO_3)_2$ is the same as the concentration of Pb^{2+}; thus, the solubility of $Pb(IO_3)_2$ is 4.0×10^{-5} M.

A More Complex Problem: The Common Ion Effect

– Calculating the solubility of $Pb(IO_3)_2$ in distilled water is a straightforward problem since the dissolution of the solid is the only source of Pb^{2+} or IO^{3-}.

How is the solubility of $Pb(IO_3)_2$ affected if we add $Pb(IO_3)_2$ to a solution of 0.10 M $Pb(NO_3)_2$. Before we set up and solve the problem algebraically, think about the chemistry occurring in this system, and decide whether the solubility of $Pb(IO_3)_2$ will increase, decrease, or remain the same. This is a good habit to develop. Knowing what answers are reasonable will help you spot errors in your calculations and give you more confidence that your solution to a problem is correct. We begin by setting up a table to help us keep track of the concentrations of Pb^{2+} and IO_3 in this system.

	$Pb_{12}(s)$	⇌	$Pb^{2+}(aq)$	+	$2IO_3^-(aq)$
Initial concentration	solid		0.10		0
Change in concentration	solid		+x		+2x
Equilibrium concentration	solid		0.10 + x = x		0 + 2x = 2x

Substititing the equilibrium concentrations into the solubility product expression

$$(0.10 + x)(2x)^2 = 2.5 \times 10^{-13}$$

and multiplying out the terms on the left leaves us with

$$4x^3 + 0.40x^2 = 2.5 \times 10^{-13}$$

This is a more difficult equation to solve than that for the solubility of $Pb(IO_3)_2$ in distilled water, and its solution is not immediately obvious. A rigorous solution to equation can be found using available computer software packages and spreadsheets.

How might we solve equation if we do not have access to a computer? One possibility is that we can apply our understanding of chemistry to simplify the algebra. From Le Châtelier's principle, we expect that the large initial concentration of Pb2+ will significantly decrease the solubility of $Pb(IO_3)_2$. In this case we can reasonably expect the equilibrium concentration of Pb^{2+} to be very close to its initial concentration; thus, the following approximation for the equilibrium concentration of Pb^{2+} seems reasonable

$$[Pb^{2+}] = 0.10 + x \approx 0.10 \text{ M}$$

Substituting into equation

$$(0.10)(2x)^2 = 2.5 \times 10^{-13}$$

and solving for x gives

$$0.40x^2 = 2.5 \times 10^{-13}$$

$$x = 7.91 \times 10^{-7}$$

Before accepting this answer, we check to see if our approximation was reasonable. In this case the approximation $0.10 + x \approx 0.10$ seems reasonable since the difference between the two values is negligible. The equilibrium concentrations of Pb^{2+} and IO^{3-}, therefore, are

$$[Pb^{2+}] = 0.10 + x \approx 0.10 \text{ M}$$

$$[I^-] = 2x = 1.6 \times 10^{-6} \text{ M}$$

The solubility of Pb(IO3)2 is equal to the additional concentration of Pb^{2+} in solution, or $7.9 \approx 10^{-7}$ mol/L. As expected, the solubility of $Pb(IO_3)_2$ decreases in the presence of a solution that already contains one of its ions. This is known as the common ion effect.As outlined in the following example, the process of making and evaluating approximations can be extended if the first approximation leads to an unacceptably large error.

Systematic Approach to Solving Equilibrium Problems

Calculating the solubility of $Pb(IO_3)_2$ in a solution of $Pb(NO_3)_2$ was more complicated than calculating its solubility in distilled water. The necessary calculations, however, were still relatively easy to organize, and the assumption used to simplify the problem was fairly obvious.

This problem was reasonably straightforward because it involved only a single equilibrium reaction, the solubility of $Pb(IO_3)_2$. Calculating the equilibrium composition of a system with multiple equilibrium reactions can become quite complicated. We will learn how to use a systematic approach to setting up and solving equilibrium problems.

As its name implies, a systematic approach involves a series of steps:

- Write all relevant equilibrium reactions and their equilibrium constant expressions.
- Count the number of species whose concentrations appear in the equilibrium constant expressions; these are your unknowns. If the number of unknowns equals the number of equilibrium constant expressions, then you have enough information to solve the problem. If not, additional equations based on the conservation of mass and charge must be written. Continue to add equations until you have the same number of equations as you have unknowns.

- Decide how accurate your final answer needs to be. This decision will influence your evaluation of any assumptions you use to simplify the problem.
- Combine your equations to solve for one unknown (usually the one you are most interested in knowing). Whenever possible, simplify the algebra by making appropriate assumptions.
- When you obtain your final answer, be sure to check your assumptions. If any of your assumptions prove invalid, then return to the previous step and continue solving. The problem is complete when you have an answer that does not violate any of your assumptions.

Besides equilibrium constant equations, two other types of equations are used in the systematic approach to solving equilibrium problems. The first of these is a mass balance equation, which is simply a statement of the conservation of matter. In a solution of a monoprotic weak acid, for example, the combined concentrations of the conjugate weak acid, HA, and the conjugate weak base, A^-, must equal the weak acid's initial concentration, *C*HA. The second type of equation is a charge balance equation. A charge balance equation is a statement of solution electroneutrality. Total positive charge from cations = total negative charge from anions Mathematically, the charge balance expression is expressed as

$$\sum_{i=1}^{n} \left|(z^+)_i\right| [M^{z+}]_i = \sum_{j=1}^{m} \left|(z^-)_j\right| \times [A^{z-}]_j$$

where $[M^{z+}]_i$ and $[A^{z-}]_j$ are, respectively, the concentrations of the *i*th cation and the *j*th anion, and $(z^+)_i$ and $(z^-)_j$ are the charges of the *i*th cation and the *j*th anion. Note that the concentration terms are multiplied by the absolute values of each ion's charge, since electroneutrality is a conservation of charge, not concentration. Every ion in solution, even those not involved in any equilibrium reactions, must be included in the charge balance equation. The charge balance equation for an aqueous solution of $Ca(NO_3)_2$ is

$$2 \times [Ca^{2+}] + [H_3O^+] = [OH^-] + [NO_3^-]$$

Note that the concentration of Ca^{2+} is multiplied by 2, and that the concentrations of H_3O^+ and OH^- are also included. Charge balance equations must be written carefully since every ion in solution must be included. This presents a problem when the concentration of one ion in solution is held constant by a reagent of unspecified composition. For example, in many situations pH is held constant using a buffer. If the composition of the buffer is not specified, then a charge balance equation cannot be written.

pH of a Monoprotic Weak Acid

To illustrate the systematic approach, let us calculate the pH of 1.0 M HF. Two equilbria affect the pH of this system. The first, and most obvious, is the acid dissociation reaction for HF

$$HF(aq) + H_2O(\ell) \rightleftharpoons H_3O^+(aq) + F^-(aq)$$

for which the equilibrium constant expression is

$$K_a = \frac{[H_3O^+][F^-]}{[HF]} = 6.8 \times 10^{-4}$$

The second equilibrium reaction is the dissociation of water, which is an obvious yet easily disregarded reaction

$$2H_2O(\ell) \rightleftharpoons H_3O^+(aq) + OH^-(aq)$$

$$K_w = [H_3O^+][OH^-] = 1.00 \times 10^{-14}$$

Counting unknowns, we find four ([HF], [F^-], [H_3O^+], and [OH^-]). To solve this problem, therefore, we need to write two additional equations involving these unknowns. These equations are a mass balance equation

$$C_{HF} = [HF] + [F^-]$$

and a charge balance equation

$$[H_3O^+] = [F^-] + [OH^-]$$

We now have four equations and four unknowns ([HF], [F^-], [H_3O^+], and [OH^-]) and are ready to solve the problem. Before doing so, however, we will simplify the algebra by making two reasonable assumptions. First, since HF is a weak acid, we expect the solution to be acidic; thus it is reasonable to assume that

$$[H_3O^+] >> [OH^-]$$

simplifkying the charge balance equation to

$$[H_3O^+] = [F^-]$$

Second, since HF is a weak acid we expect that very little dissociation occurs, and

$$[HF] >> [F^-]$$

Thus, the mass balance equation simplifies to

$$C_{HF} = [HF]$$

For this exercise we will accept our assumptions if the error introduced by each assumption is less than ±5 per cent.

Substituting equations into the equilibrium constant expression for the dissociation of HF and solving for the concentration of H_3O^+ gives us gives us

$$K_a = \frac{[H_3O^+][H_3O^+]}{C_{HF}}$$

$$[H_3O^+] = \sqrt{K_a C_{HF}} = \sqrt{(6.8 \times 10^{-4})(1.0)} = 2.6 \times 10^{-2}\,M$$

Before accepting this answer, we must verify that our assumptions are acceptable. The first assumption was that the $[OH^-]$ is significantly smaller than the $[H_3O^+]$. To calculate the concentration of OH^- we use the K_w expression (6.36)

$$[OH^-] = \frac{K_w}{[H_3O^+]} = \frac{1.00 \times 10^{-14}}{2.6 \times 10^{-2}} = 3.8 \times 10^{-13} M$$

Clearly this assumption is reasonable. The second assumption was that the $[F^-]$ is significantly smaller than the [HF]. From equation we have

$$[F^-] = 2.6 \times 10^{-2} \text{ M}$$

Since the $[F^-]$ is 2.6 per cent of CHF, this assumption is also within our limit that the error be no more than ±5 per cent. Accepting our solution for the concentration of H_3O^+, we find that the pH of 1.0 M HF is 1.59.

How does the result of this calculation change if we require our assumptions to have an error of less than ±1 per cent. In this case we can no longer assume that [HF] >> $[F^-]$. Solving the mass balance equation for [HF]

$$[HF] = C_{HF} - [F^-]$$

and substituting into the K_a expression along with equation gives

$$K_a = \frac{[H_3O^+]^2}{C_{HF} - [H_3O^+]}$$

Rearranging leaves us with a quadratic equation

$$[H_3O^+]^2 = K_a C_{HF} - K_a[H_3O^+]$$

$$[H_3O^+]^2 + K_a[H_3O^+] - K_a C_{HF} = 0$$

which we solve using the quadratic formula

$$x = \frac{-b \pm \sqrt{b^2 - 4ac}}{2a}$$

where *a, b,* and *c* are the coefficients in the quadratic equation $ax^2 + bx + c = 0$. Solving the quadratic formula gives two roots, only one of which has any chemical significance. For our problem the quadratic formula gives roots of

$$x = \frac{-6.8 \times 10^{-4} \pm \sqrt{(6.8 \times 10^{-4})^2 - (4)(1)(-6.8 \times 10^{-4})(1.0)}}{2(1)}$$

$$= \frac{-6.8 \times 10^{-4} \pm 5.22 \times 10^{-2}}{2}$$

$$= 2.57 \times 10^{-2} \text{ or } -2.63 \times 10^{-2}$$

Only the positive root has any chemical significance since the negative root implies that the concentration of H_3O^+ is negative. Thus, the $[H_3O^+]$ is 2.6×10^{-2} M, and the pH to two significant figures is still 1.59. This same approach can be extended to find the pH of a monoprotic weak base, replacing K_a with K_b, CHF with the weak base's concentration, and solving for the $[OH^-]$ in place of $[H_3O^+]$.

pH of a Polyprotic Acid or Base

A more challenging problem is to find the pH of a solution prepared from a polyprotic acid or one of its conjugate species. As an example, we will use the amino acid alanine whose structure and acid dissociation constants are shown in figure. pH of 0.10 M H_2L^+ Alanine hydrochloride is a salt consisting of the diprotic weak acid H_2L^+ and Cl^-. Because H_2L^+ has two acid dissociation reactions, a complete systematic solution to this problem will be more complicated than that for a monoprotic weak acid. Using a ladder diagram can help us simplify the problem. Since the areas of predominance for H_2L^+ and L^- are widely separated, we can assume that any solution containing an appreciable quantity of H_2L^+ will contain essentially no L^-. In this case, HL is such a weak acid that H_2L^+ behaves as if it were a monoprotic weak acid.

To find the pH of 0.10 M H_2L^+, we assume that

$$[H_3O^+] >> [OH^-]$$

Because H_2L^+ is a relatively strong weak acid, we cannot simplify the problem further, leaving us with

$$K_a = \frac{[H_3O^+]^2}{C_{H_2L^+} - [H_3O^+]}$$

Solving the resulting quadratic equation gives the $[H_3O^+]$ as 1.91×10^{-2} M or a pH of 1.72. Our assumption that $[H_3O^+]$ is significantly greater than $[OH^-]$ is acceptable.

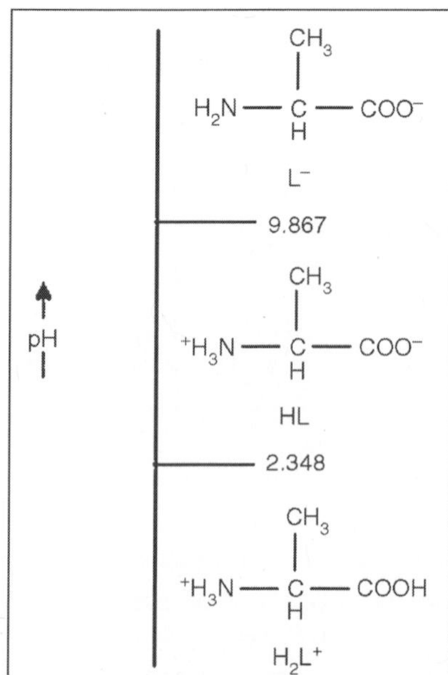

Fig. Ladder Diagram for the Amino Acid Alanine.

The alaninate ion is a diprotic weak base, but using the ladder diagram as a guide shows us that we can treat it as if it were a monoprotic weak base. we find that the pH of 0.10 M alaninate is 11.42.

pH of 0.1 M HL Finding the pH of a solution of alanine is more complicated than that for H_2L^+ or L^- because we must consider two equilibrium reactions involving HL. Alanine is an amphiprotic species, behaving as an acid

$$HL(aq) + H_2O(\ell) \rightleftharpoons H_3O^+(aq) + L^-(aq)$$

and a base

$$HL(aq) + H_2O(\ell) \rightleftharpoons OH^-(aq) + H_2L^+(aq)$$

As always, we must also consider the dissociation of water

$$2H_2O(\ell) \rightleftharpoons H_3O^+(aq) + OH^-(aq)$$

This leaves us with five unknowns $([H_2L^+],[HL],[L^-],[H_3O^+],$ and $[OH^-])$, for which we need five equations. These equations are Ka_2 and Kb_2 for HL,

$$K_{a2} = \frac{[H_3O^+][L^-]}{[HL]}$$

$$K_{b2} = \frac{K_w}{K_{al}} = \frac{[OH^-][H_2L^+]}{[HL]}$$

the K_w equation,

$$K_w = [H_3O^+][OH^-]$$

a mass balance equation on HL,

$$C_{HL} = [H_2L^+] + [HL] + [L^-]$$

and a charge balance equation

$$[H_2L^+] + [H_3O^+] = [OH^-] + [L^-]$$

From the ladder diagram it appears that we may safely assume that the concentrations of H_2L^+ and L^- are significantly smaller than that for HL, allowing us to simplify the mass balance equation to

$$C_{HL} = [HL]$$

Next we solve K_{b2} for $[H_2L^+]$

$$[H_2L^+] = \frac{K_w[HL]}{K_{al}[OH^-]} = \frac{[HL][H_3O^+]}{K_{al}} = \frac{C_{HL}[H_3O^+]}{K_{al}}$$

and K_{a2} for $[L^-]$

sabstituting these equations, along with the equation for K_w, into the charge balance equation gives us

$$\frac{C_{HL}[H_3O^+]}{K_{al}}+[H_3O^+]=\frac{K_W}{[H_3O^+]}+\frac{K_{a2}C_{HL}}{[H_3O^+]}$$

which simplifies to

$$[H_3O^+]\left(\frac{C_{HL}}{K_{al}}+1\right)=\frac{1}{[H_3O^+]}(K_W+K_{a2}C_{HL})$$

$$[H_3O^+]^2=\frac{K_{a2}C_{HL}+K_W}{(C_{HL}/K_{al})+1}$$

$$[H_3O^+]=\sqrt{\frac{K_{a1}K_{a2}C_{HL}+K_{al}K_W}{C_{HL}+K_{al}}}$$

We can simplify this equation further if $K_{al}K_W << K_{al}K_{a2}C_{HL}$, and if $K_{al} << C_{HL}$, giving

$$[H_3O^+]=\sqrt{K_{a1}K_{a2}}$$

For a solution of 0.10 M alanine, the $[H_3O^+]$ is

$$[H_3O^+]=\sqrt{(4.487\times10^{-3})(1.358\times10^{-10})}=7.807\times10^{-7}\,M$$

Verifying that the assumptions are acceptable is left as an exercise.

Triprotic Acids and Bases, and Beyond The treatment of a diprotic acid or base is easily extended to acids and bases having three or more acid–base sites.

For a triprotic weak acid such as H_3PO_4, for example, we can treat H_3PO_4 as if it was a monoprotic weak acid, $H_2PO_4{}^-$ and $HPO_4{}^{2-}$ as if they were intermediate forms of diprotic weak acids, and $PO_4{}^{3-}$ as if it was a monoprotic weak base.

WRITE EQUILIBRIUM CONSTANT EXPRESSIONS

PRESSURES CAN EXPRESS CONCENTRATIONS

Although we commonly write equilibrium quotients and equilibrium constants in terms of molar concentrations, any concentration-like term can be used, including mole fraction and molality. Sometimes the symbols K_c, K_x, and K_m are used to denote these forms of the equilibrium constant. Bear in mind that the *numerical values* of K's and Q's expressed in these different ways will not generally be the same.

Most of the equilibria we deal with in this course occur in liquid solutions and gaseous mixtures. We can express K_c values in terms of moles per litre for both, but when dealing with gases it is often more convenient to use partial pressures. These two measures of concentration are of course directly proportional:

$$C = \frac{n}{V} = \frac{\frac{PV}{RT}}{V} = \frac{P}{RT}$$

so for a reaction A*(g)* → B*(g)* we can write the equilibrium constant as,

$$Kp = \frac{P_B}{P_A}$$

All of these forms of the equilibrium constant are only approximately correct, working best at low concentrations or pressures. The only equilibrium constant that is truly "constant" (except that it still varies with the temperature!) is expressed in terms of *activities*, which you can think of as "effective concentrations" that allow for interactions between molecules. In practice, this distinction only becomes important for equilibria involving gases at very high pressures (such as are often encountered in chemical engineering) and in ionic solutions more concentrated than about 0.001 *M*. We will not deal much with activities in this course.

For a reaction such as $CO_2(g) + OH^-(aq) \rightarrow HCO_3^-(aq)$ that involves both gaseous and dissolved components, a "hybrid" equilibrium constant is commonly used:

$$K = \frac{\left[HCO_3^-\right]}{P_{CO2}\left[OH^-\right]}$$

Clearly, it is essential to be sure of the units when you see an equilibrium constant represented simply by "*K*".

Converting between K_p and K_c

It is sometimes necessary to convert between equilibrium constants expressed in different units. The most common case involves pressure- and concentration equilibrium constants. The ideal gas law relates the partial pressure of a gas to the number of moles and its volume:

$$PV = nRT$$

Concentrations are expressed in moles/unit volume *n/V*, so by rearranging the above equation we obtain the explicit relation of pressure to concentration:

$$P = (n/V)RT$$

Conversely, $c = (n/V) = P/RT$

... so a concentration [A] can be expressed as P_A(RT).

For a reaction of the form 2 A = B + 3 C, we can write

$$K_p = \frac{(P_C)^2}{(P_A)\ (P_B)^3} = \frac{([C]RT)^2}{([A]RT)\ ([B]RT)^3} = \underbrace{\boxed{\frac{[C]^2}{[A]\ [B]^3}}}_{K_c}(RT)^{\overset{\Delta n_g}{\boxed{-2}}}$$

Assuming that all of the components are gases, the difference (moles of gas in products) – (moles of gas in reactants) = Δn_g is given by

$$K_p = K_c\left(RT\right)^{\Delta ng}$$

UNCHANGING CONCENTRATIONS

Substances whose concentrations undergo no significant change in a chemical reaction do not appear in equilibrium constant expressions. How can the concentration of a reactant or product *not*change when a reaction involving that substance takes place? There are two general cases to consider.

The Substance is also the Solvent

This happens all the time in acid-base chemistry. Thus for the hydrolysis of the cyanide ion

$$CN^{-} + H_2O \rightarrow HCN + OH^{-}$$

we write,

$$\frac{\left[HCN\right]\left[OH^{-}\right]}{\left[CN^{-}\right]}$$

in which no $[H_2O]$ term appears. The justification for this omission is that water is both the solvent and reactant, but only the tiny portion that acts as a reactant would ordinarly go in the equilibrium expression. The amount of water consumed in the reaction is so minute (because K is very small) that any change in the concentration of H_2O from that of pure water (55.6 mol L^{-1}) will be negligible. Similarly, for the "dissociation" of water $H_2O = H^+ + OH^-$the equilibrium constant is expressed as the "ion product"$K_w = [H^+][OH^-]$.

But... be careful about throwing away H_2O whenever you see it. In the esterification reaction,

$$CH_3COOH + C_2H_5OH \rightarrow CH_3COOC_2H_5 + H_2O$$

that we discussed in a previous section, a $[H_2O]$ term must be present in the equilibrium expression if the reaction is assumed to be between the two liquids acetic acid and ethanol. If, on the other hand, the reaction takes place between a dilute aqueous solution of the acid and the alcohol, then the $[H_2O]$ term would not be included.

The Substance is a Solid or a Pure Liquid Phase

This is most frequently seen in solubility equilibria, but there are many other reactions in which solids are directly involved:

$$CaF_2(s) \rightarrow Ca^{2+}(aq) + 2F^{\rightarrow}(aq)$$

$$Fe_3O_4(s) + 4\ H_2(g) \rightarrow 4\ H_2O(g) + 3Fe(s)$$

These are *heterogeneous reactions* (meaning reactions in which some components are in different phases), and the argument here is that concentration is only meaningful when applied to a substance within a single phase.

Thus the term $[CaF_2]$ would refer to the "concentration of calcium fluoride within the solid CaF_2", which is a constant depending on the molar mass of CaF_2 and the density of that solid. The concentrations of the two ions will be independent of the quantity of solid CaF_2 in contact with the water; in other words, the system can be in equilibrium as long as any CaF_2 at all is present.

Throwing out the constant-concentration terms can lead to some rather sparse-looking equilibrium expressions. For example, the equilibrium expression for each of the processes shown in the following table consists solely of a single term involving the partial pressure of a gas:

1)	$CaCO_3(s) \rightarrow CaO(s) + CO_2(g)$	$K_p = P_{CO2}$	Thermal decomposition of limestone, a first step in the manufacture of cement.
2)	$Na_2SO_4 \cdot 10\ H_2O(s)$ $Na_2SO_4(s) + 10\ H_2O(g)$	$K_p = P_{H2O}{}^{10}$	Sodium sulfate decahydrate is a solid in which H_2O molecules (*"waters of hydration"*) are incorporated into the crystal structure.)
3)	$I_2(s) \rightarrow I_2(g)$	$K_p = P_{I2}$	sublimation of solid iodine; this is the source of the purple vapor you can see above solid iodine in a closed container.
4)	$H_2O(l)$! $H_2O(g)$	$K_p = P_{H2O}$	Vaporization of water. When the partial pressure of water vapor in the air is equal to *K*, the relative humidity is 100 per cent.

The last two processes 3 and 4 represent changes of state (*phase changes*) which can be treated exactly the same as chemical reactions.

In each of the heterogeneous processes shown in the table, the reactants and products can be in equilibrium (that is, permanently coexist) only when the partial pressure of the gaseous product has the value consistent with the indicated K_p. Bear in mind also that these K_p 's all increase with the temperature.

VALUES OF EQUILIBRIUM CONSTANTS

The numerical value of a quantity in terms of what it means in a practical sense is an essential part of developing a working understanding of Chemistry. This is particularly the case for equilibrium constants, whose values span the entire range of the positive numbers.

Although there is no explicit rule, for most practical purposes you can say that equilibrium constants within the range of roughly 0.01 to 100 indicate that a chemically significant amount of all components of the reaction system will be present in an equilibrium mixture and that the reaction will be *incomplete* or *"reversible"*.

As an equilibrium constant approaches the limits of zero or infinity, the reaction can be increasingly characterized as a one-way process; we say it is *"complete"* or *"irreversible"*. The latter term must of course not be taken literally;

the Le Châtelier principle still applies (especially insofar as temperature is concerned), but addition or removal of reactants or products will have less effect.

Although it is by no means a general rule, it frequently happens that reactions having very large equilibrium constants are ***kinetically hindered***, often to the extent that the reaction essentially does not take place.

The examples in the following table are intended to show that numbers (values of *K*), no matter how dull they may look, do have practical consequences!

Reaction	*K*	Remarks
$N_2(g) + O_2(g) \rightarrow 2\ NO(g)$	5×10^{-31} at 25°C, 0.0013 at 2100°C	These two very different values of *K*illustrate very nicely why reducing combustion-chamber temperatures in automobile engines is environmentally friendly.
$3\ H_2(g) + N_2(g) \rightarrow 2\ NH_3(g)$	7×10^{5} at 25°C, 56 at 1300°C	See the discussion of this reaction in the section on the Haber process.
$H_2(g) \rightarrow 2\ H(g)$	10^{-36} at 25°C, 6×10^{-5} at 5000°	Dissociation of any stable molecule into its atoms is endothermic. This means that all molecules will decompose at sufficiently high temperatures.
$H_2O(g) \rightarrow H_2(g) + \frac{1}{2}\ O_2(g)$	8×10^{-41} at 25°C	You won't find water a very good source of oxygen gas at ordinary temperatures!
$CH_3COOH(l) \rightarrow 2\ H_2O(l) + 2\ C(s)$	$K_c = 10^{13}$ at 25°C	This tells us that acetic acid has a great tendency to decompose to carbon, but nobody has ever found graphite (or diamonds!) forming in a bottle of vinegar. A good example of a super kinetically-hindered reaction!

Units Equilibrium Constants

The equilibrium expression for the synthesis of ammonia,

$$3\ H_2(g) + N_2(g) \rightarrow 2\ NH_3(g)$$

can be expressed as,

$$K_p = \frac{P^2_{NH_3}}{P_{N_2} P^3_{H_2}}$$

or,

$$K_c = \frac{[NH_3]^2}{[N_2][N_2]^3}$$

so K_p and Q_p for this process would appear to have units of atm^{-1}, and K_c and Q_c would be expressed in mol^{-2} L^2. And yet these quantities are often represented as being dimensionless. Which is correct? The answer is that both forms are acceptable.

There are some situations (which you will encounter later) in which *K*'s must be considered dimensionless, but in simply quoting the value of an equilibrium constant it is permissible to include the units, and this may even be useful in order to remove any doubt about the units of the individual terms in equilibrium expressions containing both pressure and concentration terms. In carrying out your own calculations, however, there is rarely any real need to show the units.

Strictly speaking, equilibrium expressions do not have units because the concentration or pressure terms that go into them are really *ratios* having the forms (n mol L^{-1})/(1 mol L^{-1}) or (n atm)/(1 atm) in which the unit quantity in the denominator refers to the *standard state* of the substance; thus the units always cancel out. (But first-year students are not expected to know this!)

For substances that are liquids or solids, the standard state is just the concentration of the substance within the liquid or solid, so for something like CaF*(s)*, the term going into the equilibrium expression is $[CaF_2]/[CaF_2]$ which cancels to unity; this is the reason we don't need to include terms for solid or liquid phases in equilibrium expressions. The subject of standard states would take us beyond where we need to be at this point in the course, so we will simply say that the concept is made necessary by the fact that energy, which ultimately governs chemical change, is always relative to some arbitrarily defined zero value which, for chemical substances, is the standard state.

It is important to remember that an equilibrium quotient or constant is always tied to a specific chemical equation, and if we write the equation in reverse or multiply its coefficients by a common factor, the value of *Q* or *K* will change.

The rules are very simple:

- Writing the equation in reverse will invert the equilibrium expression;
- Multiplying the coefficients by a common factor will raise *Q* or *K* to the corresponding power.

Here are some of the possibilities for the reaction involving the equilibrium between gaseous water and its elements:

$2\,H_2 + O_2 \rightarrow 2\,H_2O$	$10\,H_2 + 5\,O_2 \rightarrow 10\,H_2O$	$H_2 + ½\,O_2 \rightarrow H_2O$	$H_2O \rightarrow H_2 + ½\,O_2$
$K_p = \dfrac{P_{H_2O}^2}{P_{H_2}^2 P_{O_2}}$	$K_p = \dfrac{P_{H_2O}^{10}}{P_{H_2}^{10} P_{O_2}^5}$	$K_p = \dfrac{P_{H_2O}}{P_{H_2} P_{O_2}^{1/2}}$	$K_p = \dfrac{P_{H_2} P_{O_2}^{1/2}}{P_{H_2O}}$

Heterogeneous Reactions

Heterogeneous reactions are those involving more than one phase. Some examples:

$Fe(s) + O_2(g) \rightarrow FeO_2(s)$	air-oxidation of metallic iron (formation of rust)
$CaF_2(s) \rightarrow Ca(aq) + F^+(aq)$	dissolution of calcium fluoride in water
$H_2O(s) \rightarrow H_2O(g)$	sublimation of ice (a *phase change*)
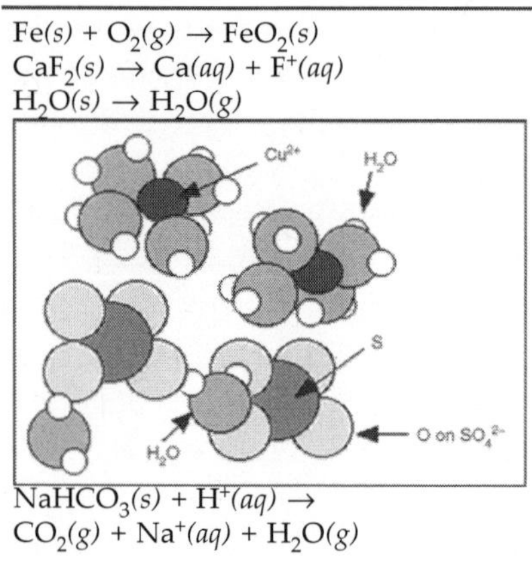	
$NaHCO_3(s) + H^+(aq) \rightarrow CO_2(g) + Na^+(aq) + H_2O(g)$	formation of carbon dioxide gas from sodium bicarbonate when water is added to baking powder (the hydrogen ions come from tartaric acid, the other component of baking powder.)

The Vapor Pressure of Solid Hydrates

A particularly interesting type of heterogeneous reaction is one in which a solid is in equilibrium with a gas. The sublimation of ice illustrated in the above table is a very common example. The equilibrium constant for this process is simply the partial pressure of water vapor in equilibrium with the solid— the *vapor pressure* of the ice.

Many common inorganic salts form solids which incorporate water molecules into their crystal structures. These water molecules are usually held rather loosely and can escape as water vapor. Copper(II) sulfate, for example forms a pentahydrate in which four of the water molecules are coordinated to the Cu^{2+} ion while the fifth is hydrogen-bonded to SO_4^{2-}. This latter water is more tightly bound, so that the pentahydrate loses water in two stages on heating:

$$CuSO_4 5H_2O \xrightarrow{140°C} CuSO_4 5H_2O \xrightarrow{400°C} CuSO_4$$

These dehydration steps are carried out at the temperatures indicated above, but at any temperature, some moisture can escape from a hydrate. For the complete dehydration of the pentahydrate we can define an equilibrium constant:

$$CuSO_4 \cdot 5H_2O(s) \rightarrow CuSO_4(s) + 5\ H_2O(g)\ K_p = 1.14\times10^{10}$$

The vapor pressure of the hydrate (for this reaction) is the partial pressure of water vapor at which the two solids can coexist indefinitely; its value is $K_p^{1/5}$ atm. If a hydrate is exposed to air in which the partial pressure of water vapor is less than its vapor pressure, the reaction will proceed to the right and the hydrate will lose moisture. Vapor pressures always increase with temperature, so any of these compounds can be dehydrated by heating.

Loss of water usually causes a breakdown in the structure of the crystal; this is commonly seen with sodium sulfate, whose vapor pressure is sufficiently large that it can exceed the partial pressure of water vapor in the air when the relative humidity is low. What one sees is that the well-formed crystals of the decahydrate undergo deterioration into a powdery form, a phenomenon known as *efflorescence*.

When a solid is able to take up moisture from the air, it is described as *hygroscopic*. A small number of anhydrous solids that have low vapor pressures not only take up atmospheric moisture on even the driest of days, but will become wet as water molecules are adsorbed onto their surfaces; this is most commonly observed with sodium hydroxide and calcium chloride. With these solids, the concentrated solution that results continues to draw in water from the air so that the entire crystal eventually dissolves into a puddle of its own making; solids exhibiting this behaviour are said to be *deliquescent*.

Name	**Formula**	**Vapor Pressure, Torr**	
		25°C	**30°C**
sodium sulfate decahydrate	$Na_2SO_4 \cdot 10H_2O$	19.2	25.3
copper(II) sulfate pentahydrate	$CuSO_4 \cdot 5H_2O$	7.8	12.5
calcium chloride monohydrate	$CaCl_2 \cdot H_2O$	3.1	5.1
(water)	H_2O	23.5	31.6

EQUILIBRIUM CONSTANT

For the hypothetical chemical reaction,

$$a\,A + b\,B \rightleftarrows c\,C + d\,D$$

the equilibrium constant is defined as,

$$K_c = \frac{[C]c[D]d}{[A]a[B]b}$$

The notation [A] signifies the molar concentration of species A. An alternative expression for the equilibrium constant involves partial pressures.

$$K_p = \frac{P_C c\ P_D d}{P_A a\ P_B b}$$

Note that the expression for the equilibrium constant includes only solutes and gases; pure solids and liquids do not appear in the expression. For example, the equilibrium expression for the reaction,

$$CaH_2\ (s) + 2\ H_2O\ (g) \rightleftarrows Ca(OH)_2\ (s) + 2\ H_2\ (g)$$

is,

$$K_c = \frac{[H_2]2}{[H_2o]2}$$

Observe that the gas-phase species H_2O and H_2 appear in the expression but the solids CaH_2 and $Ca(OH)_2$ do not appear. The equilibrium constant is most readily determined by allowing a reaction to reach equilibrium, measuring the concentrations of the various solution-phase or gas-phase reactants and products, and substituting these values into the Law of Mass Action.

Calculations Involving Reaction Stoichiometry

The one species existed in the gas phase; all other species were solids. The pressure of the system could thus be used to directly calculate the equilibrium constant, because the system pressure was the same as the partial pressure of the gas-phase species. For most reactions, however, the calculations are more complicated.

Consider the following reactions, which is an important atmospheric reaction.

$$O_2\ (g) + 2\ SO_2\ (g) \rightleftarrows 2\ SO_3\ (g)$$

Suppose the system initially contains only oxygen and sulfur dioxide. As the reaction progresses, sulfur trioxide is formed and oxygen and sulfur dioxide are consumed. The pressure of the system is the sum of the partial pressures of the components of the gas, and each of these partial pressures differs from the initial values.

$$P = P_{O2} + P_{SO2} + P_{SO3}$$

How can one take these changes into account and determine each individual partial pressure from the system pressure?

A useful concept for this purpose is the extent of reaction, which will be given the symbol x for this experiment. In reality the reaction occurs on the molecular scale, with individual molecules of oxygen, sulfur dioxide, and sulfur trioxide being involved. Chemists prefer to think in terms of moles of material, however. Each unit of x (the extent of reaction) corresponds to the number of molecules of each reactant and product indicated by the respective stoichiometric coefficients. Thus when $x = 1$, one mole of oxygen and two moles of sulfur dioxide have been consumed and two moles of sulfur trioxide have been formed. The sense of the change can be reversed. When $x = -1$, one mole of oxygen and two moles of sulfur dioxide have been formed and two moles of sulfur trioxide have been consumed.

The extent of reaction allows the change in amount of one species to be related to the change in the amount of a different species. The most straightforward way to illustrate this relationship is through the use of a table such as that shown below.

	O_2	SO_2	SO_3
Initial	P_1	P_2	P_3
Change	$-x$	$-2x$	$+2x$
Equilibrium	$P_1 - x$	$P_2 - 2x$	$P_3 + 2x$

Strictly speaking, the table should be constructed in terms of moles of each species. The equilibrium expression, however, requires partial pressures or molar concentrations. So long as each species exists in the same phase (and thus the volume is the same for each species), the moles can be replaces by partial pressure or molar concentration. In this example the extent of reaction is expressed in terms of partial pressure.

The first row of the table contains the initial partial pressures of each species. These values are called "analytical partial pressures" and are the partial pressures of each species the experimenter put in the system. The second row simply reflects the stoichiometry of the reaction and employs the extent of reaction to show how the change in amount of one reactant or product is linked to the change in the amounts of the other reactants and products.

The bottom row is simply the sum of the first two rows and contains the actual equilibrium partial pressures of each species. In this context, the extent of reaction, x, is the value necessary to reach equilibrium. At equilibrium, the values in the bottom row of the table may be substituted into the equilibrium expression to evaluate K_P.

$$P_{O2} = P_1 - x$$

$$P_{SO2} = P_2 - 2x$$

$$P_{SO3} = P_3 + 2x$$

$$K_p = \frac{P_{SO_3^2}}{P_{O2} P_{SO_2^2}} = \frac{(P_3 + 2x)^2}{(P_1 - x)(P_2 - 2x)^2}$$

Can you calculate the value of x if you are given the pressure P and the three initial partial pressures P_1, P_2, and P_3?

Can you calculate the individual partial pressures and the equilibrium constant, if you are given x and the three initial partial pressures P_1, P_2, and P_3?

Will this strategy for determining K_P from P work for the reaction shown below?

$$CO\ (g) + H_2O\ (g) \rightleftarrows CO_2\ (g) + H_2\ (g)$$

Le Châtelier's Principle

The first step in the conversion of coal into gasoline is the steam reforming reaction.

$$C\ (s) + H_2O\ (g) \rightleftarrows CO\ (g) + H_2\ (g)$$

Suppose that 0.0500 mole each of carbon, water, carbon monoxide, and hydrogen are placed in a 10.0 L glass bulb, which is then heated to 1000 K and the steam reforming reaction is allowed to reach equilibrium. This system is depicted in the experiment shown below.

The question to be answered in this experiment is:

How does a change in the analytical amount of a reactant or product affect the equilibrium amounts of reactants and products in the system?

The distinction between analytical amount and equilibrium amount is especially important in this experiment. The analytical amount of material, such as carbon, is the amount physically added to the system. The equilibrium amount is the amount of material that actually exists in the system at equilibrium. The two amounts are generally not the same, because the reaction will consume or produce the material in order to reach equilibrium.

In this experiment, for example, the analytical amount of each reactant and product is 0.0500 mole, but the equilibrium amounts of carbon and water are 0.0093 mole and the equilibrium amounts of carbon monoxide and hydrogen are 0.0907 mole. One consequence of the dynamic nature of the system is that the amount of a compound placed in a system is rarely the amount that actually exists in the system.

Effect of a Change in Temperature

When a stress is brought to bear on a system at equilibrium, the system will react in the direction that serves to relieve the stress.

A change in the analytical amount of a reactant or product produces a stoichiometric stress on the system. A change in temperature also constitutes a stress on the system, but what is the nature of this stress? Unlike changes in the analytical amounts of reactants and products, a change in temperature produces a change in the equilibrium constant itself. To understand the effect of a temperature change on the value of the equilibrium constant, it is necessary to understand how temperature affects a reaction.

An increase in temperature is the result of the flow of heat into the system. Conversely a flow of heat out of the system reduces the temperature. A reaction itself can be a source or sink for heat. The heat flow for a reaction is characterised by the Enthalpy of Reaction or the Heat of Reaction, ΔH_{rxn}.

If $\Delta H_{rxn} > 0$, the reaction is said to be endothermic, which means that the reaction draws heat from its surroundings as the reaction occurs.

If $\Delta H_{rxn} < 0$, the reaction is said to be exothermic, which means that the reaction releases heat into its surroundings as the reaction occurs.

Suppose an *exothermic* reaction is at equilibrium, and then the temperature of the system is increased. The increase in temperature corresponds to introducing heat into the system. This influx of heat pushes the system away from equilibrium and nature restores equilibrium by removing some of this additional heat.

Because the reaction is exothermic, the reaction *produces* heat when it proceeds in the forward direction. Le Châtelier's Principle states that the system will react to *remove* the added heat, thus the reaction must proceed in the reverse direction, converting products back to reactants. Conversely if the temperature of the system were decreased (heat removed from the system), the system would react in a direction that opposed the removal of the heat. The forward reaction would thus occur to release heat in an attempt to offset the heat that was removed from the system. Bear in mind that the system is being held at a specific temperature, so the net reaction (which may produce or release heat) does not actually change the temperature of the system. That extra heat is removed or replaced as necessary to maintain the temperature.

The steam reforming reaction studied in the previous experiment is an example of an endothermic reaction.

$$C\ (s) + H_2O\ (g) \rightleftarrows CO\ (g) + H_2\ (g)$$

$$\Delta H_{rxn} = +\ 131.3\ \text{kJ mole}^{-1}$$

As an endothermic reaction, the steam reforming reaction draws heat out of the system as it occurs. Although it is not strictly correct to include *heat* in a chemical equation, which shows stoichiometric relations between physical substances, one could loosely represent the endothermic nature of the reaction by writing *heat* as a reactant:

$$\textit{heat} + C\ (s) + H_2O\ (g) \rightleftarrows CO\ (g) + H_2\ (g)$$

The advantage of this construct is that it permits the appplication of Le Châtelier's Principle through the same reasoning as employed when the analytical amount of a substance is changed. The temperature-dependence of the equilibrium constant, K_P in this case, is determined by the enthalpy of reaction.

EQUILIBRIUM CONSTANT EXPRESSIONS

PRESSURES CAN EXPRESS CONCENTRATIONS

Although we commonly write equilibrium quotients and equilibrium constants in terms of molar concentrations, any concentration-like term can be used, including mole fraction and molality. Sometimes the symbols K_c, K_x, and K_m are used to denote these forms of the equilibrium constant. Bear in mind that the *numerical values* of *K*'s and *Q*'s expressed in these different ways will not generally be the same.

Most of the equilibria we deal with in this course occur in liquid solutions and gaseous mixtures. We can express K_c values in terms of moles per litre for both, but when dealing with gases it is often more convenient to use partial pressures. These two measures of concentration are of course directly proportional:

$$C = \frac{n}{V} = \frac{\frac{PV}{RT}}{V} = \frac{P}{RT}$$

so for a reaction A*(g)* → B*(g)* we can write the equilibrium constant as,

$$Kp = \frac{P_B}{P_A}$$

All of these forms of the equilibrium constant are only approximately correct, working best at low concentrations or pressures. The only equilibrium constant that is truly "constant" (except that it still varies with the temperature!) is expressed in terms of *activities*, which you can think of as "effective concentrations" that allow for interactions between molecules. In practice, this distinction only becomes important for equilibria involving gases at very high pressures (such as are often encountered in chemical engineering) and in ionic solutions more concentrated than about 0.001 *M*. We will not deal much with activities in this course. For a reaction such as $CO_2(g) + OH^-(aq) \rightarrow HCO_3^-(aq)$ that involves both gaseous and dissolved components, a "hybrid" equilibrium constant is commonly used:

$$K = \frac{\left[HCO_3^-\right]}{P_{CO2}\left[OH^-\right]}$$

Clearly, it is essential to be sure of the units when you see an equilibrium constant represented simply by "*K*".

Converting between K_p and K_c

It is sometimes necessary to convert between equilibrium constants expressed in different units. The most common case involves pressure- and concentration equilibrium constants. The ideal gas law relates the partial pressure of a gas to the number of moles and its volume:

$$PV = nRT$$

Concentrations are expressed in moles/unit volume n/V, so by rearranging the above equation we obtain the explicit relation of pressure to concentration:

$$P = (n/V)RT$$

Conversely, $c = (n/V) = P/RT$

... so a concentration [A] can be expressed as P_A(RT).

For a reaction of the form 2 A = B + 3 C, we can write

$$K_p = \frac{(P_C)^2}{(P_A)\ (P_B)^3} = \frac{([C]RT)^2}{([A]RT)\ ([B]RT)^3} = \underbrace{\frac{[C]^2}{[A]\ [B]^3}}_{K_C}(RT)^{\overset{\Delta n_g}{-2}}$$

Assuming that all of the components are gases, the difference (moles of gas in products) – (moles of gas in reactants) = Δn_g is given by,

$$K_p = K_c\left(RT\right)^{\Delta ng}$$

UNCHANGING CONCENTRATIONS

Substances whose concentrations undergo no significant change in a chemical reaction do not appear in equilibrium constant expressions.

How can the concentration of a reactant or product *not*change when a reaction involving that substance takes place? There are two general cases to consider.

The Substance is also the Solvent

This happens all the time in acid-base chemistry. Thus for the hydrolysis of the cyanide ion,

$$CN^- + H_2O \rightarrow HCN + OH^-$$

we write,

$$\frac{[HCN][OH^-]}{[CN^-]}$$

in which no $[H_2O]$ term appears. The justification for this omission is that water is both the solvent and reactant, but only the tiny portion that acts as a reactant would ordinarily go in the equilibrium expression. The amount of water consumed in the reaction is so minute (because K is very small) that any change in the concentration of H_2O from that of pure water (55.6 mol L^{-1}) will be negligible.

Similarly, for the "dissociation" of water $H_2O = H^+ + OH^-$the equilibrium constant is expressed as the "ion product"$K_w = [H^+][OH^-]$.

But... be careful about throwing away H_2O whenever you see it. In the esterification reaction,

$$CH_3COOH + C_2H_5OH \rightarrow CH_3COOC_2H_5 + H_2O$$

that we discussed in a previous section, a $[H_2O]$ term must be present in the equilibrium expression if the reaction is assumed to be between the two liquids acetic acid and ethanol. If, on the other hand, the reaction takes place between a dilute aqueous solution of the acid and the alcohol, then the $[H_2O]$ term would not be included.

The Substance is a Solid or a Pure Liquid Phase

This is most frequently seen in solubility equilibria, but there are many other reactions in which solids are directly involved:

$$CaF_2(s) \rightarrow Ca^{2+}(aq) + 2F^{\rightarrow}(aq)$$

$$Fe_3O_4(s) + 4\ H_2(g) \rightarrow 4\ H_2O(g) + 3Fe(s)$$

These are *heterogeneous reactions* (meaning reactions in which some components are in different phases), and the argument here is that concentration is only meaningful when applied to a substance within a single phase. Thus the term $[CaF_2]$ would refer to the "concentration of calcium fluoride within the solid CaF_2", which is a constant depending on the molar mass of CaF_2 and the density of that solid. The concentrations of the two ions will be independent of the quantity of solid CaF_2 in contact with the water; in other words, the system can be in equilibrium as long as any CaF_2 at all is present. Throwing out the constant-concentration terms can lead to some rather sparse-looking equilibrium expressions. For example, the equilibrium expression for each of the processes shown in the following table consists solely of a single term involving the partial pressure of a gas:

1) $CaCO_3(s) \rightarrow CaO(s) + CO_2(g)$	$K_p = P_{CO2}$	Thermal decomposition of limestone, a first step in the manufacture of cement.
2) $Na_2SO_4 \cdot 10\ H_2O(s)$ $Na_2SO_4(s) + 10\ H_2O(g)$	$K_p = P_{H2O}{}^{10}$	Sodium sulfate decahydrate is a solid in which H_2O molecules (*"waters of hydration"*) are incorporated into the crystal structure.)
3) $I_2(s) \rightarrow I_2(g)$	$K_p = P_{I2}$	sublimation of solid iodine; this is the source of the purple vapor you can see above solid iodine in a closed container.
4) $H_2O(l)$! $H_2O(g)$	$K_p = P_{H2O}$	Vaporisation of water. When the partial pressure of water vapor in the air is equal to *K*, the relative humidity is 100 per cent.

The last two processes 3 and 4 represent changes of state (*phase changes*) which can be treated exactly the same as chemical reactions.

In each of the heterogeneous processes shown in the table, the reactants and products can be in equilibrium (that is, permanently coexist) only when the partial pressure of the gaseous product has the value consistent with the indicated K_p. Bear in mind also that these K_p 's all increase with the temperature.

3

Kinetic Theory of Ideal Gas

INTRODUCTION

The Kinetic Theory picture of a gas is often called the Ideal Gas Model. It ignores interactions between molecules, and the finite size of molecules. In fact, though, these only become important when the gas is very close to the temperature at which it become liquid, or under extremely high pressure. In this lecture, we will be analyzing the behaviour of gases in the pressure and temperature range corresponding to heat engines, and in this range the Ideal Gas Model is an excellent approximation. Essentially, our programme here is to learn how gases absorb heat and turn it into work, and vice versa. This heat-work interplay is called thermodynamics.

Julius Robert Mayer was the first to appreciate that there is an equivalence between heat and mechanical work. The tortuous path that led him to this conclusion is described in an earlier lecture, but once he was there, he realized that in fact the numerical equivalence—how many Joules in one calorie in present day terminology—could be figured out easily from the results of some measurements of gas specific heat by French scientists. The key was that they had measured specific heats both at constant volume and at constant pressure. Mayer realized that in the latter case, heating the gas necessarily increased its volume, and the gas therefore did work in pushing to expand its container. Having convinced himself that mechanical work and heat were equivalent, evidently the extra heat needed to raise the temperature of the gas at constant pressure was exactly the work the gas did on its container.

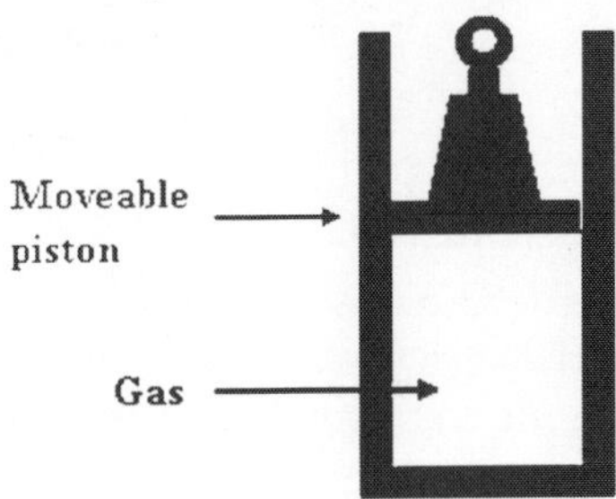

The simplest way to see what's going on is to imagine the gas in a cylinder, held in by a piston, carrying a fixed weight, able to move up and down the cylinder smoothly with negligible friction. The pressure on the gas is just the total weight pressing down divided by the area of the piston, and this total weight, of course, will not change as the piston moves slowly up or down: the gas is at constant pressure.

IDEAL GASES

An ideal gas is a substance possessing very simple thermodynamic properties to which actual gases and vapours appear to approximate indefinitely at low pressures and high temperatures. It has the characteristic equation pv=Re, and obeys Boyle's law at all temperatures. The coefficient of expansion at constant pressure is equal to the coefficient of increase of pressure at constant volume. The difference of the specific heats by equation is constant and equal to R. The isothermal elasticity – v(dp/dv) is equal to the pressure p. The adiabatic elasticity is equal to y p, where-y is the ratio S/s of the specific heats. The heat absorbed in isothermal expansion from vo to v at a temperature 0 is equal to the work done by equation (since d0 = o, and 0(dp/d0)dv =pdv), and both are given by the expression RO log e (v/vo).

The energy E and the total heat F are functions of the temperature only and their variations take the form dE = sdO, d F = Sd0. The specific heats are independent of the pressure or density by equations. If we also assume that they are constant with respect to temperature (which does not necessarily follow from the characteristic equation, but is generally assumed, and appears from Regnault's experiments to be approximately the case for simple gases), the expressions for the change of energy or total heat from 00 to 0 may be written E – E0 = s(0 – 0 0), F – Fo = S(0-00). In thiscase the ratio of the specific heats is constant as well as the difference, and the adiabatic equation takes the simple form, pv v = constant, which is at once obtained by integrating the equation for the adiabatic elasticity, – v(dp/dv) =yp.

DEVIATIONS OF ACTUAL GASES FROM THE IDEAL STATE

Since no gas is ideally perfect, it is most important for practical purposes to discuss the deviations of actual gases from the ideal state, and to consider how their properties may be thermodynamically explained and defined. The most natural method of procedure is to observe the deviations from Boyle's law by measuring the changes of pv at various constant temperatures. It is found by experiment that the change of pv with pressure at moderate pressures is nearly proportional to the change of p, in other words that the coefficient d(pv)/dp is to a first approximation a function of the temperature only.

This coefficient is sometimes called the " angular coefficient," and may be regarded as a measure of the deviations from Boyle's law, 'which may be most simply expressed at moderate pressures by formulating the variation of the angular coefficient with temperature. But this procedure in itself is not sufficient, because, although it would be highly probable that a gas obeying Boyle's law at all temperatures was practically an ideal gas, it is evident that Boyle's law would be satisfied by any substance having the characteristic equation pv = f (0), where f (0) is any arbitrary function of 0, and that the scale of temperatures given by such a substance would not necessarily coincide with the absolute scale.A sufficient test, in addition to Boyle's law, is the condition dE/dv=o at constant temperature. This gives by equation the condition Odp/d0 =p, which is satisfied by any substance possessing the characteristic equation p/0=f(v), where f(v) is any arbitrary function of v. This test was applied by Joule in the well-known experiment in which he allowed a gas to expand from one vessel to another in a calorimeter without doing external work.

Under this condition the increase of intrinsic energy would be equal to the heat absorbed, and would be indicated by fall of temperature of the calorimeter. Joule failed to observe any change of temperature in his apparatus, and was therefore justified in assuming that the increase of intrinsic energy of a gas in isothermal expansion was very small, and that the absorption of heat observed in a similar experiment in which the gas was allowed to do external work by expanding against the atmospheric pressure was equivalent to the external work done. But owing to the large thermal capacity of his calorimeter, the test, though sufficient for his immediate purpose, was not delicate enough to detect and measure the small deviations which actually exist.

JOULE–THOMSON EFFECT

In physics, the Joule–Thomson effect or Joule–Kelvin effect describes the temperature change of a gas or liquid when it is forced through a valve or porous plug while kept insulated so that no heat is exchanged with the environment.

This procedure is called a throttling process or Joule-Thomson process. At room temperature, all gases except hydrogen, helium and neon cool upon expansion by the Joule-Thomson process. The effect is named for James Prescott Joule and William Thomson, 1st Baron Kelvin who discovered it in 1852 following earlier work by Joule on Joule expansion, in which a gas undergoes free expansion in a vacuum.

Description

The adiabatic (no heat exchanged) expansion of a gas may be carried out in a number of ways. The change in temperature experienced by the gas during expansion depends on the initial and final pressure, but also on the manner in which the expansion is carried out.

- If the expansion process is reversible, meaning that the gas is in thermodynamic equilibrium at all times, it is called an isentropic expansion. In this scenario, the gas does positive work during the expansion, and its temperature decreases.

- In a free expansion, on the other hand, the gas does no work and absorbs no heat, so the internal energy is conserved. Expanded in this manner, the temperature of an ideal gas would remain constant, but the temperature of a real gas may either increase or decrease, depending on the initial temperature and pressure.
- The method of expansion in which a gas or liquid at pressure P1 flows into a region of lower pressure P2 via a valve or porous plug under steady state conditions and without change in kinetic energy, is called the Joule–Thomson process. During this process, enthalpy remains unchanged.

Temperature change of either sign can occur during the Joule-Thomson process. Each real gas has a Joule–Thomson (Kelvin) inversion temperature above which expansion at constant enthalpy causes the temperature to rise, and below which such expansion causes cooling. This inversion temperature depends on pressure; for most gases at atmospheric pressure, the inversion temperature is above room temperature, so most gases can be cooled from room temperature by isenthalpic expansion.

Physical Mechanism

As a gas expands, the average distance between molecules grows. Because of intermolecular attractive forces expansion causes an increase in the potential energy of the gas. If no external work is extracted in the process and no heat is transferred, the total energy of the gas remains the same because of the conservation of energy. The increase in potential energy thus implies a decrease in kinetic energy and therefore in temperature.

A second mechanism has the opposite effect. During gas molecule collisions, kinetic energy is temporarily converted into potential energy. As the average intermolecular distance increases, there is a drop in the number of collisions per time unit, which causes a decrease in average potential energy. Again, total energy is conserved, so this leads to an increase in kinetic energy (temperature).

Below the Joule–Thomson inversion temperature, the former effect (work done internally against intermolecular attractive forces) dominates, and free expansion causes a decrease in temperature. Above the inversion temperature, gas molecules move faster and so collide more often, and the latter effect (reduced collisions causing a decrease in the average potential energy) dominates: Joule-Thomson expansion causes a temperature increase.

The Joule–Thomson (Kelvin) Coefficient

The rate of change of temperature T with respect to pressure P in a Joule–Thomson process (that is, at constant enthalpy H) is the Joule–Thomson (Kelvin) coefficient ìJT. This coefficient can be expressed in terms of the gas's volume V, its heat capacity at constant pressure Cp, and its coefficient of thermal expansion á as:

$$\mu_{JT} \equiv \left(\frac{\partial T}{\partial P}\right)_H = \frac{V}{C_P}(\alpha T - 1).$$

The value of μJT is typically expressed in °C/bar (SI units: K/Pa) and depends on the type of gas and on the temperature and pressure of the gas before expansion. All real gases have an inversion point at which the value of μJT changes sign. The temperature of this point, the Joule–Thomson inversion temperature, depends on the pressure of the gas before expansion.

In a gas expansion the pressure decreases, so the sign of ∂P is always negative. With that in mind, the following table explains when the Joule–Thomson effect cools or warms a real gas:

If the gas temperature is	then μJT is	since ∂P is	thus∂P must be	so the gas cools
below the inversion temperature	positive	always negative	negative	
above the inversion temperature	negative	always negative	positive	warms

Helium and hydrogen are two gases whose Joule–Thomson inversion temperatures at a pressure of one atmosphere are very low (*e.g.*, about 51 K (–222 °C) for helium). Thus, helium and hydrogen warm up when expanded at constant enthalpy at typical room temperatures. On the other hand nitrogen and oxygen, the two most abundant gases in air, have inversion temperatures of 621 K (348 °C) and 764 K (491 °C) respectively: these gases can be cooled from room temperature by the Joule–Thomson effect.

For an ideal gas, ìJT is always equal to zero: ideal gases neither warm nor cool upon being expanded at constant enthalpy.In practice, the Joule–Thomson effect is achieved by allowing the gas to expand through a throttling device (usually a valve) which must be very well insulated to prevent any heat transfer to or from the gas. No external work is extracted from the gas during the expansion. The effect is applied in the Linde technique as a standard process in the petrochemical industry, where the cooling effect is used to liquefy gases, and also in many cryogenic applications (*e.g.* for the production of liquid oxygen, nitrogen, and argon). Only when the Joule–Thomson coefficient for the given gas at the given temperature is greater than zero can the gas be liquefied at that temperature by the Linde cycle. In other words, a gas must be below its inversion temperature to be liquefied by the Linde cycle. For this reason, simple Linde cycle liquefiers cannot normally be used to liquefy helium, hydrogen, or neon.

Proof that Enthalpy Remains Constant in a Joule-Thomson Process

In a Joule-Thomson process the enthalpy remains constant. To prove this, the first step is to compute the net work done by the gas that moves through the plug. Suppose that the gas has a volume of V1 in the region at pressure P1 and a volume of V2 when it appears in the region at pressure P2. Then the work done on the gas by the rest of the gas in region 1 is P1 V1. In region 2 the amount of work done by the gas is P2 V2. So, the total work done by the gas is

$$P_2V_2 - P_1V_1$$

The change in internal energy plus the work done by the gas is, by the first law of thermodynamics, the total amount of heat absorbed by the gas (here it is assumed that there is no change in kinetic energy). In the Joule-Thompson process the gas is kept insulated, so no heat is absorbed. This means that

$$E_2 - E_1 + P_2V_2 - P_1V_1 = 0$$

where E1 and E2 denote the internal energy of the gas in regions 1 and 2, respectively.

The above equation then implies that:

$$H_1 = H_2$$

where H1 and H2 denote the enthalpy of the gas in regions 1 and 2, respectively.

Derivation of the Joule-Thomson (Kelvin) coefficient

A derivation of the formula

$$m_{JT} \equiv \left(\frac{\partial T}{\partial P}\right)_H = \frac{V}{C_P}(\alpha T - 1)$$

for the Joule–Thomson (Kelvin) coefficient.

The partial derivative of T with respect to P at constant H can be computed by expressing the differential of the enthalpy dH in terms of dT and dP, and equating the resulting expression to zero and solving for the ratio of dT and dP. It follows from the fundamental thermodynamic relation that the differential of the enthalpy is given by:

$dH = TdS + VdP$ (here, S is the entropy of the gas).

Expressing dS in terms of dT and dP gives:

$$dH = T\left(\frac{\partial S}{\partial T}\right)_P dT + \left[V + T\left(\frac{\partial S}{\partial P}\right)_T\right]dP$$

Using

$C_P = T\left(\frac{\partial S}{\partial T}\right)_P$, we can write:

$$dH = C_P dT + \left[V + T\left(\frac{\partial S}{\partial P}\right)_T\right]dP.$$

The remaining partial derivative of S can be expressed in terms of the coefficient of thermal expansion via a Maxwell relation as follows. From the fundamental thermodynamic relation, it follows that the differential of the Gibbs energy is given by:

$$dG = -SdT + VdP.$$

The symmetry of partial derivatives of G with respect to T and P implies that:

$$\left(\frac{\partial S}{\partial P}\right)_T = -\left(\frac{\partial V}{\partial T}\right)_P = -V\alpha$$

where α is the coefficient of thermal expansion. Using this relation, the differential of H can be expressed as

$$dH = C_p dT + V(1 - T\alpha)dP.$$

Equating dH to zero and solving for dT/dP then gives:

$$\left(\frac{\partial T}{\partial P}\right)_H = \frac{V}{C_P}(\alpha T - 1).$$

THE GAS SPECIFIC HEATS CV AND CP

Consider now the two specific heats of this same sample of gas, let's say one mole:

Specific heat at constant volume, C_V (piston glued in place),

Specific heat at constant pressure, C_P (piston free to rise, no friction).

In fact, we already worked out C_V in the Kinetic Theory lecture: at temperature T, recall the average kinetic energy per molecule is $\frac{3}{2}kT$, so one mole of gas—Avogadro's number of molecules—will have total kinetic energy, which we'll label internal energy,

$$E_{\text{int}}\ \tfrac{3}{2}kT.N_A\ \tfrac{3}{2}RT.$$

(In this simplest case, we are ignoring the possibility of the molecules having their own internal energy: they might be spinning or vibrating—we'll include that shortly).

That the internal energy is $\frac{3}{2}kT$ per mole immediately gives us the specific heat of a mole of gas in a fixed volume,

$$C_V = \tfrac{3}{2}R$$

that being the heat which must be supplied to raise the temperature by one degree.

However, if the gas, instead of being in a fixed box, is held in a cylinder at constant pressure, experiment confirms that more heat must be supplied to raise the gas temperature by one degree. As Mayer realized, the total heat energy that must be supplied to raise the temperature of the gas one degree at constant pressure is $\frac{3}{2}k$ per molecule plus the energy required to lift the weight. The work the gas must do to raise the weight is the force the gas exerts on the piston multiplied by the distance the piston moves. If the area of piston is A, then the gas at pressure P exerts force PA. If on heating through one degree the piston rises a distance Δh, the gas does work

$$PA.\Delta h = P\Delta V.$$

Now, for one mole of gas, $PV = RT$, so at constant P

$$P\Delta V = P\Delta T.$$

Therefore, the work done by the gas in raising the weight is just $R\Delta T$, the specific heat at constant pressure, the total heat energy needed to raise the temperature of one mole by one degree,

$$C_P = C_V + R.$$

In fact, this relationship is true whether or not the molecules have rotational or vibrational internal energy. (It's known as Mayer's relationship.) For example, the specific heat of oxygen at constant volume

$$C_V(\mathrm{O}_2) = \tfrac{5}{2}R.$$

and this is understood as a contribution of $\frac{3}{2}R$ from kinetic energy, and R from the two rotational modes of a dumbbell molecule (just why there is no contribution form rotation about the third axis can only be understood using quantum mechanics). The specific heat of oxygen at constant pressure

$$C_P(O_2) = \tfrac{3}{2}R.$$

It's worth having a standard symbol for the ratio of the specific heats:

$$\frac{C_P}{C_V} = \gamma.$$

ISOTHERMS AND ADIABATS

An ideal gas in a box has three thermodynamic variables: P, V, T. But if there is a fixed mass of gas, fixing two of these variables fixes the third from $PV = nRT$ (for n moles). In a heat engine, heat can enter the gas, then leave at a different stage. The gas can expand doing work, or contract as work is done on it. To track what's going on as a gas engine transfers heat to work, say, we must follow the varying state of the gas. We do that by tracing a curve in the (P, V) plane. Supplying heat to a gas which consequently expands and does mechanical work is the key to the heat engine. But just knowing that a gas is expanding and doing work is not enough information to follow its path in the (P, V) plane.

The route it follows will depend on whether or not heat is being supplied (or taken away) at the same time. There are, however, two particular ways a gas can expand reversibly—meaning that a tiny change in the external conditions would be sufficient for the gas to retrace its path in the (P, V) plane backwards. It's important to concentrate on reversible paths, because as Carnot proved and we shall discuss later, they correspond to the most efficient engines. The two sets of reversible paths are the isotherms and the adiabats.

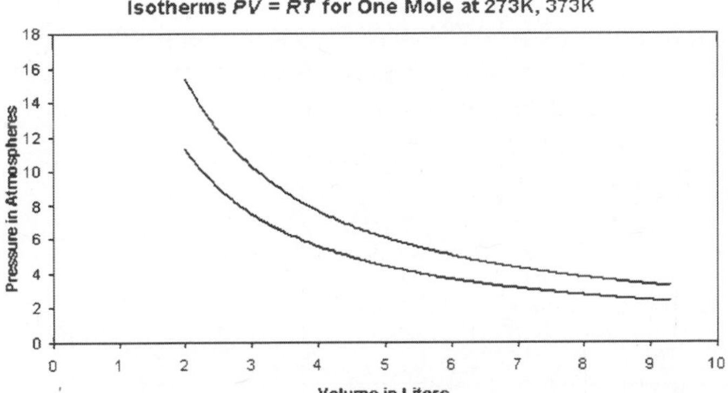

Isothermal behaviour: The gas is kept at constant temperature by allowing heat flow back and forth with a very large object (a "heat reservoir") at temperature T. From $PV = nRT$, it is evident that for a fixed mass of gas, held at constant T but subject to (slowly) varying pressure, the variables P, V will trace a hyperbolic path in the (P, V) plane. This path, $PV = nRT_1$, say is called the isotherm at temperature T1. Here are two examples of isotherms:

Adiabatic behaviour: "adiabatic" means "nothing gets through", in this case no heat gets in or out of the gas through the walls. So all the work done in compressing the gas has to go into the internal energy E_{int}.

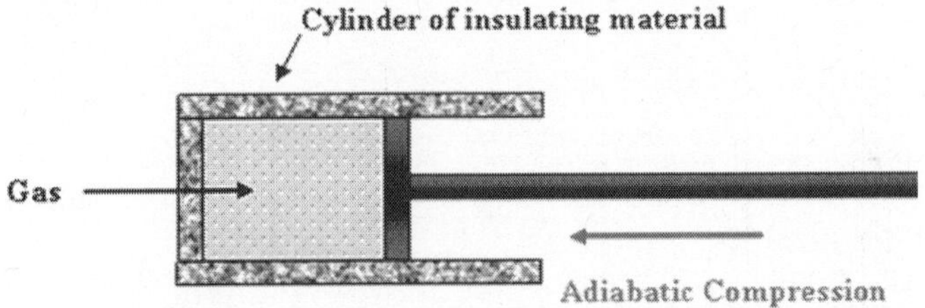

As the gas is compressed, it follows a curve in the (P, V) plane called an adiabat. To see how an adiabat differs from an isotherm, imagine beginning at some point on the blue 273K isotherm on the above graph, and applying pressure so the gas moves to higher pressure and lower volume. Since the gas's internal energy is increasing, but the number of molecules is staying the same, its temperature is necessarily rising, it will move towards the red curve, then above it.

This means the adiabats are always steeper than the isotherms.

EQUATION FOR AN ADIABAT

What equation for an adiabat corresponds to $PV = nRT_1$ for an isotherm?

On raising the gas temperature by ΔT, the change in the internal energy—the sum of molecular kinetic energy, rotational energy and vibrational energy,

$$\Delta E_{\text{int}} = C_V \Delta T.$$

This is always true: whether or not the gas is changing volume is irrelevant, all that counts in E_{int} is the sum of the energies of the individual molecules (assuming as we do here that attractive or repulsive forces between molecules are negligible). In adiabatic compression, all the work done by the external pressure goes into this internal energy, so

$$-P\Delta V = C_V \Delta T.$$

(Compressing the gas of course gives negative ΔV, positive ΔEint.)

To find the equation of an adiabat, we take the infinitesimal limit

$$-PdV = C_V dT.$$

Divide the left-hand side by PV, the right-hand side by RT (since $PV = RT$, that's OK) to find

$$-\frac{R}{C_V}\frac{dV}{V} = \frac{dT}{T}.$$

Recall now that $C_P = C_V + R$, and $C_P/C_V = \gamma$. It follows that

$$\frac{R}{C_V} = \frac{C_P - C_V}{C_V} = \gamma - 1.$$

Hence

$$-(\gamma - 1)\int\frac{dV}{V} = \int\frac{dT}{T}$$

and integrating

$$\ln T + (\gamma - 1)\ln V = \text{const.}$$

from which the equation of an adiabat is

$$TV^{-1} = \text{const}.$$

From $PV - RT$, the P, V equation for an adiabat can be found by multiplying the left-hand side of this equation by the constant PV/T, giving

$$PV^{\gamma} = \text{const. for an adiabat,}$$

where $\gamma = {}^5/_3$ for a monatomic gas, ${}^7/_5$ for a diatomic gas.

THE CARNOT CYCLE

All standard heat engines (steam, gasoline, diesel) work by supplying heat to a gas, the gas then expands in a cylinder and pushes a piston to do its work. The catch is that the heat and/or the gas must somehow then be dumped out of the cylinder to get ready for the next cycle. We examine the first step, the expansion, then go on to the full cycle—Carnot's analysis. Carnot's aim was to figure out how to maximize the efficiency of a heat engine, and then work out what that efficiency was, that is, how much of the heat supplied was actually converted into the mechanical work done by the engine.

Remember that he had in mind the analogy of the water wheel, at that time still a main driving force of industry. He knew that the most efficient water wheels were those that operated smoothly, the water went into the buckets at the top from the same level, it didn't fall through any height, and didn't splash around. In the limit of a frictionless wheel, with gentle flow on and off the wheel, the machine would be reversible—turning it in reverse to raise the water back would take the same amount of work the wheel had delivered as the water fell.

This was clearly perfect efficiency, so these were to conditions to emulate in the heat engine. The analog to having the water flow into buckets at the same height, with no wasteful drop, is to have the heat from the heat supply flow into the gas at the same temperature. There must of course be a slight drop in temperature for the heat to flow at all, but this must be minimized. This means that as the heat is supplied and the gas expands, the temperature of the gas stays the same as that of the heat supply (the "heat reservoir") and the gas is expanding isothermally.

ISOTHERMAL EXPANSION

So the first question is: how much work is done by an isothermally expanding gas? Taking the temperature of the heat reservoir to be TH (H for hot), the expanding gas follows the isothermal path $PV = nRT_H$ in the (P, V) plane.

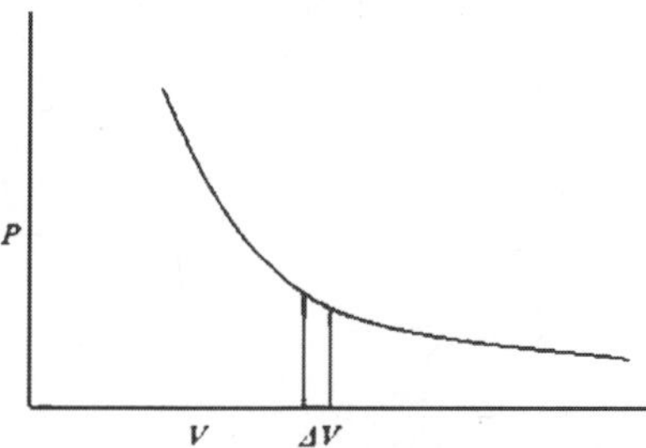

The work done by the gas in a small volume expansion ΔV is just $P\Delta V$, the area under the curve.Hence the work done in expanding isothermally from volume Va to Vb is the total area under the curve between those values,

$$\text{work done isothermally} = \int_{V_a}^{V_b} PdV = \int_{V_a}^{V_b} \frac{nRV_H}{V} dV = nRT_H \ln \frac{V_b}{V_a}.$$

Since the gas is at constant temperature TH, there is no change in its internal energy during this expansion, so the total heat supplied must be

$$nRT_H \ln \frac{V_b}{V_a},$$

the same as the external work the gas has done.

In fact, this isothermal expansion is only the first step: the gas is at the temperature of the heat reservoir, hotter than its other surroundings, and will be able to continue expanding even if the heat supply is cut off. To ensure that this further expansion is also reversible, the gas must not be losing heat to the surroundings. That is, after the heat supply is cut off, there must be no further heat exchange with the surroundings, the expansion must be adiabatic.

ADIABATIC EXPANSION

The work done in an adiabatic expansion is like that done in allowing a compressed spring to expand against a force—equal to the work needed to compress the spring in the first place, for a perfect spring, and an adiabatically enclosed gas is essentially perfect in this respect. In other words, adiabatic expansion is reversible. To find the work the gas does in expanding adiabatically from Vb to Vc, say, the above analysis is repeated with the isotherm $PV = nRT_H$ replaced by the adiabat $PV^\gamma = P_bV_b^\gamma$, work done adiabatically W_{adiabat}

$$\int_{V_a}^{V_c} PdV = P_bV_b^\gamma \int_{V_b}^{V_c} \frac{dV}{V^\gamma} = P_bV_b^\gamma \frac{V_c^{1-\gamma} - V_b^{1-\gamma}}{1-\gamma}.$$

Again, this is the area under the curve, in this case under the adiabat, from b to c in the (P, V) plane.

Since points b, c are on the same adiabat, $P_bV_b^\gamma = P_cV_c^\gamma$, and the expression can be written more neatly:

$$W_{\text{adiabat}} = \frac{P_cV_c - P_bV_b}{1-\gamma}.$$

This is a useful expression for the work done since we are plotting in the (P, V) plane, but note that from the gas law $PV = nRT$, the numerator is just $nR(Tc - T_b)$, and from this $W_{\text{adiabat}} = nC_V(T_c - T_b)$, as of course it must be—this is the loss of internal energy that has been expended by the gas on expanding against external pressure.

We've looked in detail at the work a gas does in expanding as heat is supplied (isothermally) and when there is no heat exchange (adiabatically). These are the two initial steps in a heat engine, but it is equally necessary for the engine to get back to where it began, for the next cycle. The general idea is that the piston drives a wheel, which continues to turn and pushes the gas back to the original volume.

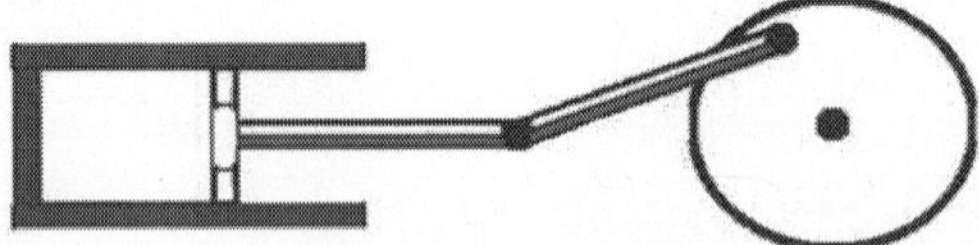

But it is also essential for the gas to be as cold as possible on this return leg, because the wheel is now having to expend work on the gas, and we want that to be as little work as possible—it's costing us. The colder the gas, the less pressure the wheel is pushing against.

To ensure that the engine is as efficient as possible, this return path to the starting point (P_a, V_a)must also be reversible. We can't just retrace the path taken in the first two legs, that would take all the work the engine did along those legs, and leave us with no net output. Now the gas cooled during the adiabatic expansion from b to c, from TH to TC, say, so we can go some distance back along the reversible colder isotherm TC. But this won't get us back to (P_a, V_a), because that's on the TH isotherm.

The simplest option—the one chosen by Carnot—is to proceed back along the cold isotherm to the point where it intersects the adiabat through a, then follow that isotherm back to a. (One could follow a more complicated path: provided it was composed of segments each being adiabatic or isothermal, it would be reversible.) Carnot's cycle is around that curved quadrilateral having these four curves as its sides.

Efficiency of the Carnot Engine

In a complete cycle of Carnot's heat engine, the gas traces the path abcd. The important question is: what fraction of the heat supplied from the hot reservoir (along the red top isotherm) is turned into mechanical work? This fraction is called the efficiency of the engine. The work output along any curve in the (P, V) plane is just $\int PdV$—the area under the curve, but it will be negative if the volume is decreasing! So the work done by the engine during the hot isothermal segment is the area abfh, then the adiabatic expansion adds the area bcef, but as the gas is compressed back, the wheel has to do work on the gas equal to the area cdge as heat is dumped into the cold reservoir, then dahg as the gas is recompressed to the starting point.

The bottom line is that the total work done by the gas is the area bounded by the four paths: the curved "parallelogram" in the picture above. We could compute this area by finding $\int PdV$ for each segment, but that is unnecessary—on completing the cycle, the gas is back to its initial temperature, so has the same internal energy. Therefore, the work done by the engine must be just the difference between the heat supplied at TH and that dumped at TC.

Now the heat supplies along the initial hot isothermal path ab, equal to the work done along that leg, is (from the paragraph above on isothermal expansion):

$$Q_H = nRT_H \ln \frac{V_b}{V_a}$$

and the heat dumped into the cold reservoir along cd is

$$Q_C = nRT_C \ln \frac{V_c}{V_d}.$$

The difference between these two is the net work output. This can be simplified using the adiabatic equations for the other two sides of the cycle:

$$T_H V_b^{\gamma-1} = T_C V_c^{\gamma-1}$$

$$T_H V_a^{\gamma-1} = T_C V_d^{\gamma-1}.$$

Dividing the first of these equations by the second,

$$\left(\frac{V_b}{V_a}\right) = \left(\frac{V_c}{V_d}\right)$$

and using that in the preceding equation for QC,

$$Q_C = nRT_C \ln\frac{V_a}{V_b} = \frac{T_C}{T_H}Q_H.$$

The work done can now be written simply:

$$W = Q_H - Q_C = \left(1 - \frac{T_C}{T_H}\right)Q_H.$$

Therefore the efficiency of the engine, defined as the fraction of the ingoing heat energy that is converted to available work, is

$$\text{efficiency} = \frac{W}{Q_H} = 1 - \frac{T_C}{T_H}.$$

These temperatures are of course in degrees Kelvin, so for example the efficiency of a Carnot engine having a hot reservoir of boiling water and a cold reservoir ice cold water will be 1– (273/373) = 0.27, just over a quarter of the heat energy is transformed into useful work.

After all the effort to construct an efficient heat engine, making it reversible to eliminate "friction" losses, etc., it is perhaps somewhat disappointing to find this figure of 27 per cent efficiency when operating between 0 and 100 degrees Celsius. Surely we can do better than that? After all, the heat energy of hot water is the kinetic energy of the moving molecules, can't we find some device to channel all that energy into useful work? Well, we can do better than 27 per cent, by having a colder cold reservoir, or a hotter hot one. But there's a limit: we can never reach 100 per cent efficiency, because we cannot have a cold reservoir at T_C= 0K, and even if we did after the first cycle the heat dumped into it would warm it up!

The Second Law of Thermodynamics states that we cannot devise an engine, working in a cycle, that simply extracts heat from a hot reservoir and delivers mechanical work.

This means any engine that takes heat and delivers work also dumps out some of the initial heat to a reservoir at a lower temperature. It's important to note that the First Law of Thermodynamics, the conservation of total energy including heat, would not be violated by an engine that powered a ship by extracting heat energy from the surrounding water.

This Second Law is saying something new. And, this Second Law does not follow from the First by logical deduction—it comes (like the First) from experiment and observation.

THE MOST EFFICIENT ENGINE

An important consequence of the second Law is that no engine can be more efficient than the Carnot cycle. Essentially, this is because a "super efficient" engine, if one existed, could be used to drive a Carnot cycle in reverse, which would pump back to the hot reservoir the heat the super efficient engine dumped in the cold reservoir, and the net effect of the two coupled engines would be to take heat from the hot reservoir and do work, contradicting the Second Law.

To see this, we plot the heat/energy flow for the Carnot cycle:

Here $Q_H = Q_C + W$ (all expressed in Joules, of course).

Since the engine is reversible, it can also be run backwards (this would be a refrigerator: outside work is supplied, and heat is extracted from a cold reservoir and dumped into a hot reservoir:

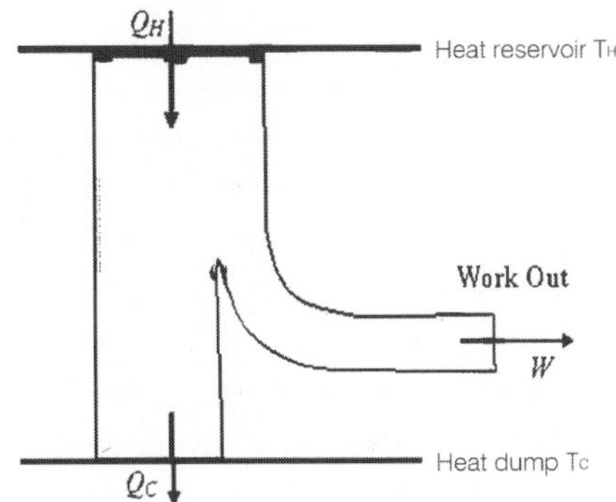

Suppose now we have a super efficient engine, represented by the first diagram above, and dumping the same heat per cycle QC into the cold reservoir, but taking in more heat energy $Q_H + \Delta$ Joules from the hot reservoir, and performing work $W + \Delta$. Now, we hook up our super efficient engine to the "Carnot refrigerator" in the diagram above. The refrigerator sucks out of the cold reservoir all the heat the super efficient engine dumped there, and needs W Joules of work per cycle to do it.

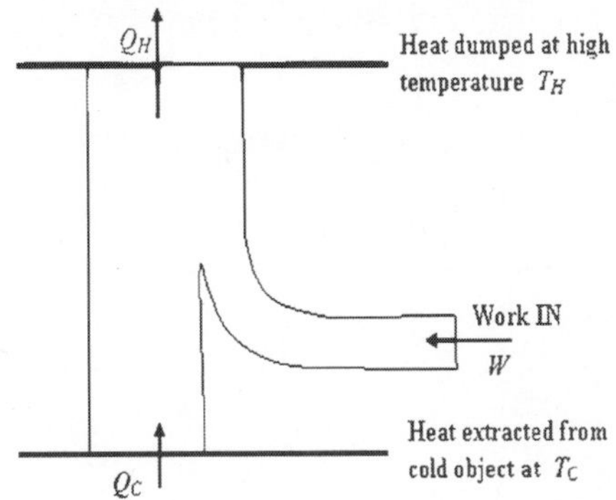

The super efficient engine can provide this, and there are still Δ Joules of work to spare. Of course, the Carnot refrigerator has also dumped QH Joules of heat in the hot reservoir. But the bottom line is that between them, the super efficient engine and the Carnot refrigerator have extracted Δ Joules of heat from the hot reservoir and performed Δ Joules of work—contradicting the Second Law.

The Second Law therefore forces the conclusion that no amount of machine design will produce an engine more efficient than the Carnot cycle. The rather low ultimate efficiencies this dictated came as a shock to nineteenth century engineers.

SPECIFIC HEAT CAPACITIES OF IDEAL GAS

If we are dealing with a gas, it is most convenient to use forms of the thermodynamics equations based on the enthalpy of the gas. From the definition of enthalpy:

$$h = e + p * v$$

where h in the specific enthalpy, p is the pressure, v is the specific volume, and e is the specific internal energy. During a process, the values of these variables change. Let's denote the change by the Greek letter delta which looks like a triangle. So "delta h" means the change of "h" from state 1 to state 2 during a process.

$$\text{delta h} = \text{delta e} + p * \text{delta v}$$

The enthalpy, internal energy, and volume are all changed, but the pressure remains the same. From our derivation of enthalpy equation, the change of specific enthalpy is equal to the heat transfer for a constant pressure process:

$$\text{delta h} = cp * \text{delta T}$$

where delta T is the change of temperature of the gas during the process, and c is the specific heat capacity. We have added a subscript "p" to the specific heat capacity to remind us that this value only applies to a constant pressure process. The equation of state of a gas relates the temperature, pressure, and volume through a gas constant R . The gas constant used by aerodynamicists is derived from the universal gas constant, but has a unique value for every gas.

$$p * v = R * T$$

If we have a constant pressure process, then:

$$p * \text{delta v} = R * \text{delta T}$$

Now let us imagine that we have a constant volume process with our gas that produces exactly the same temperature change as the constant pressure process that we have been discussing. Then the first law of thermodynamics tells us:

$$\text{delta e} = \text{delta q} - \text{delta w}$$

where q is the specific heat transfer and w is the work done by the gas. For a constant volume process, the work is equal to zero. And we can express the heat transfer as a constant times the change in temperature. This gives:

delta e = cv * delta T

where delta T is the change of temperature of the gas during the process and c is the specific heat capacity.

If we substitute the expressions for "delta e", "p * delta v", and "delta h" into the enthalpy equation we obtain:

cp * delta T = cv * delta T + R * delta T

dividing by "delta T" gives the relation:

cp = cv + R

We can define an additional variable called the specific heat ratio, which is given the Greek symbol "gamma", which is equal to cp divided by cv:

gamma = cp/cv

"Gamma" is just a number whose value depends on the state of the gas. For air, gamma = 1.4 for standard day conditions. "Gamma" appears in many fluids equations including the equation relating pressure, temperature, and volume during a simple process, the equation for the speed of sound, and all the equations for isentropic flows, and shock waves.

P – V –T SURFACE OF AN IDEAL GAS

For a fixed number of molecules the Ideal Gas Law forms the surface of a three-dimensional graph where the axis are Pressure, Volume, and Temperature.

Lines of constant pressure, constant volume, and constant temperature form a coordinate system labeling the location of an ideal gas.

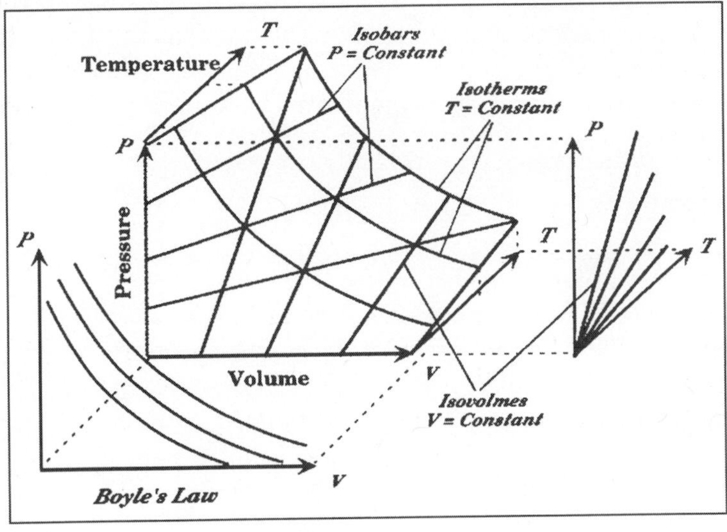

Boyle law showed that the pressure of a low-density gas is inversely proportional to its volume, when the temperature is constant.

$P \alpha 1/V$ for constant temperature.

Charles and Gay-Lussac law showed that the pressure of a low-density gas is proportional to its temperature, when the volume is constant.

$P \alpha T$ for constant volume.

The ideal gas law is only valid for low-density gas. There exists a PVT surface for high-density gases only.

PVT SURFACE FOR SUBSTANCE WHICH CONTRACTS ON FREEZING

The equilibrium state of a simple, compressible substance can be specified in terms of its pressure, volume and temperature. If any two of these state variables is specified, we can determine the third. The states of the substance can be represented as a surface in three dimensional PVT space. PVT surface above represents a substance which contracts on freezing. PVT surface image is shown in figure.

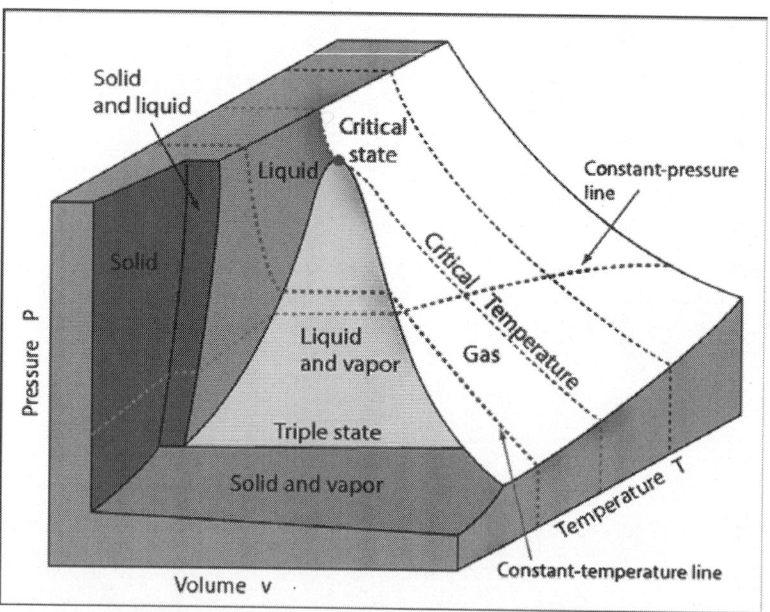

In this the solid, liquid and gas phases can be determined by regions on the surface. A point lying on a line between a single-phase and a two-phase region represents a "saturation state". The line between the liquid and liquid-vapor regions is called as liquid-saturation line. A point on the boundary between the vapor and liquid-vapor regions is called as a saturated-vapor state.

All the possible states of an ideal gas can be represented by a PVT surface as shown in figure.

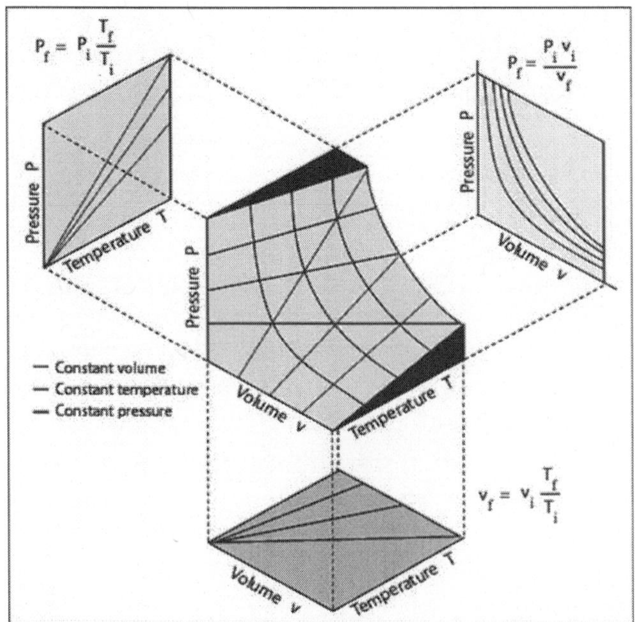

KINETIC ENERGY AND MASS FOR SLOW PARTICLES

Recall that to get momentum to be conserved in all inertial frames, we had to assume an increase of mass with speed by the factor $1/\sqrt{1-v^2/c^2}$.

This necessarily implies that even a slow-moving object has a tiny mass increase if it is put in motion.

How does this mass increase relate to the kinetic energy? Consider a mass m, moving at speed v, much less than the speed of light. Its kinetic energy

$$E = \tfrac{1}{2}mv^2,$$

as discussed above. Its mass is

$$m/\sqrt{1-v^2/c^2}$$

which we can write as $m + dm$, so dm is the tiny mass increase we know must occur. It's easy to calculate dm. For small v, we can make the approximations

$$\sqrt{1-v^2/c^2} \approx 1 + \frac{1}{2}\frac{v^2}{c^2}$$

and

$$\frac{1}{1-\frac{1}{2}\frac{v^2}{c^2}} \approx 1 + \frac{1}{2}\frac{v^2}{c^2}$$

This means the total mass at speed v is $m(1 + ½v^2/c^2)$, and writing this as $m + dm$, we see the mass increase dm equals $½ mv^2/c^2$. This means that again, the mass increase dm is related to the kinetic energy E by

$$E = (dm)c^2.$$

KINETIC ENERGY AND MASS FOR PARTICLES OF ARBITRARY SPEED

We have shown in the two sections above that when a force does work to increase the kinetic energy of a particle it also causes the mass of the particle to increase by an amount equal to the increase in energy divided by c^2. In fact this result is exactly true over the whole range of speed from zero to arbitrarily close to the speed of light, as we shall now demonstrate. For a particle of mass m accelerating along a straight line under a constant force F,

$$\text{work done} = \text{force distance}$$

so

$$K.E. = I \int R ds$$

Now

$$F = \frac{d}{dt} mv,$$

so

$$K.E. = \int \frac{d}{dt}(mv) ds = \int \frac{ds}{dt} d(mv) = \int v d(mv)$$

$$d(mv) = vdm + mdv = \left(v + m\frac{dv}{dm} \right) dm$$

We find dv/dm:

$$m = \frac{m_0}{\sqrt{1 - v^2 / c^2}} \cdot \frac{m_0}{\sqrt{1 - v^2 / c^2}} = \frac{v}{c^2 - v^2} m$$

That is,

$$\frac{dm}{dv} = \frac{v / c^2}{1 - v^2 / c^2} \cdot \frac{m_0}{\sqrt{1 - v^2 / c^2}} = \frac{v}{c^2 - v^2} m$$

so

$$\frac{dv}{dm} = \frac{c^2 - v^2}{mv}$$

and

$$d(mv) = \left(v + \frac{c^2 - v^2}{mv} \right) dm$$

Therefore

$$K.E. = \int v d(mv) = \int c^2 dm = (m - m_0)c^2$$

$$= \left(\frac{m_0}{\sqrt{1 - v^2 / c^2}} - m_0 \right) c^2$$

So we see that in the general case the work done on the body, by definition its kinetic energy, is just equal to its mass increase multiplied by c^2.

To understand why this isn't noticed in everyday life, try an example, such as a jet airplane weighing 100 tons moving at 2,000mph. 100 tons is 100,000 kilograms, 2,000mph is about 1,000 meters per second. That's a kinetic energy $\frac{1}{2}Mv^2$ of $\frac{1}{2}.10^{11}$joules, but the corresponding mass change of the airplane down by the factor c^2, 9.10^{16}, giving an actual mass increase of about half a milligram, not too easy to detect!

m and m_0

We use m_0 to denote the "rest mass" of an object, and m to denote its relativistic mass,

$$\left(\frac{m_0}{\sqrt{1 - v^2 / c^2}} \right).$$

In this notation, we follow French and Feynman. *Krane and Tipler, in contrast, use m for the rest mass*. Using m as we do gives neater formulas for momentum and energy, but is not without its dangers.

One must remember that m is *not* a constant, but a function of speed. Also, *one must remember* that the relativistic kinetic energy is $(m-m_0)c^2$, and *not* equal to $\frac{1}{2}mv^2$, even with the relativistic mass.

SPECIFIC HEAT OF AN IDEAL GAS

The equation of state for an ideal gas is

$$PV=NRT$$

where:

N is the number of moles of gas in the volume V.

Since u depends only on T,

We write the first law in terms of internal energy,

$$\delta q = du + dW$$

and assume a quasi-static process so that we can also write it in terms of enthalpy.,

$$\delta q = dh - \nu dp$$

Equating the two first law expressions above, and assuming an ideal gas, we obtain

$$C_p dT - \nu dp = C_\nu dT + p d\nu$$

Combining terms,

$$(C_p - C_\nu)dT = d(p\nu)$$

$$C_p - C_\nu = \frac{d(p\nu)}{dT}$$

Since ,

$$p\nu = RT$$

$$C_P - C_\nu = R$$

An expression that will appear often is the ratio of specific heats, which we will define as

$$\gamma = \frac{c_p}{c_\nu}$$

All ideal gases:

The specific heat at constant volume (C_v for a unit mass or Cv for one kmol) is a function of T only.

The specific heat at constant pressure ($c_p$for a unit mass or Cp for one kmol) is a function of T only.

A relation that connects the specific heats c_p , C_v , and the gas constant is

$$Cp - C_v = R$$

where the units depend on the mass considered. For a *unit mass* of gas, *e.g.*, a kilogram, Cp and C_v would be the specific heats for one kilogram of gas and R is as defined above. For *one kmol* of gas, the expression takes the form

$$Cp - C_v = \Re$$

where Cp and Cv have been used to denote the specific heats for one kmol of gas and $\Re$ is the universal gas constant.

GAS LAW EXERCISES

1. Helium takes up 5.71 litres at 0°C and 3.95 atmospheres. What is the volume of the same helium at 32°F and 800 mmHg?
2. 257 mL of oxygen in a gas tube goes from 17°C to 42°C from being out in the sun. The pressure in the tube is 39 #/in2, but it does not change as the temperature increases. What is the volume of the tube after it has heated?
3. An enormous (57,400 cubic meter) expandable helium balloon at 22°C is heated up by a fire under it and the action of the sun on the dark plastic covering on top. There will be a small increase in pressure

from 785 mmHg to 790 mmHg, but the major effect wanted is an increase in volume so the balloon can lift its cargo. To what temperature must the balloon get in order to fill out to 60,500 cubic meters?

4. What volume of air at standard pressure gets packed into an 11 ft^3 SCUBA tank at the same temperature at 15.8 atmospheres?
5. Air is 20 per cent oxygen and 80% nitrogen. What is the mass of air in an automobile tire of 19.7 L and internal pressure of 46.7 PSI at 24°C? (That pressure is the same as the 32 PSI difference you usually measure as the tire pressure 32 PSI + 14.7 PSI. You will have to use a weighted average for the molar mass of air.)
6. A constant pressure tank of gas at 1.01 Atm has propane in it at 15°C when it is at 255 cubic meters. What is its volume at 48°C?
7. A SCUBA tank is filled with air at 16.7 Atm at 24°C, but someone leaves it out in the sun to warm to 65°C. What is the tank pressure?
8. The usual partial pressure of oxygen that people get at sea level is 0.20 Atm., that is, a fifth of the usual sea level air pressure. People used to 1 Atm. air pressure begin to become "light-headed" at about 0.10 Atm oxygen. As a rule of thumb, the air pressure decreases one inch of mercury each thousand feet of altitude above sea level. At what altitude should airplane cabins be pressurized? Up to about what altitude should you be able to use unpressurized pure oxygen? (Express your answer in feet above Mean Sea Level, or MSL.)
9. Which diffuses faster, the bad smell from a cat-pan due to ammonia or an expensive French perfume with an average molecular weight of 170 g/mol? "How much faster does the faster one diffuse?
10. What is the mass of neon in a 625 mL neon tube at 3 5 7 mmHg and 25°C?
11. What is the mass of 15 litres of chlorine gas at STP?
12. How many litres of ammonia at STP are produced when 10 g of hydrogen is combined with nitrogen?
13. How many millilitres of hydrogen at 0°C and 1400 mmHg are produced if 15g of magnesium reacts with sulfuric acid?
14. What is the mass of 25 litres of fluorine gas at 2.85 atm, 450°C?
15. A nine litre tank has 150 atmospheres of bromine in it at 27°C. What is the added mass of the tank due to the gas?
16. A 250 Kg tank of liquid butane (C4H1O) burns to produce carbon dioxide at 120°C. What volume of carbon dioxide is produced at 1 Atm?
17. How many litres of product at 950 mmHg and O°C is produced by the burning of three litres of acetylene (C2H2) at 5 atm and 20°C?
18. Five grams of octane (C8H18) and enough oxygen to burn it are in an automobile cylinder compressed to 20 atm at 28°C. The mixture explodes and heats the cylinder to 150°C. What is the pressure in the (same sized) cylinder after the explosion?

19. If 0.515g of magnesium is added to HCl, it makes hydrogen gas and magnesium chloride. The hydrogen is collected at 23°C and 735mmHg. What is the volume of hydrogen?
20. What is the mass of 150 litres of propane gas (C3H8) at 37°C and 245 inHg?
21. Isopropyl alcohol, C3H7OH , makes a good fuel for cars. What volume of oxygen at 735 mmHg and 23°C is needed to burn one kilogram of isopropyl alcohol?
22. What volume does 4 Kg of nitrogen gas take up at 27°C and 3 atm?
23. The dirigible Hindenberg had 3.7E6 m^3 of hydrogen in its gas bags at 1.1 atm and 7°C. What was the weight of the hydrogen in pounds?

Answers to gas law math problems			
1. 21.4 L	2. 279 ml	3. 39.9°C	4. 174 ft^3
5. 73.9 g	6. 284 cubic meters	7. 19.0 Atm	8a. 15,000 ft. MSL
8b. 27,000 ft. MSL	9. Ammonia diffuses 3.16 times faster (Wouldn't you KNOW it?)		
10. 0.242 g	11. 47.5 g	12. 74.7 L	13. 7.51 L
14. 45.6 g	15. 8.76 Kg	16. 5.56 E5 L	17. 33.5 L
18. 35.4 Atm.	19. 532 ml	20. 2.12 Kg	21. 209 KL
22. 1.17 KL	23. 7.80 E5 L		

THE EQUATION OF STATE OF AN IDEAL GAS

Let us start our discussion by considering the simplest possible macroscopic system: *i.e.*, an ideal gas. All of the thermodynamic properties of an ideal gas are summed up in its equation of state, which determines the relationship between its pressure, volume, and temperature. Unfortunately, classical thermodynamics is unable to tell us what this equation of state is from first principles. In fact, classical thermodynamics cannot tell us anything from first principles. We always have to provide some information to begin with before classical thermodynamics can generate any new results. This initial information may come from statistical physics (*i.e.*, from our knowledge of the microscopic structure of the system under consideration), but, more usually, it is entirely empirical in nature (*i.e.*, it is the result of experiments). Of course, the ideal gas law was first discovered empirically by Robert Boyle, but, nowadays, we can justify it from statistical arguments. Recall that the number of accessible states of a monotonic ideal gas varies like

$$\Omega \alpha V^N \chi(E),$$

where is the number of atoms, and depends only on the energy of the gas (and is independent of the volume). We obtained this result by integrating over the volume of accessible phase-space. Since the energy of an ideal gas is independent of the particle coordinates (because there are no interatomic forces), the integrals over the coordinates just reduced to simultaneous volume integrals, giving the factor in the above expression. The integrals over the particle momenta were more complicated, but were clearly completely independent of , giving the V^n factor in the above expression. Now, we have a statistical rule which tells us that

$$X_\alpha = \frac{1}{\beta}\frac{\partial 1n\Omega}{\partial x_\alpha}$$

see Eq. where X_α is the mean force conjugate to the external parameter X_α (*i.e.* $đW = \sum_\alpha X\alpha dx_\alpha$,), and $\beta = 1/kT$. For an ideal gas, the only external parameter is the volume, and its conjugate force is the pressure (since $đW = pdV$). So, we can write

$$p = \frac{1}{\beta}\frac{1\partial 1n\Omega}{\partial V}$$

If we simply apply this rule to Eq. we obtain

$$p = \frac{NkT}{V}$$

However, $N = \nu N_A$, where ν is the number of moles, and N_A is Avagadro's number. Also, $kN_A = R$, where R is the ideal gas constant. This allows us to write the equation of state in its usual form

$$pV = \nu RT$$

The above derivation of the ideal gas equation of state is rather elegant. It is certainly far easier to obtain the equation of state in this manner than to treat the atoms which make up the gas as little billiard balls which continually bounce of the walls of a container. The latter derivation is difficult to perform correctly because it is necessary to average over all possible directions of atomic motion. It is clear, from the above derivation, that the crucial element needed to obtain the ideal gas equation of state is the absence of interatomic forces. This automatically gives rise to a variation of the number of accessible states with E and V of the form (6.6), which, in turn, implies the ideal gas law. So, the ideal gas law should also apply to polyatomic gases with no interatomic forces. Polyatomic gases are more complicated that monatomic gases because the molecules can rotate and vibrate, giving rise to extra degrees of freedom, in addition to the translational degrees of freedom of a monatomic gas. In other words, $X^{(E)}$, in Eq. becomes a lot more complicated in polyatomic gases. However, as long as there are no interatomic forces, the volume dependence of Ω is still V^N, and the ideal gas law should still hold true. In fact, we shall discover that the extra degrees of freedom of polyatomic gases manifest themselves by increasing the specific heat capacity. There is one other conclusion we can draw from Eq. The statistical definition of temperature is Eq.

$$\frac{1}{kT} = \frac{\partial \ln \Omega}{\partial E}$$

It follows that

$$\frac{1}{kT} = \frac{\partial \ln \chi}{\partial E}$$

We can see that since x is a function of the energy, but *not* the volume, then the temperature must be a function of the energy, but not the volume. We can turn this around and write. In other words, the internal energy of an ideal gas depends only on the temperature of the gas, and is independent of the volume. This is pretty obvious, since if there are no interatomic forces then increasing the volume, which effectively increases the mean separation between molecules, is not going to affect the molecular energies in any way. Hence, the energy of the whole gas is unaffected. The volume independence of the internal energy can also be obtained directly from the ideal gas equation of state. The internal energy of a gas can be considered as a general function of the temperature and volume, so

$$E = E(T,V)$$

It follows from mathematics that

$$dE = \left(\frac{\partial E}{\partial T}\right)_V dT + \left(\frac{\partial E}{\partial V}\right)_T dV,$$

where the subscript reminds us that the first partial derivative is taken at constant volume, and the subscript reminds us that the second partial derivative is taken at constant temperature. Thermodynamics tells us that for a quasi-static change of parameters

$$TdS = dE + pdV$$

The ideal gas law can be used to express the pressure in term of the volume and the temperature in the above expression. Thus,

$$dS = \frac{1}{T} dE + \frac{\nu R}{V} dv$$

Using Eq. (6.15), this becomes

$$dS = \frac{1}{T}\left(\frac{\partial E}{\partial T}\right)_V dT + \left[\frac{1}{T}\left(\frac{\partial E}{\partial V}\right)_T + \frac{\nu R}{V}\right] dV$$

However, dS is the exact differential of a well-defined state function, S. This means that we can consider the entropy to be a function of temperature and volume. Thus, S=S(T,V), and mathematics immediately tells us that

$$dS = \left(\frac{\partial S}{\partial T}\right)_V dT + \left(\frac{\partial S}{\partial V}\right)_T dv$$

The above expression is true for all small values of dT and dV, so a comparison with Eq. gives

$$\left(\frac{\partial S}{\partial T}\right)_V = \frac{1}{T}\left(\frac{\partial E}{\partial T}\right)_V$$

$$\left(\frac{\partial S}{\partial V}\right)_T = \frac{1}{T}\left(\frac{\partial E}{\partial V}\right)_T + \frac{\nu R}{V}$$

One well-known property of partial differentials is the equality of second derivatives, irrespective of the order of differentiation, so

$$\frac{\partial^2 S}{\partial V \partial T} = \frac{\partial^2 S}{\partial T \partial V}$$

This implies that

$$\left(\frac{\partial}{\partial V}\right)_T \left(\frac{\partial S}{\partial T}\right)_V = \left(\frac{\partial}{\partial T}\right)_V \left(\frac{\partial S}{\partial V}\right)_T$$

The above expression can be combined with Eqs. and to give

$$\frac{1}{T}\left(\frac{\partial^2 E}{\partial V \partial T}\right) = \left[-\frac{1}{T^2}\left(\frac{\partial E}{\partial V}\right)_T + \frac{1}{T}\left(\frac{\partial^2 E}{\partial T \partial V}\right)\right]$$

Since second derivatives are equivalent, irrespective of the order of differentiation, the above relation reduces to

$$\left(\frac{\partial E}{\partial V}\right)_T = 0$$

which implies that the internal energy is independent of the volume for any gas obeying the ideal equation of state. This result was confirmed experimentally by James Joule in the middle of the nineteenth century.

4

Applications Research in Analytical Chemistry

INTRODUCTION

Analytical chemistry research is largely driven by performance (sensitivity, selectivity, robustness, linear range, accuracy, precision, and speed), and cost (purchase, operation, training, time, and space). Among the main branches of contemporary analytical atomic spectrometry, the most widespread and universal are optical and mass spectrometry. In the direct elemental analysis of solid samples, the new leaders are laser-induced breakdown and laser ablation mass spectrometry, and the related techniques with transfer of the laser ablation products into inductively coupled plasma. Advances in design of diode lasers and optical parametric oscillators promote developments in fluorescence and ionization spectrometry and also in absorption techniques where uses of optical cavities for increased effective absorption pathlength are expected to expand. The use of plasma- and laser-based methods is increasing. An interest towards absolute (standardless) analysis has revived, particularly in emission spectrometry.

Great effort is put in shrinking the analysis techniques to chip size. Although there are few examples of such systems competitive with traditional analysis techniques, potential advantages include size/portability, speed, and cost. (micro Total Analysis System (μTAS) or Lab-on-a-chip). Microscale chemistry reduces the amounts of chemicals used.

Many developments improve the analysis of biological systems. Examples of rapidly expanding fields in this area are:

- Genomics - DNA sequencing and its related research. Genetic fingerprinting and DNA microarray are important tools and research fields.
- Proteomics - the analysis of protein concentrations and modifications, especially in response to various stressors, at various developmental stages, or in various parts of the body.

- Metabolomics - similar to proteomics, but dealing with metabolites.
- Transcriptomics - mRNA and its associated field
- Lipidomics - lipids and its associated field
- Peptidomics - peptides and its associated field
- Metalomics - similar to proteomics and metabolomics, but dealing with metal concentrations and especially with their binding to proteins and other molecules.

Analytical chemistry has played critical roles in the understanding of basic science to a variety of practical applications, such as biomedical applications, environmental monitoring, quality control of industrial manufacturing, forensic science and so on. The recent developments of computer automation and information technologies have extended analytical chemistry into a number of new biological fields. For example, automated DNA sequencing machines were the basis to complete human genome projects leading to the birth of genomics. Protein identification and peptide sequencing by mass spectrometry opened a new field of proteomics.

Analytical chemistry has been an indispensable area in the development of nanotechnology. Surface characterization instruments, electron microscopes and scanning probe microscopes enables scientists to visualize atomic structures with chemical characterizations.

WHAT IS ANALYTICAL CHEMISTRY?

Training in each of the five fields of chemistry provides a unique perspective to the study of chemistry. Undergraduate chemistry courses and textbooks are more than a collection of facts; they are a kind of apprenticeship. In keeping with this spirit, this chapter introduces the field of analytical chemistry and highlights the unique perspectives that analytical chemists bring to the study of chemistry.

Let's begin with a deceptively simple question. What is analytical chemistry? Like all fields of chemistry, analytical chemistry is too broad and too active a discipline for us to define completely. In this chapter, therefore, we will try to say a little about what analytical chemistry is, as well as a little about what analytical chemistry is not.

Analytical chemistry is often described as the area of chemistry responsible for characterizing the composition of matter, both qualitatively (Is there any lead in this sample?) and quantitatively (How much lead is in this sample?).

Most chemists routinely make qualitative and quantitative measurements. For this reason, some scientists suggest that analytical chemistry is not a separate branch of chemistry, but simply the application of chemical knowledge. In fact, you probably have preformed quantitative and qualitative analyses in other chemistry courses.

Defining analytical chemistry as the application of chemical knowledge ignores the unique perspective that analytical chemists bring to the study of chemistry. The craft of analytical chemistry is not in performing a routine analysis on a routine sample, which more appropriately is called chemical analysis, but in improving established analytical methods, in extending existing analytical methods to new types of samples, and in developing new analytical methods for measuring chemical phenomena.

Here is one example of this distinction between analytical chemistry and chemical analysis. Mining engineers evaluate the value of an ore by comparing the cost of removing the ore with the value of its contents. To estimate its value they analyse a sample of the ore. The challenge of developing and validating an appropriate quantitative analytical method is the analytical chemist's responsibility. After its development, the routine, daily application of the analytical method is the job of the chemical analyst.

Another distinction between analytical chemistry and chemical analysis is that analytical chemists work to improve and extend established analytical methods. For example, several factors complicate the quantitative analysis of nickel in ores, including nickel's unequal distribution within the ore, the ore's complex matrix of silicates and oxides, and the presence of other metals that may interfere with the analysis. A schematic outline of one standard analytical method in use during the late nineteenth century. The need for many reactions, digestions, and filtrations makes this analytical method both time-consuming and difficult to perform accurately.

THE ANALYTICAL PERSPECTIVE

Having noted that each field of chemistry brings a unique perspective to the study of chemistry, we now ask a second deceptively simple question. What is the analytical perspective? Many analytical chemists describe this perspective as an analytical approach to solving problems. Although there are probably as many descriptions of the analytical approach as there are analytical chemists, it is convenient for our purpose to define it as the five-step process.

Three general features of this approach deserve our attention. First, in steps 1 and 5 analytical chemists may collaborate with individuals outside the realm of analytical chemistry. In fact, many problems on which analytical chemists work originate in other fields. Second, the analytical approach includes a feedback loop (steps 2, 3, and 4) in which the result of one step may require re-evaluating the other steps. Finally, the solution to one problem often suggests a new problem.

Analytical chemistry begins with a problem, examples of which include evaluating the amount of dust and soil ingested by children as an indicator of environmental exposure to particulate based pollutants, resolving contradictory evidence regarding the toxicity of perfluoro polymers during combustion, and developing rapid and sensitive detectors for chemical and biological weapons. At this point the analytical approach may involve a collaboration between the analytical chemist and the individual or agency working on the problem.

Together they determine what information is needed. It also is important for the analytical chemist to understand how the problem relates to broader research goals or policy issues. The type of information needed and the problem's context are essential to designing an appropriate experimental procedure.

To design the experimental procedure the analytical chemist considers criteria such as the desired accuracy, precision, sensitivity, and detection limits; the urgency with which results are needed; the cost of a single analysis; the number of samples to be analysed; and the amount of sample available for analysis. Finding an appropriate balance between these parameters is frequently complicated by their interdependence.

For example, improving precision may require a larger amount of sample. Consideration is also given to collecting, storing, and preparing samples, and to whether chemical or physical interferences will affect the analysis. Finally a good experimental procedure may still yield useless information if there is no method for validating the results.

The most visible part of the analytical approach occurs in the laboratory. As part of the validation process, appropriate chemical and physical standards are used to calibrate any equipment and to standardize any reagents.

The data collected during the experiment are then analysed. Frequently the data is reduced or transformed to a more readily analyzable form. A statistical treatment of the data is used to evaluate accuracy and precision, and to validate the procedure. Results are compared to the original design criteria and the experimental design is reconsidered, additional trials are run, or a solution to the problem is proposed. When a solution is proposed, the results are subject to an external evaluation that may result in a new problem and the beginning of a new cycle.

Some scientists question whether the analytical approach is unique to analytical chemistry. Here, again, it helps to distinguish between a chemical analysis and analytical chemistry. For other analytically oriented scientists, such as a physical organic chemist or a public health officer, the primary emphasis is how the analysis supports larger research goals involving fundamental studies of chemical or physical processes, or improving access to medical care. The essence of analytical chemistry, however, is in developing new tools for solving problems, and in defining the type and quality of information available to other scientists.

QUANTITATIVE APPLICATIONS

Although not in common use, precipitation gravimetry still provides a reliable means for assessing the accuracy of other methods of analysis or for verifying the composition of standard reference materials. The general application of precipitation gravimetry to the analysis of inorganic and organic compounds.

Inorganic Analysis: The most important precipitants for inorganic cations are chromate, the halides, hydroxide, oxalate, sulfate, sulfide, and phosphate. A summary of selected methods, grouped by precipitant. Many inorganic anions can be determined using the same reactions by reversing the analyte and precipitant. For example, chromate can be determined by adding $BaCl_2$ and precipitating $BaCrO_4$. Methods for other selected inorganic anions. Methods for the homogeneous generation of precipitants.

Table. Selected Gravimetric Method for Inorganic Cations Based on Precipitation

Analyte	Precipitant	Precipitate Formed	Precipitate Weighed
Ba^{2+}	$(NH_4)_2CrO_4$	$BaCrO_4$	$BaCrO_4$
Pb^{2+}	K_2CrO_4	$PbCrO_4$	$PbCrO_4$
Ag^+	HCl	$AgCl$	$AgCl$
Hg_2^{2+}	HCl	Hg_2Cl_2	Hg_2Cl_2
Al^{3+}	NH_3	$Al(OH)_3$	Al_2O_3
Be^{2+}	NH_3	$Be(OH)_2$	BeO
Fe^{3+}	NH_3	$Fe(OH)_3$	Fe_2O_3
Ca^{2+}	$(NH_4)_2C_2O_4$	CaC_2O_4	$CaCO_3$ or CaO
Sb^{3+}	H_2S	Sb_2S_3	Sb_2S_3
As^{3+}	H_2S	As_2S_3	As_2S_3
Hg^{2+}	H_2S	HgS	HgS
Ba^{2+}	H_2SO_4	$BaSO_4$	$BaSO_4$
Pb^{2+}	H_2SO_4	$PbSO_4$	$PbSO_4$
Sr^{2+}	H_2SO_4	$SrSO_4$	$SrSO_4$
Be^{2+}	$(NH_4)_2HPO_4$	NH_4BePO_4	$Be_2P_2O_7$
Mg^{2+}	$(NH_4)_2HPO_4$	NH_4MgPO_4	$Mg_2P_2O_7$
Sr^{2+}	KH_2PO_4	$SrHPO_4$	$Sr_2P_2O_7$
Zn^{2+}	$(NH_4)_2HPO_4$	NH_4ZnPO_4	$Zn_2P_2O_7$

Table. Selected Gravimetric Methods for Inorganic Anions Based on Precipitation

Analyte	Precipitant	Precipitate Formed	Precipitate Weighed
CN^-	$AgNO_3$	$AgCN$	$AgCN$
I^-	$AgNO_3$	AgI	AgI
Br^-	$AgNO_3$	$AgBr$	$AgBr$
Cl^-	$AgNO_3$	$AgCl$	$AgCl$
ClO_3^-	$FeSO_4/AgNO_3$	$AgCl$	$AgCl$
SCN^-	$SO_2/CuSO_4$	$CuSCN$	$CuSCN$
SO_4^{2-}	$BaCl_2$	$BaSO_4$	$BaSO_4$

The majority of inorganic precipitants show poor selectivity. Most organic precipitants, however, are selective for one or two inorganic ions. Several common organic precipitants are listed in Table. Precipitation gravimetry continues to be listed as a standard method for the analysis of Mg^{2+} and SO_4^{2-} in water and wastewater analysis.

A description of the procedure for Mg^{2+} was discussed earlier in Method. Sulfate is analyzed by precipitating $BaSO_4$, using $BaCl_2$ as the precipitant. Precipitation is carried out in an acidic solution (acidified to pH 4.5–5.0 with HCl) to prevent the possible precipitation of $BaCO_3$ or $Ba_3(PO_4)_2$ and performed near the solution's boiling point.

Table. Reactions for the Homogeneous Preparation of Selected Inorganic Precipitants

Precipitant	Reaction
OH^-	$(NH_2)_2CO + 3H_2O \rightleftharpoons 2NH_4^+ + CO_2 + 2OH^-$
SO_4^{2-}	$NH_2HSO_3 + 2H_2O \rightleftharpoons NH_4^+ + H_3O^+ + SO_4^{2-}$
S^{2-}	$CH_3CSNH_2 + H_2O \rightleftharpoons CH_3CONH_2 + H_2S$
IO_3^-	$HOCH_2CH_2OH + IO_4^- \rightleftharpoons 2HCHO + H_2O + IO_3^-$
PO_4^{2-}	$(CH_3O)_3PO + 3H_2O \rightleftharpoons 3CH_3OH + H_3PO_4$
$C_2O_4^{2-}$	$(C_2H_5)_2C_2O_4 + 2H_2O \rightleftharpoons 2C_2H_5OH + H_2C_2O_4$
CO_3^{2-}	$Cl_3CCOOH + 2OH^- \rightleftharpoons CHCl_3 + CO_3^{2-} + H_2O$

Table. Selected Gravimetric Methods for Inorganic Cations Based on Precipitation with Organic Precipitants

Analyte	*Precipitant*	*Structure*	*Precipitate Formed*	*Precipitate Weighed*
Ni^{2+}	Dimethylgloxime	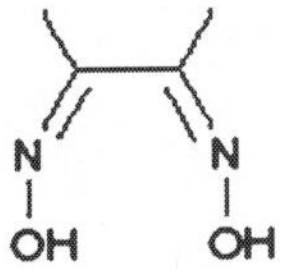 N–OH, N–OH	$Ni(C_4H_7O_2N_2)_2$	$Ni(C_4H_7O_2N_2)_2$
Fe^{3+}	Cupferron	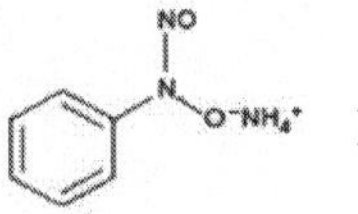 NO, N, $O^-NH_4^+$	$Fe(C_6H_5N_2O_2)_3$	Fe_2O_3
Cu2+	Cupron	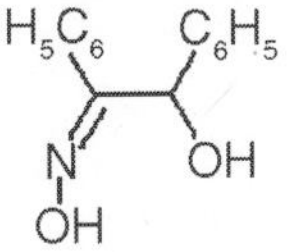 H_5C_6, C_6H_5, N, OH, OH	$CuC_{14}H_{11}O_2N$	$CuC_{14}H_{11}O_2N$
Co^{2+}	1-nitroso-2-naphthol	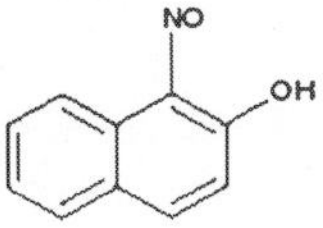 NO, OH	$Co(C_{10}H_6O_2N)_3$	Co or $CoSO_4$
K+	Sodium tetraphenylborate	$Na[B(C_6H_5)_4]$	$K[B(C_6H_5)_4]$	$K[B(C_6H_5)_4]$
NO_3^-	Nitro	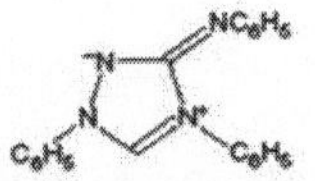 NC_6H_5, N, N, N^+, C_6H_5, C_6H_5	$C_{20}H_{16}N_4HNO_3$	$C_{20}H_{16}N_4HNO_3$

Table. Selected Gravimetric Methods for the Analysis of Organic Functional Groups and Heteroatoms Based on Precipitation

Analyte	Treatment	Precipitant	Precipitate
Organic halides	Oxidation with HNO_3 in presence	$AgNO_3$	AgX

R–X X = Cl, Br, I	of Ag^+		
Organic Halides R–X X = Cl, Br, 1	Combustion in O_2 (with Pt catalyst) in Presence of Ag^+	$AgNO_3$	AgX
Organic Sulfur	Oxidation with HNO_3 in presence of Ba^{2+}	$BaCl_2$	$BaSO_4$
Organic Sulfur	Combustion in O_2 (with Pt Catalyst) to Produce SO_2 and SO_3, which are collected in diulte H_2O_2	$BaCl_2$	$BaSO_4$
Alkoxy groups R′ – OR R = CH_3 or C_2H_5 or R′—C(=O)—OR	Reaction with Hl to produce Rl	$AgNO_3$	Agl
Alkimide group N—R (N bonded to R) R = CH_3, C_2H_5 N may be 1°, 2°, or 3°	Reaction with Hl to produce Rl	$AgNO_3$	Agl

The precipitate is digested at 80–90°C for at least 2h. Ashless filter paper pulp is added to the precipitate to aid in filtration. After filtering, the precipitate is ignited to constant weight at 800°C.

Alternatively, the precipitate can be filtered through a fineporosity fritted glass crucible (without adding filter paper pulp) and dried to constant weight at 105°C. This procedure is subject to a variety of errors, including occlusions of $Ba(NO_3)_2$,$BaCl_2$, and alkali sulfates.

Several organic functional groups or heteroatoms can be determined using gravimetric precipitation methods. The procedures for the alkoxy and alkimide functional groups are examples of indirect analyses. *Quantitative Calculations:* In precipitation gravimetry the relationship between the analyte and the precipitate is determined by the stoichiometry of the relevant reactions. The gravimetric calculations can be simplified by applying the principle of conservation of mass. The following example demonstrates the application of this approach to the direct analysis of a single analyte.

The simultaneous analysis of samples containing two analytes requires the isolation of two precipitates. The conservation of mass can be used to write separate stoichiometric equations for each precipitate. These equations can then be solved simultaneously for both analytes.

Qualitative Applications

Precipitation gravimetry can also be applied to the identification of inorganic and organic analytes, using precipitants such as those outlined in Tables. Since this does not require quantitative measurements, the analytical signal is simply the observation that a precipitate has formed. Although qualitative applications of precipitation gravimetry have been largely replaced by spectroscopic methods of analysis, they continue to find application in spot testing for the presence of specific analytes.

Evaluating Precipitation Gravimetry

Scale of Operation: The scale of operation for precipitation gravimetry is governed by the sensitivity of the balance and the availability of sample. To achieve an accuracy of ±0.1 per cent using an analytical balance with a sensitivity of ±0.1 mg, the precipitate must weigh at least 100 mg. As a consequence, precipitation gravimetry is usually limited to major or minor analytes, and macro or meso samples. The analysis of trace level analytes or micro samples usually requires a microanalytical balance.

Accuracy: For macro–major samples, relative errors of 0.1–0.2 per cent are routinely achieved. The principal limitations are solubility losses, impurities in the precipitate, and the loss of precipitate during handling. When it is difficult to obtain a precipitate free from impurities, an empirical relationship between the precipitate's mass and the mass of the analyte can be determined by an appropriate standardization.

Precision: The relative precision of precipitation gravimetry depends on the amount of sample and precipitate involved. For smaller amounts of sample or precipitate, relative precisions of 1–2 ppt are routinely obtained. When working with larger amounts of sample or precipitate, the relative precision can be extended to several parts per million. Few quantitative techniques can achieve this level of precision.

Sensitivity: For any precipitation gravimetric method, we can write the following general equation relating the signal (grams of precipitate) to the absolute amount of analyte in the sample

$$\text{Grams precipitat} = k \times \text{grams of analyte}$$

where k, the method's sensitivity, is determined by the stoichiometry between the precipitate and the analyte. Note that equation 8.13 assumes that a blank has been used to correct the signal for the reagent's contribution to the precipitate's mass. Consider, for example, the determination of Fe as Fe2O3. Using a conservation of mass for Fe we write

$$2 \times \text{moles } Fe_2O_3 = \text{moles Fe}$$

Converting moles to grams and rearranging yields an equation in the form of equation.

$$gFe_2O_3 = \frac{1}{2} \times \frac{FWFe_2O_3}{AWFe} \times gFe$$

where k is equal to

$$k = \frac{1}{2} \times \frac{FWFe_2O_3}{AWFe}$$

As can be seen from equation, we may improve a method's sensitivity in two ways. The most obvious way is to increase the ratio of the precipitate's molar mass to that of the analyte. In other words, it is desirable to form a precipitate with as large a formula weight as possible. A less obvious way to improve the calibration sensitivity is indicated by the term of 1/2 in equation, which accounts for the stoichiometry between the analyte and precipitate. Sensitivity also may be improved by forming precipitates containing fewer units of the analyte.

Selectivity: Due to the chemical nature of the precipitation process, precipitants are usually not selective for a single analyte. For example, silver is not a selective precipitant for chloride because it also forms precipitates with bromide and iodide.

Consequently, interferents are often a serious problem that must be considered if accurate results are to be obtained.

Time, Cost, and Equipment: Precipitation gravimetric procedures are time-intensive and rarely practical when analyzing a large number of samples. However, since much of the time invested in precipitation gravimetry does not require an analyst's immediate supervision, it may be a practical alternative when working with only a few samples. Equipment needs are few (beakers, filtering devices, ovens or burners, and balances), inexpensive, routinely available in most laboratories, and easy to maintain.

AN ALTERNATIVE METHOD FOR FILTERING THE PRECIPITATE

The precipitate is transferred to the filter in several steps. The first step is to decant the majority of the supernatant through the filter paper without transferring the precipitate. This is done to prevent the filter paper from becoming clogged at the beginning of the filtration process. Initial rinsing of the precipitate is done in the beaker in which the precipitation was performed.

These rinsings are also decanted through the filter paper. Finally, the precipitate is transferred onto the filter paper using a stream of rinse solution. Any precipitate clinging to the walls of the beaker is transferred using a rubber policeman (which is simply a flexible rubber spatula attached to the end of a glass stirring rod).

An alternative method for filtering the precipitate is a filtering crucible. The most common is a fritted glass crucible containing a porous glass disk filter. Fritted glass crucibles are classified by their porosity: coarse (retaining particles > 40–60 μm), medium (retaining particles > 10–15 μm), and fine (retaining particles > 4–5.5 μm). Another type of filtering crucible is the Gooch crucible, a porcelain crucible with a perforated bottom. A glass fibre mat is placed in the crucible to retain the precipitate, which is transferred to the crucible in the same manner described for filter paper. The supernatant is drawn through the crucible with the assistance of suction from a vacuum aspirator or pump.

Rinsing the Precipitate: Filtering removes most of the supernatant solution. Residual traces of the supernatant, however, must be removed to avoid a source of determinate error. Rinsing the precipitate to remove this residual material must be done carefully to avoid significant losses of the precipitate.

Of greatest concern is the potential for solubility losses. Usually the rinsing medium is selected to ensure that solubility losses are negligible. In many cases this simply involves the use of cold solvents or rinse solutions containing organic solvents such as ethanol. Precipitates containing acidic or basic ions may experience solubility losses if the rinse solution's pH is not appropriately adjusted. When coagulation plays an important role in determining particle size, a volatile inert electrolyte is often added to the rinse water to prevent the precipitate from reverting into smaller particles that may not be retained by the filtering device.

This process of reverting to smaller particles is called peptization. The volatile electrolyte is removed when drying the precipitate. When rinsing a precipitate there is a trade-off between introducing positive determinate errors due to ionic impurities from the precipitating solution and introducing negative determinate errors from solubility losses.

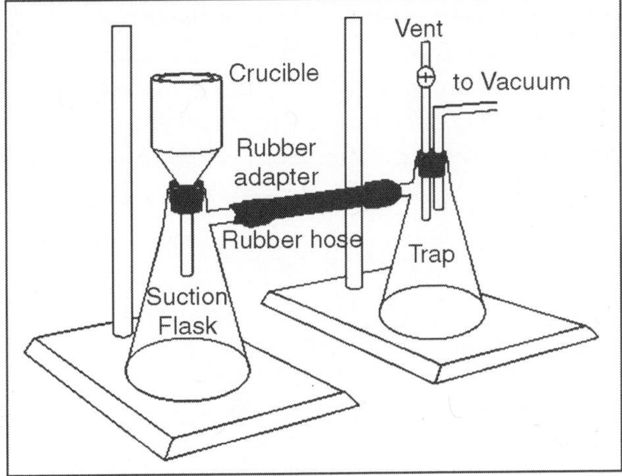

Fig. Procedure for Filtering Through a Filtering Crucible. The Trap is used to Prevent Water from a Water Aspirator from Backwashing into the Suction Flask.

In general, solubility losses are minimized by using several small portions of the rinse solution instead of a single large volume. Testing the used rinse solution for the presence of impurities is another way to ensure that the precipitate is not overrinsed. This can be done by testing for the presence of a targeted solution ion and rinsing until the ion is no longer detected in a freshly collected sample of the rinse solution. For example, when Cl^- is known to be a residual impurity, its presence can be tested for by adding a small amount of $AgNO_3$ to the collected rinse solution. A white precipitate of AgCl indicates that Cl^- is present and additional rinsing is necessary. Additional rinsing is not needed, however, if adding $AgNO_3$ does not produce a precipitate.

Drying the Precipitate: Finally, after separating the precipitate from its supernatant solution the precipitate is dried to remove any residual traces of rinse solution and any volatile impurities. The temperature and method of drying depend on the method of filtration, and the precipitate's desired chemical form.

A temperature of 110°C is usually sufficient when removing water and other easily volatilized impurities. A conventional laboratory oven is sufficient for this purpose. Higher temperatures require the use of a muffle furnace, or a Bunsen or Meker burner, and are necessary when the precipitate must be thermally decomposed before weighing or when using filter paper. To ensure that drying is complete the precipitate is repeatedly dried and weighed until a constant weight is obtained.

Filter paper's ability to absorb moisture makes its removal necessary before weighing the precipitate. This is accomplished by folding the filter paper over the precipitate and transferring both the filter paper and the precipitate to a porcelain or platinum crucible. Gentle heating is used to first dry and then to char the filter paper. Once the paper begins to char, the temperature is slowly increased. Although the paper will often show traces of smoke, it is not allowed to catch fire as any precipitate retained by soot particles will be lost. After the paper is completely charred the temperature is slowly raised to a higher temperature. At this stage any carbon left after charring is oxidized to CO_2.

Fritted glass crucibles cannot withstand high temperatures and, therefore, should only be dried in an oven at temperatures below 200°C. The glass fibre mats used in Gooch crucibles can be heated to a maximum temperature of approximately 500°C.

Composition of Final Precipitate: The quantitative application of precipitation gravimetry, which is based on a conservation of mass, requires that the final precipitate have a well-defined composition. Precipitates containing volatile ions or substantial amounts of hydrated water are usually dried at a temperature that is sufficient to completely remove the volatile species.

For example, one standard gravimetric method for the determination of magnesium involves the precipitation of $MgNH_4PO_4.6H_2O$. Unfortunately, this precipitate is difficult to dry at lower temperatures without losing an inconsistent amount of hydrated water and ammonia. Instead, the precipitate is dried at temperatures above 1000 °C, where it decomposes to magnesium pyrophosphate, Mg_2P_2O7.

An additional problem is encountered when the isolated solid is nonstoichiometric. For example, precipitating Mn^{2+} as $Mn(OH)_2$, followed by heating to produce the oxide, frequently produces a solid with a stoichiometry of MnOx, where x varies between 1 and 2. In this case the non-stoichiometric product results from the formation of a mixture of several oxides that differ in the oxidation state of manganese. Other non-stoichiometric compounds form as a result of lattice defects in the crystal structure.

Representative Method: The best way to appreciate the importance of the theoretical and practical is to carefully examine the procedure for a typical precipitation gravimetric method. Although each method has its own unique considerations, the determination of Mg^{2+} in water and wastewater by precipitating $MgNH_4PO_4.6H_2O$ and isolating $Mg_2P_2O_7$ provides an instructive example of a typical procedure.

CALIBRATION AND STANDARDS

CALIBRATION

With the exception of absolute methods of analysis that involve chemical reactions of known stoichiometry (*e.g.*, gravimetric and titrimetric determinations), a calibration or standardization procedure is required to establish the relation between a measured physico-chemical response to an analyte and the amount or concentration of the analyte producing the response.

Techniques and methods where calibration is necessary are frequently instrumental, and the detector response is in the form of an electrical signal. An important consideration is the effect of matrix components on the analyte detector signal, which may be supressed or enhanced, this being known as the matrix effect. When this is known to occur, matrix matching of the calibration standards to simulate the gross composition expected in the samples is essential (*i.e.* matrix components are added to all the analyte standards in the same amounts as are expected in the samples).

There are several methods of calibration, the choice of the most suitable depending on the characteristics of the analytical technique to be employed, the nature of the sample and the level of analyte(s) expected. These include:

- *External standardization:* A series of at least four calibration standards containing known amounts or concentrations of the analyte and matrix components, if required, is either prepared from laboratory chemicals of guaranteed purity (AnalaR or an equivalent grade) or purchased as a concentrated standard ready to use. The response of the detection system is recorded for each standard under specified and stable conditions and additionally for a blank, sometimes called a reagent blank (a standard prepared in an identical fashion to the other standards but omitting the analyte). The data is either plotted as a calibration graph or used to calculate a factor to convert detector responses measured for the analyte in samples into corresponding masses or concentrations.
- Standard addition.
- Internal standardization.

The last two methods of calibration are described in Topic. Instruments and apparatus used for analytical work must be correctly maintained and calibrated against reference values to ensure that measurements are accurate and reliable.

Performance should be checked regularly and records kept so that any deterioration can be quickly detected and remedied. Microcomputer and microprocessor controlled instrumentation often has built-in performance checks that are automatically initiated each time an instrument is turned on.

Some examples of instrument or apparatus calibration are

- Manual calibration of an electronic balance with certified weights;
- Calibration of volumetric glassware by weighing volumes of pure water.
- Calibration of the wavelength and absorbance scales of spectrophotometers with certified emission or absorption characteristics.
- Calibration of temperature scales and electrical voltage or current readouts with certified measurement equipment.

CHEMICAL STANDARD

Materials or substances suitable for use as chemical standards are generally single compounds or elements. They must be of known composition, and high purity and stability. Many are available commercially under the name AnalaR.

Primary standards, which are used principally in titrimetry to standardize a reagent (titrant) (*i.e.* to establish its exact concentration) must be internationally recognized and should fulfil the following requirements:

- Be easy to obtain and preserve in a high state of purity and of known chemical composition.
- Be non-hygroscopic and stable in air allowing accurate weighing.
- Have impurities not normally exceeding 0.02 per cent by weight.
- Be readily soluble in water or another suitable solvent.
- React rapidly with an analyte in solution.
- Other than pure elements, to have a high relative molar mass to minimize weighing errors.

Primary standards are used directly in titrimetric methods or to standardize solutions of secondary or working standards (*i.e.* materials or substances that do not fulfill all of the above criteria, that are to be used subsequently as the titrant in a particular method). Chemical standards are also used as reagents to effect reactions with analytes before completing the analysis by techniques other than titrimetry.

Some approved primary standards for titrimetric analysis are given in *Table*.

REFERENCE MATERIAL

Reference materials are used to demonstrate the accuracy, reliability and comparability of analytical results. A certified or standard reference material (CRM or SRM) is a reference material, the values of one or more properties of which have been certified by a technically valid procedure and accompanied by a traceable certificate or other documentation issued by a certifying body such as the

Some primary standards used in titrimetric analysis

Type of titration	Primary standard
Acid-base	Sodium carbonate, Na_2CO_3
	Sodium tetraborate, $Na_2B_4O_7.10H_2O$
	Potassium hydrogen phthalate, $KH(C_8H_4O_4)$
	Benzoic acid, C_6H_5COOH
Redox	Potassium dichromate, $K_2Cr_2O_7$
	Potassium iodate, KIO_3
	Sodium oxalate, $Na_2C_2O_4$
Precipitation (silver halide)	Silver nitrate, $AgNO_3$
	Sodium chloride, NaCl
Complexometric (EDTA)	Zinc, Zn
	Magnesium, Mg
	EDTA (disodium salt), $C_{10}H_{14}N_2O_8Na_2$

Bureau of Analytical Standards. CRMs or SRMs are produced in various forms and for different purposes and they may contain one or more certified components, such as

- Pure substances or solutions for calibration or identification.
- Materials of known matrix composition to facilitate comparisons of analytical data.
- Materials with approximately known matrix composition and specified components.

They have a number of principal uses, including

- Validation of new methods of analysis.
- Standardization/calibration of other reference materials.
- Confirmation of the validity of standardized methods.
- Support of quality control and quality assurance schemes.

EXPERIMENTAL FACTOR IN TESTING

Significance tests involve a comparison between a calculated experimental factor and a tabulated factor determined by the number of values in the set(s) of experimental data and a selected probability level that the conclusion is correct.

They are used for several purposes, such as:

- To check individual values in a set of data for the presence of determinate errors (bias).
- To compare the precision of two or more sets of data using their variances.
- To compare the means of two or more sets of data with one another or with known values to establish levels of accuracy.

Tests are based on a null hypothesis - an assumption that there is no significant difference between the values being compared. The hypothesis is accepted if the calculated experimental factor is less than the corresponding tabulated factor, otherwise it is rejected and there is said to be a significant difference between the values at the selected probability level. The conclusion should always be stated clearly and unambiguously.

Probability levels of 90 per cent, 95 per cent and 99 per cent are generally considered appropriate for most purposes, but it should be remembered that there are also corresponding 10 per cent, 5 per cent or 1 per cent probabilities, respectively, of the opposite conclusion being valid. For example, if a test indicates that the null hypothesis is correct and that there is no significant difference between two values at the 95 per cent probability level, it also allows the possibility that there is a significant difference at the 5 per cent level.

Separate tabular values for some significance test factors have been compiled for what are described as one-tailed and two-tailed tests. The exact purpose of the comparison that is to be made determines which table to use.

- The one-tailed test is used either to establish whether one experimental value is significantly greater than the other OR the other way around.
- The two-tailed test is used to establish whether there is a significant difference between the two values being compared, whether one is higher or lower than the other not being specified. The two-tailed test is by far the most widely used. Examples are given below.

OUTLIERS

Inspection of a set of replicate measurements or results may reveal that one or more is considerably higher or lower than the remainder and appears to be outside the range expected from the inherent effects of indeterminate (random) errors alone. Such values are termed outliers, or suspect values, because it is possible that they may have a bias due to a determinate error.

On occasions, the source of error may already be known or it is discovered on investigation, and the outlier(s) can be rejected without recourse to a statistical test. Frequently, however, this is not the case, and a test of significance such as the Q-test should be applied to a suspect value to determine whether it should be rejected and therefore not included in any further computations and statistical assessments of the data.

Q-TEST

Also known as Dixon's Q-test, this is one of several that have been devised to test suspected outliers in a set of replicates. It involves the calculation of a ratio, *Qexptl*, defined as the absolute difference between a suspect value and the value closest to it divided by the spread of all the values in the set:

Qexptl = suspect value – nearest value/(largest value – smallest value)

Qexptl is then compared with a tabulated value, *Qtab*, at a selected level of probability, usually 90 per cent or 95 per cent, for a set of *n* values. If *Qexptl* is less than *Qtab*, then the null hypothesis that there is no significant difference between the suspect value and the other values in the set is accepted, and the suspect value is retained for further data processing. However, if *Qexptl* is greater than *Qtab*, then the suspect value is regarded as an outlier and is rejected. A rejected value should NOT be used in the remaining calculations.

Table. Critical Values of Q at the 95% (P = 0.05) Level for a Two-tailed Test

Sample size	*Critical value*
4	0.831
5	0.717
6	0.621
7	0.570
8	0.524

Example 1. Four replicate values were obtained for the determination of a pesticide in river water

0.403, 0.410, 0.401, 0.380 mg dm^{-3}

Inspection of the data suggests that 0.380 μg dm^{-3} is a possible outlier.

Q_{expel}] |0.380 – 0.401 |/(0.410 – 0.380) = 0.021/0.03 = 0.70

Q_{tab} = 0.83 for four values at the 95 per cent probability level

As Q_{expel} is less than Q_{tab}, 0.380 μg dm^{-3} is not an outlier at the 95 per cent level and should be retained.

Example 2. If, in Example 1, three additional values of 0.400, 0.413 and 0.411 μg dm^{-3} were included, 0.380 μg dm^{-3} is still a possible outlier.

Q_{expei} = |0.080 – 0.400 |/(0.413 – 0.380) = 0.020/0.033 = 0.61

Q_{tab} = 0.57 for seven values at the 95 per cent probability level

Now, as *Qexptl* is greater than *Qtab*, 0.380 mg dm^{-3} is an outlier at the 95 per cent level and should be rejected. Note that because the three additional values are all around 0.4 mg dm^{-3}, the suspect value of 0.380 mg dm^{-3} appears even more anomalous.

F-TEST

This test is used to compare the precisions of two sets of data which may originate from two analysts in the same laboratory, two different methods of analysis for the same analyte or results from two different laboratories. A statistic, F, is defined as the ratio of the population variances, s_1 2/s_2 2, or the sample variances, s_1 2/s_2 2, of the two sets of data where the larger variance is always placed in the numerator so that $F \geq 1$.

If the null hypothesis is true, the variances are equal and the value of F will be one or very close to it. As for the Q-test, an experimental value, F_{exptl}, is calculated and compared with a tabulated value, F_{tab}, at a defined probability level, usually 90 per cent or 95 per cent, and for the number of degrees of freedom, $N - 1$, for each set of data. If F_{exptl} is less than F_{tab}, then the null hypothesis that there is no significant difference between the two variances and hence between the precision of the two sets of data, is accepted. However, if F_{exptl} is greater than F_{tab}, there is a significant difference between the two variances and hence between the precisions of the two sets of data.

Some values of F_{tab} at the 95 per cent probability level are given in *Table.* The columns in the table correspond to the numbers of degrees of freedom for the numerator set of data, while the rows correspond to the number of degrees of freedom for the denominator set. Two versions of the table are available, depending on the exact purpose of the comparison to be made: a one-tailed Ftest will show whether the precision of one set of data is significantly better than the other, while a two-tailed F-test will show whether the two precisions are significantly different.

Table. Critical Values of Fat the 95% (P = 0.05) Level for a Two-tailed Test

v_1	5	7	9
v_2			
5	7.146	6.853	6.681
7	5.285	4.995	4.823
9	4.484	4.197	4.026

v_1 = number of degrees of freedom of the fumerator, v_2 = number of freedom of the denominator

The application of a two-tailed F-test is demonstrated by the following example.

Example 3. A proposed new method for the determination of sulfate in an industrial waste effluent is compared with an existing method, giving the following results:

Method	*Mean/g dm^{-3} replicates*	*No. of of freedom*	*No. of degree*	*s/mg dm^{-3}*
Existing	72	8	7	3.38
New	72	8	7	1.50

Is there a significant difference between the precisions of the two methods

$$F_{exqel} = s^2_{existing} \Big/ s^2_{new} = \frac{(3.38)^2}{(1.50)^2} = 5.08$$

The two-tailed tabular value for F with 7 degrees of freedom for both the numerator and the denominator is

$$F_{7,7} = 5.00 \text{ at the 95\% probability level}$$

As *Fexptl* is greater than *Ftab*, the null hypothesis is rejected; the two methods are giving significantly different precisions.

T-TEST

This test is used to compare the experimental means of two sets of data or to compare the experimental mean of one set of data with a known or reference value. A statistic, *t*, is defined, depending on the circumstances, by one of three alternative equations.

Comparison of two experimental means, $\overline{x}_A$ and $\overline{x}_B$

$$t = \frac{(\overline{x}_A - \overline{x}_B)}{S_{pooled}} \times \left(\frac{NM}{N+M}\right)^{1/2}$$

where *spooled* is the pooled estimated standard deviation for sets A and B, and N and M are the numbers of values in sets A and B respectively. If $N = M$, then the second term reduces to $(N/2)^{1/2}$. A simplified version of equation (4), it can be used to calculate *spooled* as there are only two sets of data.

$$s_{pooled} = \left\{\left[(N-1)s_A^2 + (M-1)s_B^2\right]/[N+M-2]\right\}^{1/2}$$

In some circumstances, the use of equation (1) may not be appropriate for the comparison of two experimental means. Examples of when this may be the case are if

- The amount of sample is so restricted as to allow only one determination by each of the two methods.
- The methods are to be compared for a series of samples containing different levels of analyte rather than replicating the analysis at one level only.
- Samples are to be analyzed over a long period of time when the same experimental conditions cannot be guaranteed.

It may therefore be essential or convenient to pair the results (one from each method) and use a paired t-test where t is defined by

$$t = \frac{\overline{x}_4}{s_4} \times 4^{1/2}$$

d being the mean difference between paired values and *sd* the estimated standard deviation of the differences.

Comparison of one experimental mean with a known value, *m*

$$t = \frac{(\overline{x} - \mu)}{s} \times N^{1/2}$$

Using the appropriate equation, an experimental value, t_{exptl}, is calculated and compared with a tabulated value, t_{tab}, at a defined probability level, usually between 90 and 99 per cent, and for $N - 1$ degrees of freedom $(N + M - 2)$ degrees of freedom (equation).

If t_{exptl} is less than t_{tab}, then the null hypothesis that there is no significant difference between the two experimental means or between the experimental mean and a known value is accepted, *i.e.* there is no evidence of a bias. However, if t_{exptl} is greater than t_{tab}, there is a significant difference indicating a bias.

Both one-tailed and two-tailed t-tests can be used, depending on circumstances, but two-tailed are often preferred. The application of all three t-test equations is demonstrated by the following examples.

Table. Critical Values of t at the 95% and 99% (P = 0.05 and 0.01) Levels for a Two-tailed Test

Number of degrees of freedom	*95 per cent level*	*99 per cent level*
2	4.30	9.92
5	2.57	4.03
10	2.23	3.10
18	2.10	2.88

Example 1. Two methods for the determination of polyaromatic hydrocarbons in soils were compared by analyzing a standard with the following results:

No. of determinations by each method: 10

No. of degrees of freedom: 18

UV spectrophotometry: $\overline{x}$ = 28.00 mg kg^{-1} s = 0.30 mg kg^{-1}

Fluorimetry: $\overline{x}$ = 26.25 mg kg^{-1} s = 0.23 mg kg^{-1}

Do the mean results for the two methods differ significantly?

Equation (2) is first used to calculate a pooled standard deviation:

$$s_{pecial} = \left\{\left[(N-1)s_A^2 + (M-1)s_B^2\right]/[N+M-2]\right\}^{1/2} = \{(9\times 0.3^2 + 9\times 0.23^2)/18\}^{1/2}$$

s_{pecial} = 0.267 mg kg^{-1}

The equation (1) is used to evaluate t_{expei}

$$t_{expel} = \frac{(\overline{x}_A - \overline{x}_B)}{s_{pecial}} \times \left(\frac{NM}{N+M}\right)^{1/2}$$

$= \{(28.0 - 26.25)/0.267\} \times 5^{1/2} = 14.7$

For 18 degrees of freedom, the two-tailed value of t_{tab} at the 95 per cent probability level is 2.10, and the 99 per cent level it is 2.88.

As t_{exptl} is greater than t_{tab} at both the 95 and 99 per cent probability levels, there is a significant difference between the means of the two methods.

Example 2. A new high performance liquid chromatographic method for the determination of pseudoephedrine in a pharmaceutical product at two different levels was compared with an established method with the following results:

Table. Pseudoephedrine per dose(mg)

Method 1	*Method 2*
59.9	58.6
59.3	58.3
60.4	60.5
30.7	29.4
30.2	30.4
30.1	28.9

Do the means of the two methods differ significantly? Because the two levels of pseudoephedrine differ considerably, equation for a paired t-test is used to calculate t_{exptl}. The differences between the pairs of values are 1.3, 1.0, –0.1, 1.3, –0.2 and 1.2 mg per dose, and the estimated standard deviation of the differences from their mean of 0.750 mg per dose is 0.706 mg per dose. Substitution of these values into the equation gives

$$t_{\exp el} = \frac{\overline{x}_A}{s_A} \times N^{1/2} = (0.750/0.706) \times 6^{1/2} = 2.60$$

For 5 degrees of freedom, the two-tailed value of t_{tab} at the 95 per cent probability level is 2.57. As t_{exptl} is greater than t_{tab}, there is a significant difference between the means of the two methods. (*Note:* using equation would give a t_{exptl} value of 0.08 and an incorrect conclusion.)

ANALYSIS OF VARIANCE

Analysis of variance, also known as ANOVA, is a statistical technique for investigating different sources of variability associated with a series of results.

It enables the effect of each source to be assessed separately and compared with the other(s) using F-tests. Indeterminate or random errors affect all measurements, but additional sources of variability may also arise. The additional sources can be divided into two types:

- Additional random effects, described as random-effect factors;
- Specific effects from determinate sources, described as controlled or fixedeffect factors.

Where one additional effect may be present, a one-way ANOVA is used, whilst for two additional effects, two-way ANOVA is appropriate. Both involve much lengthier calculations than the simpler tests of significance, but facilities for these are available with computer packages such as Microsoft Excel and Minitab.

Typical examples of the use of ANOVA are:

- Analysis of a heterogeneous material where variation in composition is an additional random factor.
- Analysis of samples by several laboratories, methods or analysts where the laboratories, methods or analysts are additional fixed-effect factors.
- Analysis of a material stored under different conditions to investigate stability where the storage conditions provide an additional fixed-effect factor.

5

Method of Classifying Analytical Techniques

Analyzing a sample generates a chemical or physical signal whose magnitude is proportional to the amount of analyte in the sample. The signal may be anything we can measure; common examples are mass, volume, and absorbance. For our purposes it is convenient to divide analytical techniques into two general classes based on whether this signal is proportional to an absolute amount of analyte or a relative amount of analyte.

Consider two graduated cylinders, each containing 0.01 M $Cu(NO_3)^2$. Cylinder 1 contains 10 mL, or 0.0001 mol, of Cu^{2+}; cylinder 2 contains 20 mL, or 0.0002 mol, of Cu^{2+}. If a technique responds to the absolute amount of analyte in the sample, then the signal due to the analyte, *SA*, can be expressed as

$$S_A = kn_A$$

where *n*A is the moles or grams of analyte in the sample, and *k* is a proportionality constant. Since cylinder 2 contains twice as many moles of $Cu_2{}^+$ as cylinder 1, analyzing the contents of cylinder 2 gives a signal that is twice that of cylinder. A second class of analytical techniques are those that respond to the relative amount of analyte; thus

$$S_A = kC_A$$

where *C*A is the concentration of analyte in the sample. Since the solutions in both cylinders have the same concentration of Cu^{2+}, their analysis yields identical signals. Techniques responding to the absolute amount of analyte are called total analysis techniques. Historically, most early analytical methods used total analysis techniques, hence they are often referred to as "classical" techniques. Mass, volume, and charge are the most common signals for total analysis techniques, and the corresponding techniques are gravimetry, titrimetry, and coulometry.

With a few exceptions, the signal in a total analysis technique results from one or more chemical reactions involving the analyte. These reactions may involve any combination of precipitation, acid–base, complexation, or redox chemistry. The stoichiometry of each reaction, however, must be known to solve equation for the moles of analyte. Techniques, such as spectroscopy, potentiometry, and voltammetry, in which the signal is proportional to the relative amount of analyte in a sample are called concentration techniques. Since most concentration techniques rely on measuring an optical or electrical signal, they also are known as "instrumental" techniques. For a concentration technique, the relationship between the signal and the analyte is a theoretical function that depends on experimental conditions and the instrumentation used to measure the signal. For this reason the value of k in equation must be determined experimentally.

ELECTROCHEMICAL SENSORS: A POWERFUL TOOL IN ANALYTICAL CHEMISTRY

An overview of analytical chemistry development demonstrates that electrochemical sensors represent the most rapidly growing class of chemical sensors. A chemical sensor can be defined as a device that provides continuous information about its environment. Ideally, a chemical sensor provides a certain type of response directly related to the quantity of a specific chemical species. All chemical sensors consist of a transducer, with transforms the response into a detectable signal on modern instrumentation, and a chemically selective layer, which isolates the response of the analyte from its immediate environment. They can be classified according to the property to be determined as: electrical, optical, mass or thermal sensors and they are designed to detect and respond to an analyte in the gaseous, liquid or solid state.

Compared to optical, mass and thermal sensors, electrochemical sensors are especially attractive because of their remarkable detectability, experimental simplicity and low cost. They have a leading position among the presently available sensors that have reached the commercial stage and which have found a vast range of important applications in the fields of clinical, industrial, environmental and agricultural analyses.

There are three main types of electrochemical sensors: potentiometric, amperometric and conductometric. For potentiometric sensors, a local equilibrium is established at the sensor interface, where either the electrode or membrane potential is measured, and information about the composition of a sample is obtained from the potential difference between two electrodes. Amperometric sensors exploit the use of a potential applied between a reference and a working electrode, to cause the oxidation or reduction of an electroactive species; the resultant current is measured. On the other hand, conductometric sensors are involved with the measurement of conductivity at a series of frequencies.

Given the impressive progress in the electrochemical sensor area, and their growing impact on analytical chemistry, it would be impossible in the context of this review to mention all the advances in electrochemical sensor research. Other review articles on electrochemical sensors are found in the litreature. The main emphasis of this work is to provide a general overview of electrochemical sensors in a presentation of fundamental aspects and developments in a field in which a significant number of Brazilian researchers have been involved.

POTENTIOMETRIC SENSORS

Potentiometric sensors have found the most widespread practical applicability since the early 1930's, due to their simplicity, familiarity and cost. There are three basic types of potentiometric devices: ion- selective electrodes (IES), coated wire electrodes (CWES) and field effect transistors (FETS).

The ion selective electrode is an indicator electrode capable of selectively measuring the activity of a particular ionic species. In the classic configuration, such electrodes are mainly membrane-based devices, consisting of permselective ion-conducting materials, which separate the sample from the inside of the electrode. One electrode is the working electrode whose potential is determined by its environment.

The second electrode is a reference electrode whose potential is fixed by a solution containing the ion of interest at a constant activity. Since the potential of the reference electrode is constant, the value of the potential difference (cell potential) can be related to the concentration of the dissolved ion. The detailed theory of the processes at the membrane interface, which generate the potential, is available elsewhere.

Different strategies for producing an electrode that is selective to one species are based primarily on the nature and composition of the membrane material. Research in this area has opened up a whole series of applications to an almost unlimited number of analytes, where the only restriction is the selection of dopant and ionophore matrix of the membrane. Depending on the nature of the membrane, ISEs can be divided into three groups: glass, liquid or solid electrodes. More than two dozen ISEs are commercially available from Orion, Radiometer, Corning, Beckman, Hitachi and others, and they are widely used for the analysis of organic ions and of cationic or anionic species from various effluents, as well as in the manufacture and monitoring of drug, using selected response membrane electrodes.

The most widely used potentiometric device is the pH electrode, which has been used for several decades. Its success is attributed to a series of undisputed advantages, such as simplicity, rapidity, non-destructiveness, low cost, applicability to a wide concentration range and, particularly, to its extremely high selectivity for hydrogen ions. Glass electrodes, based on a thin ion-sensitive glass membrane, are the most common and they are available in many shapes and sizes. Nevertheless, measurements of pH can also be performed using other types of potentiometric sensors. Application of glass electrodes for other monovalent cations, including sodium, lithium, amonium and potassium sensors based on new glass compositions, have also been reported. Although the use of glass membrane electrode to measure the pH of solutions, has had enormous success, its utilization is limited to measurements in aqueous media. Determinations of hydrogen ion in non aqueous solutions require corrections.

Liquid–membrane-electrode ISE, based on water-immiscible liquid substances impregnated in a polymeric membrane, are widely used for direct potentiometric measurements of several polyvalent cations as well as certain anions. The polymeric membrane is used to separate the test solution from the inner compartment containing a solution of the target ion. The membrane–active recognition can be by a liquid ion exchanger or by a neutral macrocyclic compound having molecule–sized dimensions containing cavities to surround the target ions.

Recent advances in the field can be illustrated by the work of some Brazilian research groups. A review about analytical applications of conducting polymers in potentiometric sensors has been published by Kubota and co-workers. The construction of a hydrogen ion-selective potentiometric electrode based on a tridodecylamine ionophore dispersed in a poly(vinylchloride) (PVC) membrane, or poly(1-aminoanthracene) films has been described. In addition, Rodwedder *et al.* and Fatibello and co-workers have shown the use of coated graphite epoxy ion selective electrodes for determinaton of cations using ion-pair formation with tricaprylylmethylammonium cation in a PVC matrix. Using a similar system with incorporation of saccharinate anion and toluidine, Rover *et al.* have described the construction of a tubular ion selective electrode useful for determination of saccharin.

A more sensitive system for saccharin determination has been described by Alfaya *et al.* using a thin film of silsesquioxane 3-n-propylpyridinium chloride polymer coated on a graphite rod. The successful use of thin film electrodes modified, by nickel(II) hexacyanoferrate, for potassium determination has been described by Stradiotto and co-workers. The construction and application of ion selective electrodes applied for determination of pharmaceutical compounds, such as acetylsalicylic acid and vitamin B-6, has also been described. The development of homogenous membrane tubular ion-selective electrode for cyanide determination was investigated by Queiroz and co-workers. Also a tubular ion selective electrode based on the ionophore nanoctin on PVC membrane is described for ammonium determination.

There is a growing interest devoted to the development of potentiometric sensors based on solid-state membranes. They can be made of single cristals, polycrystalline pellets, or mixed crystals that are selective to anions (*e.g.*; F^-, Cl^-, Br^-, I^-, SCN^-, S^{2-}) or cations (*e.g.*;Cd^{2+}, Cu^{2+}, Pb^{2+}).

Coated-wire electrodes (CWEs) were first introduced in the mid of 1970's by Freiser. In the classical CWE design, a conductor is directly coated with an appropriate ion-selective polymer membrane (usually poly(vinyl chloride, poly(vinylbenzyl chloride) or poly(acrylic acid) to form an electrode system that is sensitive to electrolyte concentrations. The CWE response is similar to that of classical ISE, with regard to detectability and range of concentration. The great advantage is that the design eliminates the need for an internal reference electrode, resulting in benefits during miniaturization, for example. This is particularly useful for the *in vitro* and *in vivo* biomedical and clinical monitoring of different kind of analytes.

Ion-selective field effect transistors (ISFET) work as an extension of CWE. ISFET incorporate the ion-sensing membrane directly on the gate area of a field effect transistor (FET). The FET is a solid–state device that exhibits high-imput impedance and low-output impedance and therefore is capable of monitoring charge buildup on the ion-sensing membrane. The construction is based on the technology used to fabricate microelectronic chips, and the great contribution is that it is possible to prepare small multisensor systems with multiple gates, for sensing several ions simultaneously, while their small size permits the *in vivo* determination of analytes.

Alteration of the polymer matrix of potentiometric sensors by the immobilizaton of biological/biomedical materials has been pursued to alter the selectivity patterns of matrix ISE. Typical applications of these systems include enzyme determination and immunoassays. Enzyme electrodes are fabricated by immobilizing or covalenty binding an enzyme to an ion or gas-selective electrode or to an ISFET. The electrode will respond after the substrate diffuses to the immobilized enzyme and reacts. This reaction results in the release of a specific by product that can be detected directly by the ion-selective electrode. The success of the enzyme electrode depends, in part, on the immobilization of the enzyme layer, whose technology has experienced phenomenal growth in the recent years. An example is the use of laccase (enzyme label) for oxygen measurements.

The immobilization of intact microorganisms on the surface of an ISE has also been described in the literature. In addition, potentiometric immunosensors have been developed, which are based on the change in potential when either an antibody or an antigen is bound to its specific partner that has been immobilized on the electrode.

Recent advances can be illustrated by the use of carbon working and silver reference electrodes, supported in ceramic plates, for the analysis of human serum albumin (HSA). The anti-HSA IgG is adsorbed onto the surface of a thick-film electrode. After a competition between the bi-labeled analyte-tracer and the analyte for the Ab sites, and removal of unbound tracer, Bi^{3+} is released, which can be detected by potentiometric stripping analysis after a pre-concentration step.

Various on-line monitoring systems can benefit from the inherent specificity, wide scope, dynamic behaviour and simplicity of potentiometric sensors. They have become widely used as detectors in high speed automated flow analysers, such as air-segmented and flow-injection systems. In addition, the coupling of modern ion chromatography with potentiometric detection has been used with significant success. Miniaturization of ISE has also permitted their use as on-column detectors for capillary electrophoresis. Thus it is possible to conclude that potentiometric sensors have been important since 1930, when the commercialization of a glass electrode resulted in the foundation of one the most successful analytical instrument companies (Beckman Instruments). History also shown that, since 1960, when ion-selective electrodes revolutionized the approach to the difficult analysis of inorganic ions, up to now, the growth of patents for different formulations of glass, for different membrane types and for diverse shapes and sizes of electrodes testify to interest in the area. Therefore, there are many commercially available ion-sensing potentiometric devices.

These systems tend to be low in cost, simple to use, easily automated for rapid sampling, with low interferences from the matrix, and can be applied to small volumes. These advantages make potentiometric sensors an ideal choice for both clinical and industrial measurements where speed, simplicity, and accuracy are essential.

AMPEROMETRIC SENSORS

"Amperometric sensor" is a term that is an anomaly in itself. In the context of electroanalytical techniques, amperometric measurements are made by recording the current flow in the cell at a single applied potential. On the other hand, a voltammetric measurement is made when the potential difference across an electrochemical cell is scanned from one preset value to another and the cell current is recorded as a function of the applied potential. In both cases, the essential operational feature of voltammetric or amperometric devices is the transfer of electrons to or from the analyte The basic instrumentation requires controlled-potential equipment and the electrochemical cell consists of two electrodes immersed in a suitable electrolyte.

A more complex and usual arrangement involves the use of a three-electrode cell, one of the electrodes serving as a reference electrode. While the working electrode is the electrode at which the reaction of interest occurs, the reference electrode (*e.g.*; Ag/AgCl, Hg/Hg_2Cl_2) provides a stable potential compared to the working electrode. An inert conducting material (*e.g.*; platinum, graphite) is usually used as auxiliary electrode. A supporting electrolyte is required in controlled-potential experiments to eliminate electromigration effects, decrease the resistance of the solution and maintain the ionic strength constant. The theoretical aspects and experimental procedures have been well documented.

In order to limit the scope of the definition of amperometric sensor to a reasonable range, this paper will not include studies on the mechanistic aspects involving electron-transfer reactions and electrode process. Although these studies are extremely important to amperometric sensor evolution, our discussion will concern itself only with the recent advances in quantitative amperometric sensors.

The performance of amperometric sensors is strongly influenced by the working electrode material. Consequently, much effort has been devoted to electrode fabrication and maintenance. Although classical electrochemical measurements of anaytes started in 1922, when Heyrovsky invented the dropping mercury electrode, for which he received a Nobel prize, solid electrodes constructed of noble metals and various forms of carbon have been the sensors of choice in recent years. The impressive progress in this area, and its growing impact on electroanalytical chemistry, is more recent. Research into electrochemical sensors is proceeding in a number of directions.

Mercury was very attractive as an electrode material for many years because it has an extended cathodic potential range window, high reproducibility and a renewable surface. The hanging mercury drop electrode or mercury film electrode was the most popular working electrode for stripping analysis. Numerous methods have been developed for determinations of metals, anions, organometallics and organic compounds by stripping analysis at concentration levels down to $1x10^{-10}$ mol L^{-1} using a simple pre-concentration step. The limited anodic potential of mercury electrodes and its toxicity are the principal disadvantages of the method.

Solid electrodes (carbon, platinum, gold, silver, nickel, copper, dimensionally stable anions) have been very popular as electrode materials because of their versatile potential window, low background current, low cost, chemical inertness, and suitability for various sensing and detection applications. The proliferation of chemically modified electrodes (QME) generates a modern approach to electrode systems, where deliberate alteration of the electrode surface is introduced by incorporation of an appropriate surface modifier. While there was some progress in the development of amperometric sensors in the early 1970's, most amperometric sensors developed were applicable only to a controlled, stringent set of laboratory conditions.

Fortunately the miniaturization of the working electrode has gained much attention and microeletrodes (ME) were developed with dimensions not greater than 2 mm, increasing the possibility of *in vivo* and *in vitro* measurements with small apparatus. This resulted in a predictable boon for the development of amperometric sensors for real sample analysis. An example of this advance was the development of biossensors.

They are capable of permitting a biospecific reagent, immobilized or retained at a suitable electrode, to convert the biological recognition process into a quantitative amperometric response. Particularly relevant is the coupling of sensitive amperometric sensors with liquid chromatography and flow injection systems. Finally, with the growing need to carry out decentralized analytical determinations remote from the laboratory, the development of screen-printed electrodes has provided modern amperometric sensors that are potentially portable. The new possibilities of design and fabrication of the electrodes, with the incorporation of microelectrodes, microelectrodes arrays, and chemically modified electrodes into various highly sensitive sensor systems (biosensors), has increased industrial and clinical interest. An overview of each of these contributions to electroanalysis.

CHEMICALLY MODIFIED ELECTRODES

Immobilisation of chemical microstrutures onto electrode surfaces has been a major growth area in electrochemistry in recent years. Chemically modified electrodes (CME) result from a deliberate immobilization of a modifier agent onto the electrode surface though chemical reactions, chemisorption, composite formation or polymer coating. Compared to conventional electrodes, greater control of electrode characteristics and reactivity is achieved by surface modificaton, since the immobilization transfers the physicochemical properties of the modifier to the electrode surface. Its application to electroanalysis is well documented in the literature.

The interest in this area is motivated by the many potential applications of such a system. Examples include development of electrocatalytic systems with high chemical selectivity and activity, coating of semiconducting electrodes with photosensitising and anticorrosive properties, electrochromic displays, microelectrochemical devices for the field of molecular electronics and electrochemical sensors. The main benefits to analytical applications include acceleration of electron-transfer reactions, preferential accumulation, or selective membrane permeation and interferent exclusion. Such steps can impart higher selectivity, detectability and stability to amperometric devices, which have been extensively reviewed.

One of the common approaches for incorporating a modifier onto the surface has been coverage with an appropriate polymer film. Most polymers are applied to electrode surfaces by a combination of adsorptive attraction and low solubility in the electrolyte solution, using pre-formed polymers or electrochemical polymerisation.

An early approach in the use of pre-formed polymers was their use as anchoring groups for coordinating metal complexes to pyrolytic graphite electrodes. Examples include poly(4-vinylpyridine) (PVP), poly(vinylferrocene) (PVF), poly(p-nitrostyrene), metalopolymers and others, used as redox monomers in polymer modification schemes after synthetic procedures.

The advantages of preconcentrating CME were obtained by coating the electrode surface with a thin film of an ion-exchange polymer. The strategy of coating an anion-exchange polymer onto an electrode surface is similar to the stripping voltammetric method principle. This technique uses solid electrodes coated with a thin layer of an ion-exchange polymer, which allows the quick pre-concentration and simultaneous amperometric detection of an ion redox analyte. The response depends on the concentration of electroactive species incorporated by ion exchange inside the polymeric layer. If an ion exchanger characterized by proper selectivity is used, it is possible to determine trace levels of ionic electroactive analytes (metals, pharmaceutical compounds, biological compounds) at submicromolar concentration levels. The most popular cation exchanging system is Nafion, although poly(styrene sulphonate), poly(vinyl sulphate), deprotonated poly(acrylic acid) and poly(-L-lysine), Tosflex, poly(vinylpyridine) and cellulose acetate have received significant attention. In addition, the preconcentrating agent also may acts as a permeselective coating, offering selectivity by exclusion from the surface of unwanted matrix constituents, while allowing transport of the target analyte. Pratical examples are obtained by use of size-exclusion coatings such as poly(diaminobenzene), hydrophobic lipid or charged-exclusion ionomeric Nafion coatings. Electronically conducting polymers [such as poly(pyrrole), poly(thiophene), and poly(aniline)] have attracted considerable attention due to their ability to incorporate or expel ionic species during oxidative electro-polymerization of their monomers. The vast literature concerning fundamental investigations and proposed applications has been reviewed.

Differents types of inorganic films, such as metal oxide, clay, zeolite, and metal ferrocyanide, can also be formed on electrode surfaces. These films are of interest because they frequently show well-defined structures, are thermally and chemically stable, and are usually inexpensive and readily available.

The improvements in selectivity and detectability due modifications of the electrode surface have been investigated by several Brazilian researchers. Serrano and co-workers have described the catalytic detection of NADH by using carbon paste electrodes coated by electropolymerized films of 3,4-dihidroxybenzaldehyde. Toma and co-workers also have shown the use of electrostatically assembled tetraruthenated cobalt and zinc porphyrin multi-layer films applied for the development of amperometric sensors. Kominsky and Bertotti have developed amperometric sensors based on electrodes modified by an electrochemically deposited molybdenum oxide layer and used them to determine some inorganic compounds. Kubota and co-workers have studied the construction and application of an amperometric sensor for dopamine using a glassy carbon electrode with a Nafion membrane doped with a (2,2´-bipyridil) copper (II) choride complex. Stradiotto and co-workers have reported a new way of modifying an electrode by combining vanadium pentoxide xerogel and a cationic surfactant able to detect anions, such as penatcyanonitrosylferrate (II) and hexacyanoferrate (III). Thin films of materials such as Prussian blue and related materials can be formed on the electrode surface and show interesting properties. Mattos and Gorton have shown that glucose biosensors can be constructed using Prussian blue films, with excellent performance.

These numerous studies, testify to the reliability of amperometric sensing devices as an electroanaytical technique ready to be used for solving real analytical problems. The extensive possibilities of application, as well as high selectivity in preconcentration and detection, makes these devices appropriate for analysis in complex samples.

Amperometric Biosensors

Biosensors consist of a biological component (biochemical receptor) coupled to a transducer that will convert the biological into an electrical signal. A biological molecule, like an enzyme, cell, tissue slice, organelle, peptide, antibody, nucleic acid *etc.*, ensures molecular recognition and may transform the analyte in some way. The electrochemical tranducer (potentiometric, amperometric or conductimetric), monitors the change in properties. The choice of transducer depends on the biological reaction. Molecular recognition could be accompanied by chemical conversion of the analyte to its respective products, which are determined by the biocatalytic sensor. In other cases, when an antibody is used, the biospecific recognition systems and interactions take place without analyte conversion, resulting in an affinity sensor.

As the electrochemical biosensor is a self-contained integrated device, the biochemical receptor should be retained in direct spatial contact with an electrochemical transduction element.

Clark and Lyons were the pioneers in indicating the possibility of using enzyme-containing membranes coupled to eletrodes. Updike and Hicks prepared the first biosensor by polymerizing a gel containing glucose oxidase onto a Clark oxygen electrode. Besides the entrapment of the biochemical receptor behind a membrane and within a polymeric matrix, the receptors could be immobilized within self-assembled monolayers (SAM) or bilayer lipid membranes (BLM), covalently bonded on membranes or previously activated surfaces or obtained through bulk modification of the entire electrode material. The biosensor lifetime depends on the stability of the immobilized material.

The first biosensor was proposed to oxidize glucose to gluconic acid using an electrode to amperometrically detect the consumption of oxygen, since its consumption is proportional to glucose concentration. Glucose oxidation actually uses a prosthetic group (flavin adenine dinucleotide - FAD) to transfer the electrons from the substrate. After substrate oxidation, $FADH_2$ returns to FAD in the presence of oxygen. The electron acceptor (mediator-M_{ox}) is reduced to M_{red} by the cofator and then oxidized to M_{ox} when in contact with the anode, which is polarized at the appropriate potential. This is the second-generation biosensor.

A more direct method is to have no mediator but a electrode on which the reduced enzyme can be directly oxidized. Kulys *et al.* suggested organic conducting salts to transfer electrons directly and are responsible for initiating third generation biosensors.

Investigations about construction, characterization and application of amperometric biosensors were initiated a long time ago in Brazil. Currently, several groups have introduced wide improvements in this field. An example is the amperometric biosensor for fructose detection developed by Kubota and co-workers using the immobilization of D-fructose 5-dehidrogenase on an electrode coated with polypyrrole film. Yamanaka and co-workers have demonstrated the development of an amperometric biosensor based on chlolinesterase for the determination of pesticides.

The use of a sol-gel matrix to develop new biosensors has been reviewed by Alfaya and Kubota. Kubota and co-workers have also described a flow system method for continuous determination of phenolic compounds in environmental matrices employing lacase and tyrosinase-based biosensors as detectors. The use of a DNA biosensor has been investigated by Serrano and co-workers. The authors have demonstrated that the use of a DNA biosensor is a very promising tool for the investigation and study of the action of selected drugs. Mattos *et al.* have also developed glucose biosensors using the enzyme glucose oxidase immobilized on Prussian blue films electrodeposited on glassy carbon electrodes. The analytical use of vegetable tissue or crude extracts as enzymatic sources has been extensively studied by Fatibello and co-workers and by Angnes and co-workers. Examples of many biosensors and/or flow injection procedures for analysis of compounds of biological, environmental, food, pharmaceutical and industrial interest have been described using vegetable tissues or crude extracts from sweet potato, avocado, zucchini, peach, yam, cassava, artichoke, turnip, horseradish, and tissues from fruits of the palm tree *latania sp.*

A wide variety of compounds can be analysed with an amperometric biosensor, according to the bioreceptor chosen. Instead of the pure redox enzymes, organelles (chloroplasts, mitochondria), animal and vegetable tissues or microorganisms can be immobilized onto the transducer. Recently, mimetic molecules have also been suggested, instead of enzymes or antibodies.

The interaction between the antibody (A_b) and antigen (A_g) is characterised by its affinity constant, K, defined by the equilibrium concentrations of the A_b-A_g complex, the free analyte and the free antibody binding sites (where $K= [A_b\text{-}A_g]/[A_b]\,[A_g]$). Antibody reversibly binds antigen or hapten with high affinity constants between 5×10^4 and 10^{12} L mol^{-1}. The extreme affinity of antigen-antbody interactions results in great detectability using these immunoassay methods.

As the physicochemical changes caused by the binding of the analyte to the biomolecule are usually very slight, in many cases auxiliary reactions have to be coupled. Indirect methods are based on labelling one of the immunochemical reaction partners with an enzyme, catalyst, fluorophore, electroactive substance or liposome. In indirect sensors, the free conjugate must be separated from the immunoreagent-bound conjugate before the marker activity can be measured.

There are different schemes of immunoassay and the most useful for immunosensors are the sandwich and the competitive binding reactions. A sandwich format uses an A_b covalently immobilized onto the surface of the sensor, the test sample is incubated and, after washing, the complex is further exposed to a second marked A_b. Good detectability is reached with this format but it can not be applied to small molecules, because it is difficult for them to have more than one determinant epitope.

In the competitive case, antigen determination by competitive binding assays with labeled antigen is used. A known amount of labeled antigen is added to a sample with an unknown antigen concentration. When this mixture reacts with the A_b immobilized at the transducer surface labeled antigen and antigen from the sample (unlabeled antigen) compete for the binding sites of the antibody. The affinity of antibodies for labeled antigen is usually lower than their affinity for unlabeled (free) antigen, owing to steric factors. After removal of all unbound antigen in the washing step, the amount of bound labeled antigen is determined through an enzyme-catalyzed reaction. The higher the quantity of A_g in the sample, the lower is the fraction of labeled antigen in the A_g-A_b complex. The final stage of this immunoasssay format is in fact the quantitative detection of the label.

Yamanaka and co-workers has been contributed to the field of immunoassay development in Brazil, where investigations of biosensors based on the reaction between antigen and antibody coupled to an amperometric transducer are currently in progress.

Overviews about enzyme labels suitable for amperometric, potentiometric and conductimetric detection are found in the literature. Examples of applications of immuno devices in clinical medicine and in environmental monitoring are also given, but virtually unlimited areas of application exist in food analysis and various other industrial processes.

Other Versatile Amperometric Devices

Microelectrodes. The need for amperometric sensors with small dimensions has induced extensive investigations on the use of microelectrode systems. The miniaturization of the working electrode was first proposed by Wightman for the *in vivo* and *in vitro* determination of neurotransmitters. Nevertheless, microelectrodes exhibit several other attractive possibilities, including the exploration of microscopic domains, detection in microflow system, time-resolved probing of processes in single cells, the *in vivo* monitoring of neurochemical events (*e.g.* dopamine release) and analyses of very small sample volumes.

Electrodes of different materials have been miniaturized in many geometrical shapes, with the common characteristic that the electrode dimension is significantly smaller than the diffusion layer at their surface. Due to the greatly reduced double-layer capacitance of microelectrodes, associated with their small area, and radial diffusion to the edges of microelectrodes, the signal–to-background characteristics are much better than with conventional electrodes. Because of their geometries and low current intensities, it is possible to work in highly resistive situations, including low dielectric solvents, at low temperature, in the gas phase, with ionically conductive polymers, with oil-based lubrificants and with solutions without addition of a supporting electrolyte.

Such dimensions offer obvious analytical advantages. Fine metal (Pt, Au, Ir) wires, carbon fibres or thin metal films are commonly used for these preparations. The geometric characteristics of the microeletrodes lead to analytical methods developed using a simple droplet (mL).

In Brazil, some groups have extensively contributed to the development of this area. Machado, Avacca and co-workers have reviewed some theoretical aspects and studied the utilization of Pt, Co, or Ni-Co microelectrodes (UME) to characterize copper electrocrystallization and hydrogen evolution. Mori and Bertotti have investigated the chemical reaction between nitrite and iodide by using microelectrodes and Cavalheiro and co-workers have shown the application of nanostructured carbon fibre disk microelectrodes for determination of adenosine and uric acid in physiological buffers. Nevertheless, the most successful works in these area are published by Angnes and co-workers. These authors have described a notable performance of gold electrodes constructed from recordable CDs. The versatile, economic and very sensitive electrodes are applied for quantification of several organic and inorganic compounds. Stradiotto and co-workers have also used microelectrodes applied to small volumes.

The potential for use of UME in biological fluids has been dramatically increased with chemical modifications, such as ion-exchange polymers that have made their applicability in biological fluids more viable. The charged membranes are selective for specific ions and reduce the interference of proteins and other materials indigenous to the physiological system.

Recently, arrangements of surfaces containing several microelectrodes with small dimensions, has gained attention. The surface is obtained with uniform (array) or random (emsemble) dispersion of a conductor in an insulating matrix. Such surfaces consist of spaced microdisks or ensembles fabricated by deposition of a metal conductor into the pores of a microporous host membrane. Nanoelectrode emsembles (NEE) have improved selectivity because the Faradaic current is proportional to the total geometric area while the background current depends on the sum of areas of the electrode elements in the NEE. As a result, they present a large collective current and have became a very promising area. In Brazil, Angnes and co-workers, Pasquini and co-workers, Fungaro and Brett and Fertonani *et al.* have contributed to the design and application of these devices.

Screen printed electrodes. The demand for easy to use, portable electroanalytical sensors, providing the opportunity to perform clinical, environmental or industrial analyses away from a centralized laboratory, seems to have been fulfilled with screen printing technology.

Screen printing is a simple process for reproducibly and inexpensively depositing electrode substrates onto inert backing supports, usually PVC or ceramic materials. In general, the support is coated with a conducting ink, which is recovered by a second film with isolating characteristics with the aim to define an electrical contact at the extremities and other useful areas related to the electrode surfaces (working, reference and auxiliary electrodes). They are potentially portable, simple to operate, reliable, and inexpensive to manufacture. Many ink-type substrates have been used for sensor construction, the most successful have included carbon and the noble metals. Taking into consideration that carbon is inexpensive and conductive, it is the most used substrate for the economic fabrication of electrodes. Screen–printing equipment and inks for screen printed preparations are commercially available and applications in a variety of situations are described in the literature.

Chemically modified electrodes may also be produced by incorporating specific reagents into the screen-printing inks, increasing the selectivity and detectability of measurements. The modest cost of screen printed electrodes, coupled with their versatility allow the production of disposable devices, which are relevant to industrial laboratories.

Amperometric sensors based on screen printed electrodes have shown promise and successful applications for the determination of several naturally occurring biomolecular drugs, potential environmental toxins and industrial pollutants. Angnes and co-workers and Hart *et al.* have published excellent reviews describing in more depth some theoretical aspects and methods involving screen printed electrodes, unmodified or modified, for determination of such important organic compounds as ascorbic acid, uric acid, glucose, paracetamol, salicylic acid, ethanol, cholesterol, DNA, and environmental pollutants such as formaldehyde, hydrazines, nitrite, phenol, pesticides and herbicides and some metals. Recently, Kubota and co-workers have also described the use of a simple laccase-based screen-printed electrode for monitoring phenolic compounds in environmental samples.

In conclusion, screen printed electrodes are widely accepted in analytical chemistry as disposable electrochemical sensors, providing rapid and simple analytical determinations. Numerous devices based on screen-printing have matured to commercial success. For example blood glucose determinations using screen printed electrochemical sensors may now be viewed as the technique of choice for personal monitoring.

Automated Flow Systems

Automation within the clinical or industrial laboratory has seen major improvements in speed, reproducibility, cost, simplicity, and versatility. Applications of electrochemical detectors in automated flow systems have gained popularity in recent years. Such detectors have been particularly useful in connection with liquid chromatography and flow injection systems.

Amperometric detection is usually performed by controlling the potential of the working electrode at a fixed value and monitoring the current as a function of time. The magnitude of the peak current serves as a measure of the analyte concentration that passed through the detector. The potentiometric detector was discussed previously.

The coupling of electrochemical sensors to automated flow methods; segmented flow analysis (discrete segments divided by air bubbles) or flow injection analysis (FIA), where there is an unsegmented flowing stream into which reproducible sample volumes are injected, have combined the possibility of on-line sample pre-treatment aimed to isolate the analyte with the high good detectabillity of amperometric detection.

A wide range of cell designs and flow-injections systems using amperometric detection have been investigated by Gutz and co-workers and Angnes and co-workers. These authors have used amperometric detectors based on both thin-layer and wall-jet configurations. It is possible to employ many types, from mercury drop electrodes to carbon-fibre microelectrodes, unmodified or with a modified surface capable of having electrocatalytic or preconcentration properties. A typical flow injection analysis of ascorbic acid in pharmaceutical compounds using amperometric detection with electrodes modified with Prussian blue films was developed by Stradiotto and co-workers.

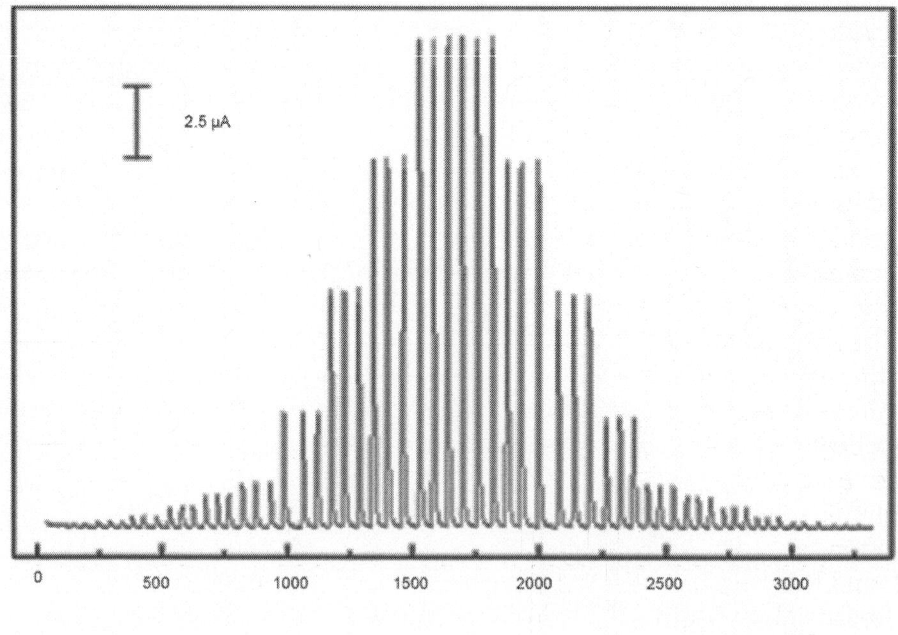

Fig. Flow injection amperometry for increasing and decreasing concentration of ascorbic acid (0.005, 0.013, 0.025, 0.050, 0.075, 0.10, 0.25, 0.50 and 1.0 mmol L^{-1}) in a 1.0 mol 1^{-1} KCl (pH 3.7) solution. Injection volume. 90μL; flow-rate mL min^{-1}.

The expansion of this research to incorporate the amperometric detector (highly sensitive and selective) into liquid chromatograph, adds the great advantage of a separation technique. Electrochemical detection is usually performed by controlling the potential of the working electrode, capable of reducing or oxidizing an electroactive analyte and monitoring the current as a function of the elution time. Taking into consideration that many organic compounds of biological, environmental or pharmaceutical interest present oxidable or reducible sites in their molecule, the applicability of electrochemical detection is enormous. The analytical capability of these detectors can be enhanced using surface modification and/or electrochemical procedures to increase the selectivity of the amperometric sensor or by using more than one working electrode operating simultaneously at different potentials using parallel detection. The power of electrochemical detection applied to capillary electrophoresis is also reviewed by Silva.

Conductometric Sensors

Sensors in this group rely on changes of electric conductivity of a film or a bulk material, whose conductivity is affected by the analyte present. Conductimetric methods are fundamentally non-selective. Only with the advent of modified surfaces for selectivity and much improved instrumentation have these become more viable methods for designing sensors. There are some very practical considerations that make conductimetric methods attractive, such as low cost and simplicity, since no reference electrodes are needed. Improved instrumentation has contributed to rapid and easy determination of analytes, based only on the measurement of conductivity. The most predominant materials used in these sensors will be examined first. The methods of numerical processing of the analytical signal will then be discussed, followed by certain phenomena or properties which lend themselves to the purpose of sensing.

Thin films are used mostly as gas sensors, due to their conductivity changes following surface chemisorption. For example, due to oxygen chemisorption, CdS films can be used as oxygen sensors. Porous films of $MnWO_4$ can work as a humidity sensor, oxides doped with copper or copper oxide can be very sensitive to gases containing H_2S and semiconducting Ga_2O_3 thin films can detect CH_4. Polymers, either conductive themselves or with a modifier, are also often used. Polypyrrole can detect volatile amines, or when doped with ClO_4^- and tosylate, it can be made into a sensor for NH_3.

Resistance measured from a DC current is typical. Often the measurement is done with AC current (impedance), which also allows one to obtain changes in capacitive impedance. For example zeolite layers can change the capacitance in response to the presence of combustion gases. Discrimination between conductivity and capacitance through impedance can extend the use of sensing materials. Impedance sensors were described for NO_2 and tobacco smoke, odor detection or determination of water in an oil-in-water emulsion. Intelligent materials are frequently developed for this purpose.

Humidity is very often a parameter for which mixed oxide conductivity sensors are used although it can also be an unwanted interference. A gas sensor for ethanol and acetone vapors, insensitive to humidity, can be based on sintered bismuth tungstate. Reports on the development of a multichannel aroma sensor are also found. Impedance measurements at various frequencies allow one to detect antibody-antigen binding.

In Brazil, the advantages of the use of conductometric detection coupled to flow injection analysis was described by Jardim and co-workers to measure CO_2 in the atmosphere. Fatibello and Borges have also shown a flow injection conductometric system appropriate to evaluate acidity in industrial hydrated ethyl alcohol.

In addition, an automatic rain sensors based on a conductometric sensor was proposed by Gutz and co-workers, and Krug and co-workers. Tubino and Barros have also described a conductometric sensor coupled to a flow injection system to measure soil extracts and acidity in vinegars. Recently, the development of electrochemical sensors based on measurement of electrochemical admittance spectroscopy were also investigated by Stradiotto and co-workers.

SELECTING AN ANALYTICAL METHOD

A method is the application of a technique to a specific analyte in a specific matrix. Methods for determining the concentration of lead in drinking water can be developed using any of the techniques mentioned in the previous section.

Insoluble lead salts such as $PbSO_4$ and $PbCrO_4$ can form the basis for a gravimetric method. Lead forms several soluble complexes that can be used in a complexation titrimetric method or, if the complexes are highly absorbing, in a spectrophotometric method.

Lead in the gaseous free-atom state can be measured by an atomic absorption spectroscopic method. Finally, the availability of multiple oxidation states (Pb, Pb^{2+}, Pb^{4+}) makes coulometric, potentiometric, and voltammetric methods feasible.

The requirements of the analysis determine the best method. In choosing a method, consideration is given to some or all the following design criteria: accuracy, precision, sensitivity, selectivity, robustness, ruggedness, scale of operation, analysis time, availability of equipment, and cost. Each of these criteria is considered in more detail in the following sections.

Accuracy

Accuracy is a measure of how closely the result of an experiment agrees with the expected result. The difference between the obtained result and the expected result is usually divided by the expected result and reported as a percent relative error.

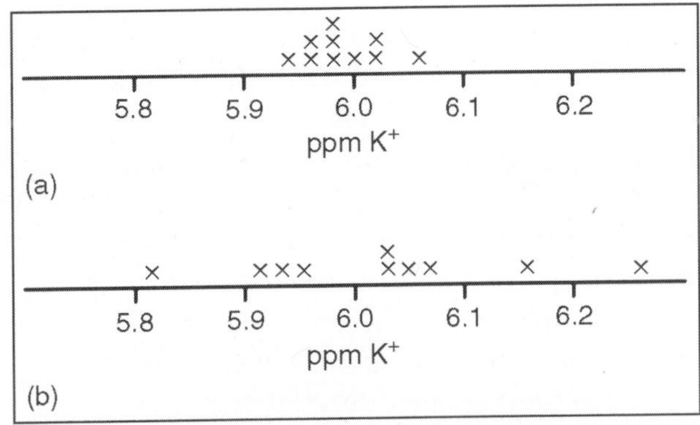

Fig. Two Determinations of the Concentration of K+ in Serum, Showing the Effect of Precision.

$$\%\text{Error} = \frac{\text{obtained result} - \text{expected result}}{\text{expected result}} \times 100$$

Analytical methods may be divided into three groups based on the magnitude of their relative errors.3 When an experimental result is within 1 per cent of the correct result, the analytical method is highly accurate. Methods resulting in relative errors between 1 per cent and 5 per cent are moderately accurate, but methods of low accuracy produce relative errors greater than 5 per cent.

The magnitude of a method's relative error depends on how accurately the signal is measured, how accurately the value of k in equations is known, and the ease of handling the sample without loss or contamination. In general, total analysis methods produce results of high accuracy, and concentration methods range from high to low accuracy.

Precision

When a sample is analyzed several times, the individual results are rarely the same. Instead, the results are randomly scattered. Precision is a measure of this variability. The closer the agreement between individual analyses, the more precise the results. For example, in determining the concentration of K^+ in serum, the results shown in Figure are more precise than those in Figure. It is important to realise that precision does not imply accuracy. That the data in Figure are more precise does not mean that the first set of results is more accurate. In fact, both sets of results may be very inaccurate.

As with accuracy, precision depends on those factors affecting the relationship between the signal and the analyte. Of particular importance are the uncertainty in measuring the signal and the ease of handling samples reproducibly. In most cases the signal for a total analysis method can be measured with a higher precision than the corresponding signal for a concentration method.

Sensitivity

The ability to demonstrate that two samples have different amounts of analyte is an essential part of many analyses. A method's sensitivity is a measure of its ability to establish that such differences are significant. Sensitivity is often confused with a method's detection limit. The detection limit is the smallest amount of analyte that can be determined with confidence.

Sensitivity is the change in signal per unit change in the amount of analyte and is equivalent to the proportionality constant, *k*, in equations. If *S*A is the smallest increment in signal that can be measured, then the smallest difference in the amount of analyte that can be detected is

$$\Delta n_A = \frac{\Delta S_A}{k} \text{(total analysis method)}$$

$$\Delta C_A = \frac{\Delta S_A}{k} \text{(concentration method)}$$

Suppose that for a particular total analysis method the signal is a measurement of mass using a balance whose smallest increment is ±0.0001 g. If the method's sensitivity is 0.200, then the method can conceivably detect a difference of as little as

$$\Delta n_A = \frac{\pm 0.0001 g}{0.200} = \pm 0.0005 g$$

in the absolute amount of analyte in two samples. For methods with the same *S*A, the method with the greatest sensitivity is best able to discriminate among smaller amounts of analyte.

Selectivity

An analytical method is selective if its signal is a function of only the amount of analyte present in the sample. In the presence of an interferent, equations can be expanded to include a term corresponding to the interferent's contribution to the signal, *S*I,

$$S_{samp} = S_A + S_1 = k_A n_A + k_1 n_1 \quad \text{(total analysis method)}$$

$$S_{samp} = S_A + S_1 = k_A C_A + k_1 C_1 \quad \text{(concentration method)}$$

where *S*samp is the total signal due to constituents in the sample; *k*A and *k*I are the sensitivities for the analyte and the interferent, respectively; and *n*I and *C*I are the moles (or grams) and concentration of the interferent in the sample.

The selectivity of the method for the interferent relative to the analyte is defined by a selectivity coefficient, *K*A,I

$$K_{A,I} = \frac{k_1}{k_A}$$

which may be positive or negative depending on whether the interferent's effect on the signal is opposite that of the analyte.* A selectivity coefficient greater than +1 or less than –1 indicates that the method is more selective for the interferent than for the analyte. Solving equation for *k*I

$$k_I = K_{A,I} \times k_A$$

substituting into equations, and simplifying gives

$$S_{samp} = k_A(n_A + K_{A,I} \times n_1) \qquad \text{(total analysis method)}$$

$$S_{samp} = k_A(C_A + K_{A,I} \times C_1) \qquad \text{(concentration method)}$$

The selectivity coefficient is easy to calculate if *kA* and *kI* can be independently determined. It is also possible to calculate *KA,I* by measuring *S*samp in the presence and absence of known amounts of analyte and interferent.

Knowing the selectivity coefficient provides a useful way to evaluate an interferent's potential effect on an analysis. An interferent will not pose a problem as long as the term *KA,I* × *nI* in equation is significantly smaller than *nA*, or *KA,I* × *CI* in equation is significantly smaller than *C*A. Not surprisingly, methods whose signals depend on chemical reactivity are often less selective and, therefore, more susceptible to interferences. Problems with selectivity become even greater when the analyte is present at a very low concentration.

Robustness and Ruggedness

For a method to be useful it must provide reliable results. Unfortunately, methods are subject to a variety of chemical and physical interferences that contribute uncertainty to the analysis. When a method is relatively free from chemical interferences, it can be applied to the determination of analytes in a wide variety of sample matrices. Such methods are considered robust.

Random variations in experimental conditions also introduce uncertainty. If a method's sensitivity is highly dependent on experimental conditions, such as temperature, acidity, or reaction time, then slight changes in those conditions may lead to significantly different results. A rugged method is relatively insensitive to changes in experimental conditions.

Scale of Operation

Another way to narrow the choice of methods is to consider the scale on which the analysis must be conducted. Three limitations of particular importance are the amount of sample available for the analysis, the concentration of analyte in the sample, and the absolute amount of analyte needed to obtain a measurable signal. The first and second limitations define the scale of operations shown in Figure; the last limitation positions a method within the scale of operations.

The scale of operations in Figure shows the analyte's concentration in weight percent on the *y*-axis and the sample's size on the *x*-axis. For convenience, we divide analytes into major (1 per cent w/w), minor (0.01 per cent w/w – 1% w/w), trace (10–7% w/w – 0.01 per cent w/w) and ultratrace

(10–7 per cent w/w) components, and we divide samples into macro (0.1 g), meso (10 mg – 100 mg), micro (0.1 mg – 10 mg) and ultramicro (0.1 mg) sample sizes. Note that both the *x*-axis and the *y*-axis use a logarithmic scale. The analyte's concentration and the amount of sample used provide a characteristic description for an analysis. For example, samples in a macro–major analysis weigh more than 0.1 g and contain more than 1 per cent analyte.

Diagonal lines connecting the two axes show combinations of sample size and concentration of analyte containing the same absolute amount of analyte. As shown in Figure, for example, a 1-g sample containing 1 per cent analyte has the same amount of analyte (0.010 g) as a 100-mg sample containing 10 per cent analyte or a 10-mg sample containing 100 per cent analyte.

Since total analysis methods respond to the absolute amount of analyte in a sample, the diagonal lines provide an easy way to define their limitations. Consider, for example, a hypothetical total analysis method for which the minimum detectable signal requires 100 mg of analyte.

Using figure, the diagonal line representing 100 mg suggests that this method is best suited for macro samples and major analytes. Applying the method to a minor analyte with a concentration of 0.1 per cent w/w requires a sample of at least 100 g. Working with a sample of this size is rarely practical, however, due to the complications of carrying such a large amount of material through the analysis.

Alternatively, the minimum amount of required analyte can be decreased by improving the limitations associated with measuring the signal. For example, if the signal is a measurement of mass, a decrease in the minimum amount of analyte can be accomplished by switching from a conventional analytical balance, which weighs samples to ±0.1 mg, to a semimicro (±0.01 mg) or microbalance (±0.001 mg). Concentration methods frequently have both lower and upper limits for the amount of analyte that can be determined. The lower limit is dictated by the smallest concentration of analyte producing a useful signal and typically is in the parts per million or parts per billion concentration range. Upper concentration limits exist when the sensitivity of the analysis decreases at higher concentrations.

An upper concentration level is important because it determines how a sample with a high concentration of analyte must be treated before the analysis. Consider, for example, a method with an upper concentration limit of 1 ppm (micrograms per milliliter). If the method requires a sample of 1 mL, then the upper limit on the amount of analyte that can be handled is 1 mg.

Using figure, and following the diagonal line for 1 g of analyte, we find that the analysis of an analyte present at a concentration of 10 per cent w/w requires a sample of only 10 mg! Extending such an analysis to a major analyte, therefore, requires the ability to obtain and work with very small samples or the ability to dilute the original sample accurately.

Using this example, analyzing a sample for an analyte whose concentration is 10 per cent w/w requires a 10,000-fold dilution. Not surprisingly, concentration methods are most commonly used for minor, trace, and ultratrace analytes, in macro and meso samples.

Equipment, Time, and Cost

Finally, analytical methods can be compared in terms of their need for equipment, the time required to complete an analysis, and the cost per sample. Methods relying on instrumentation are equipment-intensive and may require significant operator training. For example, the graphite furnace atomic absorption spectroscopic method for determining lead levels in water requires a significant capital investment in the instrument and an experienced operator to obtain reliable results.

Other methods, such as titrimetry, require only simple equipment and reagents and can be learned quickly.

The time needed to complete an analysis for a single sample is often fairly similar from method to method. This is somewhat misleading, however, because much of this time is spent preparing the solutions and equipment needed for the analysis. Once the solutions and equipment are in place, the number of samples that can be analyzed per hour differs substantially from method to method.

This is a significant factor in selecting a method for laboratories that handle a high volume of samples. The cost of an analysis is determined by many factors, including the cost of necessary equipment and reagents, the cost of hiring analysts, and the number of samples that can be processed per hour. In general, methods relying on instruments cost more per sample than other methods.

Making the Final Choice

Unfortunately, the design criteria discussed earlier are not mutually independent.8 Working with smaller amounts of analyte or sample, or improving selectivity, often comes at the expense of precision. Attempts to minimize cost and analysis time may decrease accuracy. Selecting a specific method requires a careful balance among these design criteria. Usually, the most important design criterion is accuracy, and the best method is that capable of producing the most accurate results. When the need for results is urgent, as is often the case in clinical labs, analysis time may become the critical factor.

The best method is often dictated by the sample's properties. Analyzing a sample with a complex matrix may require a method with excellent selectivity to avoid interferences. Samples in which the analyte is present at a trace or ultratrace concentration usually must be analyzed by a concentration method. If the quantity of sample is limited, then the method must not require large amounts of sample.

Determining the concentration of lead in drinking water requires a method that can detect lead at the parts per billion concentrations. Selectivity is also important because other metal ions are present at significantly higher concentrations. Graphite furnace atomic absorption spectroscopy is a commonly used method for determining lead levels in drinking water because it meets these specifications. The same method is also used in determining lead levels in blood, where its ability to detect low concentrations of lead using a few microliters of sample are important considerations.

ANALYTICAL TECHNIQUES AND METHODS

ANALYTICAL TECHNIQUES

There are numerous chemical or physico-chemical processes that can be used to provide analytical information. The processes are related to a wide range of atomic and molecular properties and phenomena that enable elements and compounds to be detected and/or quantitatively measured under controlled conditions. The underlying processes define the various analytical techniques. The more important of these are listed in *Table*, together with their suitability for qualitative, quantitative or structural analysis and the levels of analyte(s) in a sample that can be measured.

Atomic and molecular spectrometry and chromatography, which together comprise the largest and most widely used groups of techniques, can be further subdivided according to their physico-chemical basis.

Spectrometric techniques may involve either the emission or absorption of electromagnetic radiation over a very wide range of energies, and can provide qualitative, quantitative and structural information for analytes from major components of a sample down to ultra-trace levels.

Analytical techniques and principal applications

Technique	Property measured	Principal areas of application
Gravimetry	Weight of pure analyte or compound of known stoichiometry	Quantitative for major or minor components
Titrimetry	Volume of standard reagent solution reacting with the analyte	Quantitative for major or minor components
Atomic and molecular spectrometry	Wavelength and intensity of electromagnetic radiation emitted or absorbed by the analyte	Qualitative, quantitative or structural for major down to trace level components
Mass spectrometry	Mass of analyte or fragments of it	Qualitative or structural for major down to trace level components isotope ratios
Chromatography and electrophoresis	Various physico-chemical properties of separated analytes	Qualitative and quantitative separations of mixtures at major to trace levels
Thermal analysis	Chemical/physical changes in the analyte when heated or cooled	Characterization of single or mixed major/minor components
Electrochemical analysis	Electrical properties of the analyte in solution	Qualitative and quantitative for major to trace level components
Radiochemical analysis	Characteristic ionizing nuclear radiation emitted by the analyte	Qualitative and quantitative at major to trace levels

Spectrometric techniques and principal applications

Technique	Basis	Principal applications
Plasma emission spectrometry	Atomic emission after excitation in high temperature gas plasma	Determination of metals and some non-metals mainly at trace levels
Flame emission spectrometry	Atomic emission after flame excitation	Determination of alkali and alkaline earth metals
Atomic absorption spectrometry	Atomic absorption after atomization by flame or electrothermal means	Determination of trace metals and some non-metals
Atomic fluorescence spectrometry	Atomic fluorescence emission after flame excitation	Determination of mercury and hydrides of non-metals at trace levels
X-ray emission spectrometry	Atomic or atomic fluorescence emission after excitation by electrons or radiation	Determination of major and minor elemental components of metallurgical and geological samples
γ-spectrometry	γ-ray emission after nuclear excitation	Monitoring of radioactive elements in environmental samples
Ultraviolet/visible spectrometry	Electronic molecular absorption in solution	Quantitative determination of unsaturated organic compounds
Infrared spectrometry	Vibrational molecular absorption	Identification of organic compounds
Nuclear magnetic resonance spectrometry	Nuclear absorption (change of spin states)	Identification and structural analysis of organic compounds
Mass spectrometry	Ionization and fragmentation of molecules	Identification and structural analysis of organic compounds

The most important atomic and molecular spectrometric techniques and their principal applications are listed in *Table*.

Chromatographic techniques provide the means of separating the components of mixtures and simultaneous qualitative and quantitative analysis, as required. The linking of chromatographic and spectrometric techniques, called hyphenation, provides a powerful means of separating and identifying unknown compounds. Electrophoresis is another separation technique with similarities to chromatography that is particularly useful for the separation of charged species. The principal separation techniques and their applications are listed in *Table*.

Separation techniques and principal applications

Technique	Basis	Principal applications
Thin-layer chromatography	Differential rates of migration of analytes through a stationary phase by movement of a liquid or gaseous mobile phase	Qualitative analysis of mixtures
Gas chromatography		Quantitative and qualitative determination of volatile compounds
High-performance liquid chromatography		Quantitative and qualitative determination of nonvolatile compounds
Electrophoresis	Differential rates of migration of analytes through a buffered medium	Quantitative and qualitative determination of ionic compounds

ANALYTICAL METHODS

An analytical method consists of a detailed, stepwise list of instructions to be followed in the qualitative, quantitative or structural analysis of a sample for one or more analytes and using a specified technique. It will include a summary and lists of chemicals and reagents to be used, laboratory apparatus and glassware, and appropriate instrumentation.

The quality and sources of chemicals, including solvents, and the required performance characteristics of instruments will also be specified as will the procedure for obtaining a representative sample of the material to be analysed. This is of crucial importance in obtaining meaningful results. The preparation or pre-treatment of the sample will be followed by any necessary standardization of reagents and/or calibration of instruments under specified conditions.

Qualitative tests for the analyte(s) or quantitative measurements under the same conditions as those used for standards complete the practical part of the method. The remaining steps will be concerned with data processing, computational methods for quantitative analysis and the formatting of the analytical report. The statistical assessment of quantitative data is vital in establishing the reliability and value of the data, and the use of various statistical parameters and tests is widespread.

Many standard analytical methods have been published as papers in analytical journals and other scientific literature, and in textbook form. Collections by trades associations representing, for example, the cosmetics, food, iron and steel, pharmaceutical, polymer plastics and paint, and water industries are available. Standards organizations and statutory authorities, instrument manufacturers' applications notes, the Royal Society of Chemistry and the US Environmental Protection Agency are also valuable sources of standard methods.

Often, laboratories will develop their own in-house methods or adapt existing ones for specific purposes. Method development forms a significant part of the work of most analytical laboratories, and method validation and periodic revalidation is a necessity. Selection of the most appropriate analytical method should take into account the following factors:

- The purpose of the analysis, the required time scale and any cost constraints.
- The level of analyte(s) expected and the detection limit required.
- The nature of the sample, the amount available and the necessary sample preparation procedure.
- The accuracy required for a quantitative analysis.
- The availability of reference materials, standards, chemicals and solvents, instrumentation and any special facilities.
- Possible interference with the detection or quantitative measurement of the analyte(s) and the possible need for sample clean-up to avoid matrix interference.
- The degree of selectivity available "methods may be selective for a small number of analytes or specific for only one.
- Quality control and safety factors.

METHOD VALIDATION

Analytical methods must be shown to give reliable data, free from bias and suitable for the intended use. Most methods are multi-step procedures, and the process of validation generally involves a stepwise approach in which optimised experimental parameters are tested for robustness (ruggedness), that is sensitivity to variations in the conditions, and sources of errors investigated. A common approach is to start with the final measurement stage, using calibration standards of known high purity for each analyte to establish the performance characteristics of the detection system (*i.e.* specificity, range, quantitative response (linearity), sensitivity, stability and reproducibility).

Robustness in terms of temperature, humidity and pressure variations would be included at this stage, and a statistical assessment made of the reproducibility of repeated identical measurements (replicates). The process is then extended backwards in sequence through the preceding stages of the method, checking that the optimum conditions and performance established for the final measurement on analyte calibration standards remain valid throughout. Where this is not the case, new conditions must be investigated by modification of the procedure and the process repeated. A summary of this approach is shown in *Figure* in the form of a flow diagram.

Step 1 — Performance characteristics of detector for single analyte calibration standards

Step 2 — Process repeated for mixed analyse calibration standards

Step 3 — Process repeated for analyte calibration standards with possible interfering substances and for reagent blanks

Step 4 — Process repeated for analyte calibration standards with anticipated matrix components to evaluate matrix interference

Step 5 — Analysis of 'spiked' simulated sample matrix, i.e., matrix with added known amounts of analyte(s), to test recoveries

Step 6 — Field trials in routine laboratory with more junior personnel to test ruggedness

Flow chart for method validation

At each stage, the results are assessed using appropriate statistical tests and compared for consistency with those of the previous stage. Where unacceptable variations arise, changes to the procedure are implemented and the assessment process repeated. The performance and robustness of the overall method are finally tested with field trials in one or more routine analytical laboratories before the method is considered to be fully validated.

THE DISTRIBUTION OF MEASUREMENTS AND RESULTS

An analysis, particularly a quantitative analysis, is usually performed on several replicate samples. How do we report the result for such an experiment when results for the replicates are scattered around a central value? To complicate matters further, the analysis of each replicate usually requires multiple measurements that, themselves, are scattered around a central value.

Consider, for example, the data in Table for the mass of a penny. Reporting only the mean is insufficient because it fails to indicate the uncertainty in measuring a penny's mass. Including the standard deviation, or other measure of spread, provides the necessary information about the uncertainty in measuring mass.

Nevertheless, the central tendency and spread together do not provide a definitive statement about a penny's true mass. If you are not convinced that this is true, ask yourself how obtaining the mass of an additional penny will change the mean and standard deviation.

How we report the result of an experiment is further complicated by the need to compare the results of different experiments. For example, Table shows results for a second, independent experiment to determine the mass of a U.S. penny in circulation.

Although the results shown in Tables are similar, they are not identical; thus, we are justified in asking whether the results are in agreement. Unfortunately, a definitive comparison between these two sets of data is not possible based solely on their respective means and standard deviations.

Developing a meaningful method for reporting an experiment's result requires the ability to predict the true central value and true spread of the population under investigation from a limited sampling of that population. In this section we will take a quantitative look at how individual measurements and results are distributed around a central value.

Table. Results for a Second Determination of the Mass of a United States Penny in Circulation.

Penny	Mass (g)
1	3.052
2	3.141
3	3.083
4	3.083
5	3.048
$\overline{X}$	3.081
s	0.037

Populations and Samples

In the previous section we introduced the terms "population" and "sample" in the context of reporting the result of an experiment. Before continuing, we need to understand the difference between a population and a sample. A population is the set of all objects in the system being investigated. These objects, which also are members of the population, possess qualitative or quantitative characteristics, or values, that can be measured. If we analyse every member of a population, we can determine the population's true central value and spread. The probability of occurrence for a particular value, *P*(*V*), is given as

$$P(V) = \frac{M}{N}$$

where *V* is the value of interest, *M* is the value's frequency of occurrence in the population, and *N* is the size of the population. In determining the mass of a circulating United States penny, for instance, the members of the population are all United States pennies currently in circulation, while the values are the possible masses that a penny may have.

In most circumstances, populations are so large that it is not feasible to analyse every member of the population. This is certainly true for the population of circulating U.S. pennies. Instead, we select and analyse a limited subset, or sample, of the population. The data in Tables 4.1 and 4.10, for example, give results for two samples drawn at random from the larger population of all U.S. pennies currently in circulation.

Probability Distributions for Populations

To predict the properties of a population on the basis of a sample, it is necessary to know something about the population's expected distribution around its central value. The distribution of a population can be represented by plotting the frequency of occurrence of individual values as a function of the values themselves. Such plots are called probability distributions. Unfortunately, we are rarely able to calculate the exact probability distribution for a chemical system. In fact, the probability distribution can take any shape, depending on the nature of the chemical system being investigated. Fortunately many chemical systems display one of several common probability distributions. Two of these distributions, the binomial distribution and the normal distribution, are discussed next.

Binomial Distribution The binomial distribution describes a population in which the values are the number of times a particular outcome occurs during a fixed number of trials. Mathematically, the binomial distribution is given as

$$P(X,N) = \frac{N!}{X!(N-X)!} \times p^X \times (1-p)^{N-X}$$

where $P(X,N)$ is the probability that a given outcome will occur X times during N trials, and p is the probability that the outcome will occur in a single trial.* If you flip a coin five times, $P(2,5)$ gives the probability that two of the five trials will turn up "heads." A binomial distribution has well-defined measures of central tendency and spread. The true mean value, for example, is given as

$$\mu = Np$$

and the true spread is given by the variance

$$\sigma^2 = Np(1-p)$$

or the standard deviation

$$\sigma = \sqrt{Np(1-p)}$$

The binomial distribution describes a population whose members have only certain, discrete values. A good example of a population obeying the binomial distribution is the sampling of homogeneous materials. The binomial distribution can be used to calculate the probability of finding a particular isotope in a molecule. Normal Distribution The binomial distribution describes a population whose members have only certain, discrete values. This is the case with the number of 13C atoms in a molecule, which must be an integer number no greater then the number of carbon atoms in the molecule. A molecule, for example, cannot have 2.5 atoms of 13C. Other populations are considered continuous, in that members of the population may take on any value.

The most commonly encountered continuous distribution is the Gaussian, or normal distribution, where the frequency of occurrence for a value, X, is given by

$$f(X) = \frac{1}{\sqrt{2\pi\sigma^2}} \exp\left[\frac{-(X-\mu)^2}{2\sigma^2}\right]$$

The shape of a normal distribution is determined by two parameters, the first of which is the population's central, or true mean value, ?, given as

$$\mu = \frac{\sum_{i=1}^{N} X_i}{n}$$

where n is the number of members in the population. The second parameter is the population's variance, μ^2, which is calculated using the following equation

$$\sigma^2 = \frac{\sum_{i=1}^{N} (X_i - \mu)^2}{n}$$

Examples of normal distributions with $\mu = 0$ and $\sigma^2 = 25$, 100 or 400, are shown in figure. Several features of these normal distributions deserve attention. First, note that each normal distribution contains a single maximum corresponding and that the distribution is symmetrical about this value. Second, increasing the population's variance increases the distribution's spread while decreasing its height. Finally, because the normal distribution depends solely on μ and σ^2, the area, or probability of occurrence between any two limits defined in terms of these parameters is the same for all normal distribution curves.

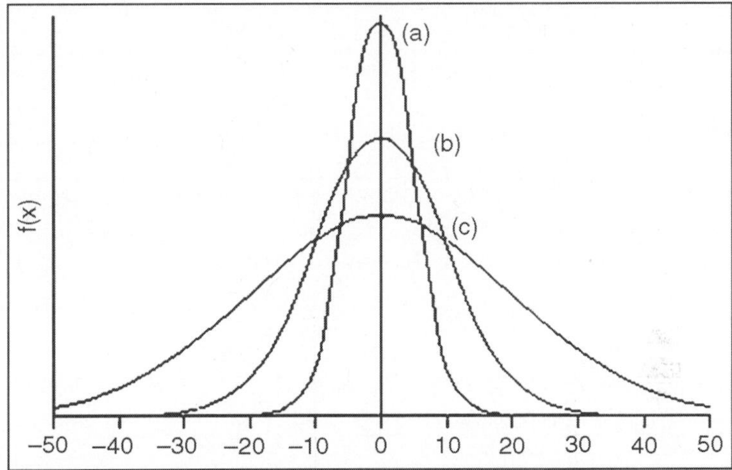

Fig. Normal Distributions for (a) $\mu = 0$ and $\sigma^2 = 25$; (b) $\mu = 0$ and $\sigma^2 = 100$; and (c) $\mu = 0$ and $\sigma^2 = 400$.

For example, 68.26 per cent of the members in a normally distributed population have values within the range $\mu \pm 1\sigma$, regardless of the actual values of μ and σ. As shown in Example, probability tables can be used to determine the probability of occurrence between any defined limits.

Confidence Intervals for Populations

If we randomly select a single member from a population, what will be its most likely value? This is an important question, and, in one form or another, it is the fundamental problem for any analysis. One of the most important features of a population's probability distribution is that it provides a way to answer this question. Earlier we noted that 68.26 per cent of a normally distributed population is found within the range of $\mu \pm 1\sigma$. Stating this another way, there is a 68.26 per cent probability that a member selected at random from a normally distributed population will have a value in the interval of $\mu \pm 1\sigma$. In general, we can write

$$X_i = \mu \pm z\sigma$$

where the factor *z* accounts for the desired level of confidence. Values reported in this fashion are called confidence intervals. Equation, for example, is the confidence interval for a single member of a population. Confidence intervals can be quoted for any desired probability level, several examples of which are shown in Table. For reasons that will be discussed later in the chapter, a 95 per cent confidence interval frequently is reported.

Table. Confidence Intervals for Normal Distribution Curves Between the Limits $\mu \pm z\sigma$

z	Confidence Interval (%)
0.50	38.30
1.00	68.26
1.50	86.64
1.96	95.00
2.00	95.44
2.50	98.76
3.00	99.73
3.50	99.95

Alternatively, a confidence interval can be expressed in terms of the population's standard deviation and the value of a single member drawn from the population. Thus, equation can be rewritten as a confidence interval for the population mean

$$\mu = X_i \pm z\sigma$$

Confidence intervals also can be reported using the mean for a sample of size *n,* drawn from a population of known σ. The standard deviation for the mean value, $\mu\bar{x}$, which also is known as the standard error of the mean, is

$$\sigma\bar{X} = \frac{z\sigma}{\sqrt{n}}$$

The confidence interval for the population's mean, therefore, is

$$\mu = \bar{X} \pm \frac{z\sigma}{\sqrt{n}}$$

Probability Distributions for Samples

We introduced two probability distributions commonly encountered when studying populations. We have yet to address, however, how we can identify the probability distribution for a given population. we assumed that the amount of aspirin in analgesic tablets is normally distributed.

We are justified in asking how this can be determined without analyzing every member of the population. When we cannot study the whole population, or when we cannot predict the mathematical form of a population's probability distribution, we must deduce the distribution from a limited sampling of its members.

Sample Distributions and the Central Limit Theorem Let's return to the problem of determining a penny's mass to explore the relationship between a population's distribution and the distribution of samples drawn from that population. The data shown in Tables are insufficient for our purpose because they are not large enough to give a useful picture of their respective probability distributions.

EXTERNAL METHODS OF QUALITY ASSESSMENT

Internal methods of quality assessment should always be viewed with some level of skepticism because of the potential for bias in their execution and interpretation. For this reason, external methods of quality assessment also play an important role in quality assurance programmes. One external method of quality assessment is the certification of a laboratory by a sponsoring agency. Certification is based on the successful analysis of a set of proficiency standards prepared by the sponsoring agency. For example, laboratories involved in environmental analyses may be required to analyze standard samples prepared by the Environmental Protection Agency.

A second example of an external method of quality assessment is the voluntary participation of the laboratory in a collaborative test sponsored by a professional organization such as the Association of Official Analytical Chemists. Finally, individuals contracting with a laboratory can perform their own external quality assessment by submitting blind duplicate samples and blind standard samples to the laboratory for analysis. If the results for the quality assessment samples are unacceptable, then there is good reason to consider the results suspect for other samples provided by the laboratory.

Evaluating Quality Assurance Data

In the previous section we described several internal methods of quality assessment that provide quantitative estimates of the systematic and random errors present in an analytical system. Now we turn our attention to how this numerical information is incorporated into the written directives of a complete quality assurance programme. Two approaches to developing quality assurance programmes have been described9: a prescriptive approach, in which an exact method of quality assessment is prescribed; and a performance-based approach, in which any form of quality assessment is acceptable, provided that an acceptable level of statistical control can be demonstrated.

Prescriptive Approach

With a prescriptive approach to quality assessment, duplicate samples, blanks, standards, and spike recoveries are measured following a specific protocol. The result for each analysis is then compared with a single predetermined limit. If this limit is exceeded, an appropriate corrective action is taken. Prescriptive approaches to quality assurance are common for programmes and laboratories subject to federal regulation. For example, the Food and Drug Administration (FDA) specifies quality assurance practices that must be followed by laboratories analyzing products regulated by the FDA.

A good example of a prescriptive approach to quality assessment is the protocol outlined in figure, published by the Environmental Protection Agency (EPA) for laboratories involved in monitoring studies of water and wastewater. Independent samples *A* and *B* are collected simultaneously at the sample site.

Sample A is split into two equal-volume samples, and labeled A_1 and A_2. Sample B is also split into two equal-volume samples, one of which, BSF, is spiked with a known amount of analyte. A field blank, DF, also is spiked with the same amount of analyte. All five samples (A_1, A_2, B, BSF, and DF) are preserved if necessary and transported to the laboratory for analysis.

The first sample to be analyzed is the field blank. If its spike recovery is unacceptable, indicating that a systematic error is present, then a laboratory method blank, DL, is prepared and analyzed. If the spike recovery for the method blank is also unsatisfactory, then the systematic error originated in the laboratory. An acceptable spike recovery for the method blank, however, indicates that the systematic error occurred in the field or during transport to the laboratory. Systematic errors in the laboratory can be corrected, and the analysis continued. Any systematic errors occurring in the field, however, cast uncertainty on the quality of the samples, making it necessary to collect new samples. If the field blank is satisfactory, then sample B is analyzed. If the result for B is above the method's detection limit, or if it is within the range of 0.1 to 10 times the amount of analyte spiked into BSF, then a spike recovery for BSF is determined. An unacceptable spike recovery for BSF indicates the presence of a systematic error involving the sample. To determine the source of the systematic error, a laboratory spike, BSL, is prepared using sample B and analyzed.

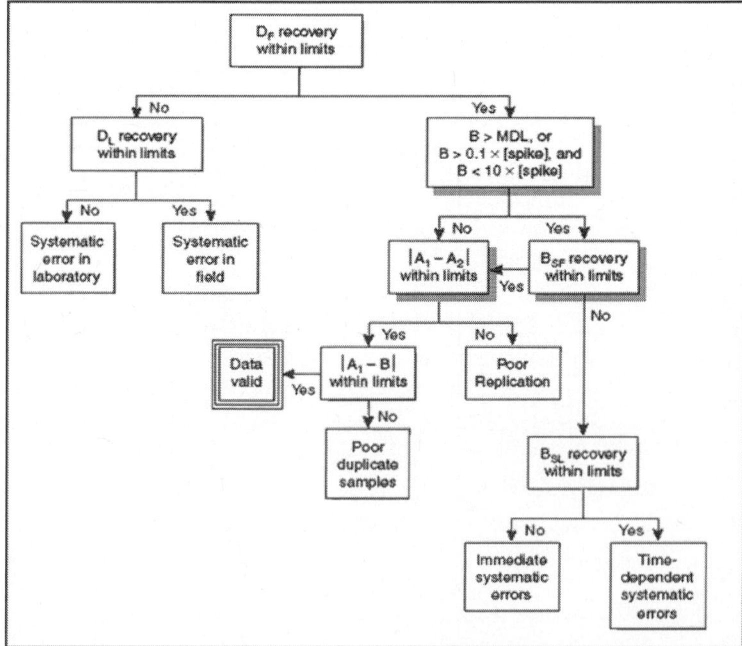

If the spike recovery for BSL is acceptable, then the systematic error requires a long time to have a noticeable effect on the spike recovery. One possible explanation is that the analyte has not been properly preserved or has been held beyond the acceptable holding time. An unacceptable spike recovery for BSL suggests an immediate systematic error, such as that due to the influence of the sample's matrix. In either case, the systematic errors are fatal and must be corrected before the sample is reanalyzed.

If the spike recovery for BSF is acceptable, or if the result for sample B is below the method's detection limit or outside the range of 0.1 to 10 times the amount of analyte spiked in BSF, then the duplicate samples A_1 and A_2 are analyzed. The results for A_1 and A_2 are discarded if the difference between their values is excessive. If the difference between the results for A_1 and A_2 is within the accepted limits, then the results for samples A_1 and B are compared. Since samples collected from the same sampling site at the same time should be identical in composition, the results are discarded if the difference between their values is unsatisfactory, and accepted if the difference is satisfactory.

This protocol requires four to five evaluations of quality assessment data before the result for a single sample can be accepted; a process that must be repeated for each analyte and for each sample. Other prescriptive protocols are equally demanding. For example, figure shows a portion of the quality assurance protocol used for the graphite furnace atomic absorption analysis of trace metals in aqueous solutions. This protocol involves the analysis of an initial calibration verification standard and an initial calibration blank, followed by the analysis of samples in groups of ten.

Each group of samples is preceded and followed by continuing calibration verification (CCV) and continuing calibration blank (CCB) quality assessment samples. Results for each group of ten samples can be accepted only if both sets of CCV and CCB quality assessment samples are acceptable.

The advantage to a prescriptive approach to quality assurance is that a single consistent set of guidelines is used by all laboratories to control the quality of analytical results. A significant disadvantage, however, is that the ability of a laboratory to produce quality results is not taken into account when determining the frequency of collecting and analyzing quality assessment data.

Laboratories with a record of producing high-quality results are forced to spend more time and money on quality assessment than is perhaps necessary. At the same time, the frequency of quality assessment may be insufficient for laboratories with a history of producing results of poor quality.

Performance-Based Approach

In a performance-based approach to quality assurance, a laboratory is free to use its experience to determine the best way to gather and monitor quality assessment data. The quality assessment methods remain the same (duplicate samples, blanks, standards, and spike recoveries) since they provide the necessary information about precision and bias.

What the laboratory can control, however, is the frequency with which quality assessment samples are analyzed, and the conditions indicating when an analytical system is no longer in a state of statistical control.

Furthermore, a performance-based approach to quality assessment allows a laboratory to determine if an analytical system is in danger of drifting out of statistical control. Corrective measures are then taken before further problems develop. The principal tool for performance-based quality assessment is the control chart. In a control chart the results from the analysis of quality assessment samples are plotted in the order in which they are collected, providing a continuous record of the statistical state of the analytical system.

Quality assessment data collected over time can be summarized by a mean value and a standard deviation. The fundamental assumption behind the use of a control chart is that quality assessment data will show only random variations around the mean value when the analytical system is in statistical control.

When an analytical system moves out of statistical control, the quality assessment data is influenced by additional sources of error, increasing the standard deviation or changing the mean value.

Control charts were originally developed in the 1920s as a quality assurance tool for the control of manufactured products.

Two types of control charts are commonly used in quality assurance: a property control chart in which results for single measurements, or the means for several replicate measurements, are plotted sequentially; and a precision control chart in which ranges or standard deviations are plotted sequentially.

In either case, the control chart consists of a line representing the mean value for the measured property or the precision, and two or more boundary lines whose positions are determined by the precision of the measurement process.

The position of the data points about the boundary lines determines whether the system is in statistical control.

$$CL = \overline{X} = \frac{\Sigma X_i}{n}$$

The positions of the boundary lines are determined by the standard deviation, S, of the points used to determine the central line

$$S = \sqrt{\frac{\Sigma(X_i - \overline{X})^2}{n-1}}$$

with the upper and lower warning limits (*UWL* and *LWL*), and the upper and lower control limits (*UCL* and *LCL*) given by

$$UWL = CL + 2S$$
$$LWL = CL - 2S$$
$$UCL = CL + 3S$$
$$LCL = CL - 3S$$

Property control charts can also be constructed using points that are the mean value, – Xi, for a set of r replicate determinations on a single sample. The mean for the ith sample is given by

$$\bar{X}_i = \frac{\sum_{j=1}^{r} X_{ij}}{j}$$

where X_{ij} is the jth replicate. The centre line for the control chart, therefore, is

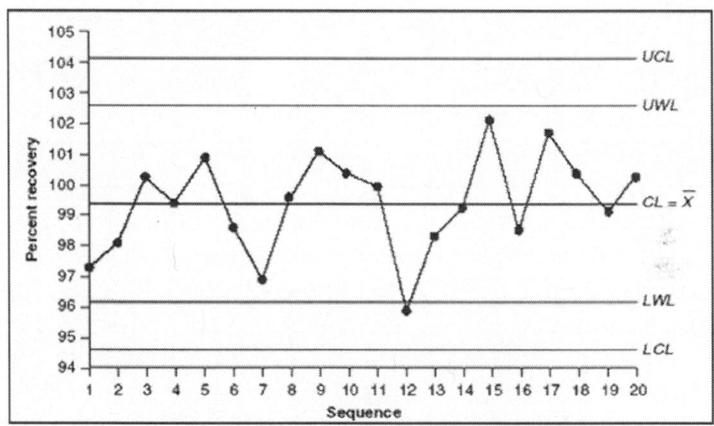

Fig. Property Control Chart for Example

$$CL = \bar{X} = \frac{\sum \bar{X}_i}{n}.$$

To determine the standard deviation for the warning and control limits, it is necessary to calculate the variance for each sample, si_2.

$$s_i^2 = \frac{\sum_{j=1}^{r} \left(X_{ij} - \bar{X}_i\right)^2}{r-1}$$

The overall standard deviation, S, is the square root of the average variance for the samples used to establish the control plot.

$$S = \sqrt{\frac{\sum s_i^2}{n}}$$

Finally, the resulting warning and control limits are

$$UWL = CL + \frac{2S}{\sqrt{r}}$$

$$LWL = CL - \frac{2S}{\sqrt{r}}$$

$$UCL = CL + \frac{2S}{\sqrt{r}}$$

Table. Statistical Factors for the Upper Warning Limit and Upper Control Limit

Replicates	*fUWL*	*fUCL*
2	2.512	3.267
3	2.050	2.575
4	1.855	2.282
5	1.743	2.115
6.	1.669	2.004

$$LCL = CL - \frac{3S}{\sqrt{r}}$$

Constructing a Precision Control Chart: The most common measure of precision used in constructing a precision control chart is the range, R, between the largest and smallest results for a set of j replicate analyses on a sample.

$$R = X_{large} - X_{small}$$

To construct the control chart, ranges for a minimum of 15–20 samples (preferably 30 or more samples) are obtained while the system is known to be in statistical control. The line for the average range, R –, is determined by the mean of these n samples

$$\overline{R} = \frac{\Sigma R_i}{n}$$

The upper control line and the upper warning line are given by

$$UCL = f_{UCL} \times \overline{R}$$

$$UWL = f_{UWL} \times \overline{R}$$

where f_{UCL} and f_{UWL} are statistical factors determined by the number of replicates used to determine the range. Because the range always is greater than or equal to zero, there is no lower control limit or lower warning limit.

The precision control chart is strictly valid only for the replicate analysis of identical samples, such as a calibration standard or a standard reference material. Its use for the analysis of non-identical samples, such as a series of clinical or environmental samples, is complicated by the fact that the range usually is not independent of the magnitude of X_{large} and X_{small}.

For example, table shows the relationship between –R and the concentration of chromium in water. Clearly the significant difference in the average range for these concentrations of Cr makes a single precision control chart impossible. One solution to this problem is to prepare separate precision control charts, each of which covers a range of concentrations for which –R is approximately constant.

Interpreting Control Charts: The purpose of a control chart is to determine if a system is in statistical control. This determination is made by examining the location of individual points in relation to the warning limits and the control limits, and the distribution of the points around the central line. If we assume that the data are normally distributed, then the probability of finding a point at any distance from the mean value can be determined from the normal distribution curve. The upper and lower control limits for a property control chart, for example, are set to ?3S, which, if S is a good approximation for σ, includes 99.74 per cent of the data.

The probability that a point will fall outside the UCL or LCL, therefore, is only 0.26 per cent. The most likely explanation when a point exceeds a control limit is that a systematic error has occurred or that the precision of the measurement process has deteriorated. In either case the system is assumed to be out of statistical control.

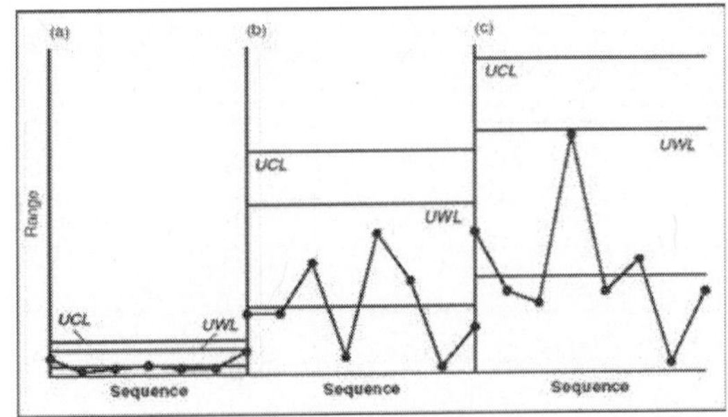

Fig. Example of the Use of Subrange Precision Control Charts for Samples that Span a Range of Analyte Concentrations. The Precision Control Charts are used for (a) Low Concentrations of Analyte; (b) Intermediate Concentrations of Analyte; and (c) High Concentrations of Analyte.

Rule 1. A system is considered to be out of statistical control if any single point exceeds either the *UCL* or the *LCL*. The upper and lower warning limits, which are located at 2S, should only be exceeded by 5 per cent of the data.

Rule 2. A system is considered to be out of statistical control if two out of three consecutive points are between the *UWL* and *UCL* or between the *LWL* and *LCL*. When a system is in statistical control, the data points should be randomly distributed about the centre line. The presence of an unlikely pattern in the data is another indication that a system is no longer in statistical control.

Rule 3. A system is considered to be out of statistical control if a run of seven consecutive points is completely above or completely below the centre line.

Rule 4. A system is considered to be out of statistical control if six consecutive points are all increasing in value or all decreasing in value. The points may be on either side of the centre line.

Rule 5. A system is considered to be out of statistical control if 14 consecutive points alternate up and down in value. The points may be on either side of the centre line.

Rule 6. A system is considered to be out of statistical control if any obvious "nonrandom" pattern is observed. The same rules apply to precision control charts with the exception that there are no lower warning and lower control limits.

Using Control Charts for Quality Assurance: Control charts play an important role in a performance-based programme of quality assurance because they provide an easily interpreted picture of the statistical state of an analytical system. Quality assessment samples such as blanks, standards, and spike recoveries can be monitored with property control charts. A precision control chart can be used to monitor duplicate samples. The first step in using a control chart for quality assurance is to determine the mean value and the standard deviation (except when using the range) for the quality assessment data while the system is under statistical control.

These values must be established under the same conditions that will be present during the normal use of the control chart. Thus, preliminary data should be randomly collected throughout the day, as well as over several days, to account for short-term and long-term variability.

The preliminary data are used to construct an initial control chart, and discrepant points are determined using the rules discussed in the previous section. Questionable points are dropped, and the control chart is replotted.

As the control chart is used, it may become apparent that the original limits need adjusting. Control limits can be recalculated if the number of new data points is at least equivalent to the amount of data used to construct the original control chart. For example, if 15 points were initially used, the limits can be reevaluated after 15 additional points are collected. The 30 points are pooled together to calculate the new limits. A second modification can be made after a further 30 points have been collected. Another indication that a control chart needs to be modified is when points rarely exceed the warning limits. In this case the new limits can be recalculated using the last 20 points.

Once a control chart is in use, new quality assessment data should be added at a rate sufficient to ensure that the system remains in statistical control. As with prescriptive approaches to quality assurance, when a quality assessment sample is found to be out of statistical control, all samples analyzed since the last successful verification of statistical control must be reanalyzed.

The advantage of a performance-based approach to quality assurance is that a laboratory may use its experience, guided by control charts, to determine the frequency for collecting quality assessment samples. When the system is stable, quality assessment samples can be acquired less frequently.

6

Numbers Analysis in Analytical Chemistry

INTRODUCTION

Analytical chemistry is inherently a quantitative science. Whether determining the concentration of a species in a solution, evaluating an equilibrium constant, measuring a reaction rate, or drawing a correlation between a compound's structure and its reactivity, analytical chemists make measurements and perform calculations. In this section we briefly review several important topics involving the use of numbers in analytical chemistry.

UNITS FOR EXPRESSING CONCENTRATION

Concentration is a general measurement unit stating the amount of solute present in a known amount of solution

$$\text{Concentration} = \frac{\text{amount of solute}}{\text{amount of solution}}$$

Although the terms "solute" and "solution" are often associated with liquid samples, they can be extended to gas-phase and solid-phase samples as well. The actual units for reporting concentration depend on how the amounts of solute and solution are measured. Table lists the most common units of concentration.

Molarity and Formality

Both molarity and formality express concentration as moles of solute per liter of solution. There is, however, a subtle difference between molarity and formality. Molarity is the concentration of a particular chemical species in solution.

Formality, on the other hand, is a substance's total concentration in solution without regard to its specific chemical form. There is no difference between a substance's molarity and formality if it dissolves without dissociating into ions. The molar concentration of a solution of glucose, for example, is the same as its formality.

For substances that ionize in solution, such as NaCl, molarity and formality are different. For example, dissolving 0.1 mol of NaCl in 1 L of water gives a solution containing 0.1 mol of $Na^+$and 0.1 mol of Cl^-. The molarity of NaCl, therefore, is zero since there is essentially no undissociated NaCl in solution.

The solution, instead, is 0.1 M in $Na^+$and 0.1 M in Cl^-. The formality of NaCl, however, is 0.1 F because it represents the total amount of NaCl in solution. The rigorous definition of molarity, for better or worse, is largely ignored in the current literature, as it is in this text. When we state that a solution is 0.1 M NaCl we understand it to consist of Na^+ and Cl^- ions. The unit of formality is used only when it provides a clearer description of solution chemistry.

Table. Common Units for Reporting Concentration

Name	Units[a]	Symbol
molarity	$\frac{\text{moles solute}}{\text{liters solution}}$	M
formality	$\frac{\text{number FWs solute}}{\text{liters solution}}$	F
normality	$\frac{\text{number FWs solute}}{\text{liters solution}}$	N
molality	$\frac{\text{moles solutge}}{\text{kg solvent}}$	%w/w
weight %	$\frac{\text{g solute}}{\text{100g solvent}}$	%w/w
volume %	$\frac{\text{mL solute}}{\text{100 mL solution}}$	%v/N
weight-to-volume %	$\frac{\text{g solute}}{\text{100 mL solution}}$	ppm
parts per million	$\frac{\text{gsolute}}{10^6\text{g solution}}$	ppb

Molar concentrations are used so frequently that a symbolic notation is often used to simplify its expression in equations and writing. The use of square brackets around a species indicates that we are referring to that species' molar concentration. Thus, [Na] is read as the "molar concentration of sodium ions."

Normality

Normality is an older unit of concentration that, although once commonly used, is frequently ignored in today's laboratories. Normality is still used in some handbooks of analytical methods, and, for this reason, it is helpful to understand its meaning. For example, normality is the concentration unit used in *Standard Methods for the Examination of Water and Wastewater,* 1 a commonly used source of analytical methods for environmental laboratories.

Normality makes use of the chemical equivalent, which is the amount of one chemical species reacting stoichiometrically with another chemical species. Note that this definition makes an equivalent, and thus normality, a function of the chemical reaction in which the species participates. Although a solution of H_2SO_4 has a fixed molarity, its normality depends on how it reacts.

The number of equivalents, *n,* is based on a reaction unit, which is that part of a chemical species involved in a reaction. In a precipitation reaction, for example, the reaction unit is the charge of the cation or anion involved in the reaction; thus for the reaction

$$Pb^{2+}(aq) + 2I^-(aq) \rightleftharpoons pbI_2(s)$$

n =2 for $Pb_2{}^+$ and n =1 for I–. In an acid–base reaction, the reaction unit is the number of H^+ ions donated by an acid or accepted by a base. For the reaction between sulfuric acid and ammonia

$$H_2SO_4(aq) + 2NH_3(aq) \rightleftharpoons 2H_4{}^+(aq) + SO_4{}^{2-}(aq)$$

we find that n = 2 for H_2SO_4 and n =1 for NH_3. For a complexation reaction, the reaction unit is the number of electron pairs that can be accepted by the metal or donated by the ligand. In the reaction between $Ag^+$and NH_3

$$Ag^+(aq) + 2NH_3(aq) \rightleftharpoons Ag(NH_3)_2{}^+(aq)$$

the value of n for Ag^+ is 2 and that for NH3 is 1. Finally, in an oxidation–reduction reaction the reaction unit is the number of electrons released by the reducing agent or accepted by the oxidizing agent; thus, for the reaction

$$2Fe^{3+}(aq) + Sn^{2+}(aq) \rightleftharpoons Sn^{4+}(aq) + 2Fe^{2+}(aq)$$

Clearly, determining the number of equivalents for a chemical species requires an understanding of how it reacts. Normality is the number of equivalent weights (EW) per unit volume and, like formality, is independent of speciation. An equivalent weight is defined as the ratio of a chemical species' formula weight (FW) to the number of its equivalents

$$EW = \frac{FW}{n}$$

Consequently, the following simple relationship exists between normality and molarity.

$$N = n \times M$$

Molality

Molality is used in thermodynamic calculations where a temperature independent unit of concentration is needed. Molarity, formality and normality are based on the volume of solution in which the solute is dissolved. Since density is a temperature dependent property a solution's volume, and thus its molar, formal and normal concentrations, will change as a function of its temperature. By using the solvent's mass in place of its volume, the resulting concentration becomes independent of temperature.

Weight, Volume, and Weight-to-Volume Ratios

Weight percent (per cent w/w), volume percent (per cent v/v) and weight-to-volume percent (per cent w/v) express concentration as units of solute per 100 units of sample. A solution in which a solute has a concentration of 23 per cent w/v contains 23 g of solute per 100 mL of solution. Parts per million (ppm) and parts per billion (ppb) are mass ratios of grams of solute to one million or one billion grams of sample, respectively. For example, a steel that is 450 ppm in Mn contains 450 μg of Mn for every gram of steel. If we approximate the density of an aqueous solution as 1.00 g/mL, then solution concentrations can be expressed in parts per million or parts per billion using the following relationships.

$$\text{ppm} = \frac{\text{mg}}{\text{liter}} = \frac{\mu\text{g}}{\text{mL}}$$

$$\text{ppb} = \frac{\mu\text{g}}{\text{liter}} = \frac{\text{ng}}{\text{mL}}$$

For gases a part per million usually is a volume ratio. Thus, a helium concentration of 6.3 ppm means that one litre of air contains 6.3 μL of He.

Converting Between Concentration Units

The units of concentration most frequently encountered in analytical chemistry are molarity, weight percent, volume percent, weight-to-volume percent, parts per million, and parts per billion. By recognizing the general definition of concentration given in equation, it is easy to convert between concentration units.

p-Functions

Sometimes it is inconvenient to use the concentration units in Table. For example, during a reaction a reactant's concentration may change by many orders of magnitude. If we are interested in viewing the progress of the reaction graphically, we might wish to plot the reactant's concentration as a function of time or as a function of the volume of a reagent being added to the reaction.

Such is the case in Figure, where the molar concentration of H^+ is plotted (*y*-axis on left side of figure) as a function of the volume of NaOH added to a solution of HCl.

The initial $[H^+]$ is 0.10 M, and its concentration after adding 75 mL of NaOH is 5.0×10^{-13} M. We can easily follow changes in the $[H^+]$ over the first 14 additions of NaOH. For the last ten additions of NaOH, however, changes in the $[H^+]$ are too small to be seen. When working with concentrations that span many orders of magnitude, it is often more convenient to express the concentration as a p-function. The p-function of a number X is written as pX and is defined as

$$pX = -\log(X)$$

Thus, the pH of a solution that is 0.10 M H^+ is

$$pH = -\log[H^+] = -\log(0.10) = 1.00$$

and the pH 5.0×10^{-13} M H^+ is

$$pH = -\log[H^+] = -\log(5.0 \times 10^{-13}) = 12.30$$

Figure shows how plotting pH in place of $[H^+]$ provides more detail about how the concentration of H^+ changes following the addition of NaOH.

FUNDAMENTAL UNITS OF MEASURE

Imagine that you find the following instructions in a laboratory procedure: "Transfer 1.5 of your sample to a 100 volumetric flask, and dilute to volume." How do you do this? Clearly these instructions are incomplete since the units of measurement are not stated.

Compare this with a complete instruction: "Transfer 1.5 g of your sample to a 100-mL volumetric flask, and dilute to volume." This is an instruction that you can easily follow.

Measurements usually consist of a unit and a number expressing the quantity of that unit. Unfortunately, many different units may be used to express the same physical measurement. For example, the mass of a sample weighing 1.5 g also may be expressed as 0.0033 lb or 0.053 oz. For consistency, and to avoid confusion, scientists use a common set of fundamental units, several of which are listed in Table.

These units are called SI units after the *Système International d'Unités.* Other measurements are defined using these fundamental SI units. For example, we measure the quantity of heat produced during a chemical reaction in joules, (J), where

$$1J = 1\frac{m^2kg}{s^2}$$

Table provides a list of other important derived SI units, as well as a few commonly used non-SI units.

Chemists frequently work with measurements that are very large or very small. A mole, for example, contains 602,213,670,000,000, 000,000,000 particles, and some analytical techniques can detect as little as 0.000000000000001 g of a compound. For simplicity, we express these measurements using scientific notation; thus, a mole contains 6.0221367×10^{23} particles, and the stated mass is 1 m $\times 10^{-15}$ g. Sometimes it is preferable to express measurements without the exponential term, replacing it with a prefix. A mass of 1×10^{-15} g is the same as 1 femtogram. Table lists other common prefixes.

Table. Fundamental SI Units

Measurement	Unit	kg
mass	kilogram	kg
volume	litre	L
distance	meter	m
temperature	kelvin	K
time	second	s
current	ampere	A
amount of substance	mole	mol

Table. Other Sl and Non-Sl units

Measurement	Unit	Symbol	Equivalent Sl units
length	angstrom	A	1Å = 1 × 10^{-10}m
force	newton	N	1N = 1 m.kg/s^2
pressure	pascal	Pa	1Pa = 1N/m^2 = 1 kg/(m.s^2)
	atmosphere	atm	1atm = 101,325 Pa
energy, work, heat	joule	J	1j = 1N.m = 1m^2.kg/s^2
power	watt	W	1W = 1J/s = 1m^2.kg/s^3
charge	coulomb	C	1C = 1A.s
potential	volt	V	1V = 1W/A = 1m^2.kg/(s^3.A)
temperature	degree Celsius	°C	°C = K – 273.15
	degree Fahrenheit	°F	°F = 1.8(K – 273.15)+32

Table. Common Prefixes for Exponential Notation.

Exponential	Prefix	Symbol
10^{12}	tera	T
10^{9}	giga	G
10^{6}	mega	M
10^{3}	kilo	k
10^{-1}	deci	d
10^{-2}	centi	c
10^{-3}	milli	m
10^{-6}	micro	m
10^{-9}	nano	n
10^{-12}	pico	p
10^{-15}	femto	f
10^{-18}	atto	a

Significant Figures

Recording a measurement provides information about both its magnitude and uncertainty.

For example, if we weigh a sample on a balance and record its mass as 1.2637 g, we assume that all digits, except the last, are known exactly. We assume that the last digit has an uncertainty of at least ±1, giving an absolute uncertainty of at least ±0.0001 g, or a relative uncertainty of at least

$$\frac{\pm 0.0001\text{g}}{1.2637\text{g}} \times 100 = \pm 0.0079\%$$

Significant figures are a reflection of a measurement's uncertainty. The number of significant figures is equal to the number of digits in the measurement, with the exception that a zero (0) used to fix the location of a decimal point is not considered significant. This definition can be ambiguous.

For example, how many significant figures are in the number 100? If measured to the nearest hundred, then there is one significant figure. If measured to the nearest ten, however, then two significant figures are included. To avoid ambiguity we use scientific notation. Thus, 1×10^2 has one significant figure, whereas 1.0×10^2 has two significant figures. For measurements using logarithms, such as pH, the number of significant figures is equal to the number of digits to the right of the decimal, including all zeros.

Digits to the left of the decimal are not included as significant figures since they only indicate the power of 10. A pH of 2.45, therefore, contains two significant figures. Exact numbers, such as the stoichiometric coefficients in a chemical formula or reaction, and unit conversion factors, have an infinite number of significant figures. A mole of $CaCl_2$, for example, contains exactly two moles of chloride and one mole of calcium. In the equality

$$1000 \text{ mL} = 1 \text{ L}$$

both numbers have an infinite number of significant figures.

Recording a measurement to the correct number of significant figures is important because it tells others about how precisely you made your measurement. For example, suppose you weigh an object on a balance capable of measuring mass to the nearest ± 0.1 mg, but record its mass as 1.762 g instead of 1.7620 g. By failing to record the trailing zero, which is a significant figure, you suggest to others that the mass was determined using a balance capable of weighing to only the nearest ± 1 mg. Similarly, a buret with scale markings every 0.1 mL can be read to the nearest ± 0.01 mL. The digit in the hundredth's place is the least significant figure since we must estimate its value. Reporting a volume of 12.241 mL implies that your buret's scale is more precise than it actually is, with divisions every 0.01 mL.

Significant figures are also important because they guide us in reporting the result of an analysis. When using a measurement in a calculation, the result of that calculation can never be more certain than that measurement's uncertainty. Simply put, the result of an analysis can never be more certain than the least certain measurement included in the analysis.

As a general rule, mathematical operations involving addition and subtraction are carried out to the last digit that is significant for all numbers included in the calculation. When multiplying and dividing, the general rule is that the answer contains the same number of significant figures as that number in the calculation having the fewest significant figures. Thus,

$$\frac{22.91 \times 1.152}{16.302} = 0.21361 = 0.214$$

It is important to remember, however, that these rules are generalizations. What is conserved is not the number of significant figures, but absolute uncertainty when adding or subtracting, and relative uncertainty when multiplying or dividing. For example, the following calculation reports the answer to the correct number of significant figures, even though it violates the general rules outlined earlier.

$$\frac{101}{99} = 1.02$$

Since the relative uncertainty in both measurements is roughly 1 per cent (101 ± 1, 99 ± 1), the relative uncertainty in the final answer also must be roughly 1 per cent. Reporting the answer to only two significant figures, as required by the general rules, implies a relative uncertainty of 10 per cent. The correct answer, with three significant figures, yields the expected relative uncertainty.

Finally, to avoid "round-off " errors in calculations, it is a good idea to retain at least one extra significant figure throughout the calculation. This is the practice adopted in this textbook. Better yet, invest in a good scientific calculator that allows you to perform lengthy calculations without recording intermediate values. When the calculation is complete, the final answer can be rounded to the correct number of significant figures using the following simple rules.

1. Retain the least significant figure if it and the digits that follow are less than halfway to the next higher digit; thus, rounding 12.442 to the nearest tenth gives 12.4 since 0.442 is less than halfway between 0.400 and 0.500.
2. Increase the least significant figure by 1 if it and the digits that follow are more than halfway to the next higher digit; thus, rounding 12.476 to the nearest tenth gives 12.5 since 0.476 is more than halfway between 0.400 and 0.500.
3. If the least significant figure and the digits that follow are exactly halfway to the next higher digit, then round the least significant figure to the nearest even number; thus, rounding 12.450 to the nearest tenth gives 12.4, but rounding 12.550 to the nearest tenth gives 12.6. Rounding in this manner prevents us from introducing a bias by always rounding up or down.

BASIC EQUIPMENT AND INSTRUMENTATION

Measurements are made using appropriate equipment or instruments. The array of equipment and instrumentation used in analytical chemistry is impressive, ranging from the simple and inexpensive, to the complex and costly. With two exceptions, we will postpone the discussion of equipment and instrumentation to those chapters where they are used. The instrumentation used to measure mass and much of the equipment used to measure volume are important to all analytical techniques and are therefore discussed in this section.

Instrumentation for Measuring Mass

An object's mass is measured using a balance. The most common type of balance is an electronic balance in which the balance pan is placed over an electromagnet.

The sample to be weighed is placed on the sample pan, displacing the pan downward by a force equal to the product of the sample's mass and the acceleration due to gravity.

The balance detects this downward movement and generates a counterbalancing force using an electromagnet. The current needed to produce this force is proportional to the object's mass. A typical electronic balance has a capacity of 100–200 g and can measure mass to the nearest ±0.01 to ±1 mg.

Another type of balance is the single-pan, unequal arm balance. In this mechanical balance the balance pan and a set of removable standard weights on one side of a beam are balanced against a fixed counterweight on the beam's other side. The beam itself is balanced on a fulcrum consisting of a sharp knife edge.

Adding a sample to the balance pan tilts the beam away from its balance point. Selected standard weights are then removed until the beam is brought back into balance. The combined mass of the removed weights equals the sample's mass. The capacities and measurement limits of these balances are comparable to an electronic balance.

(a) Photo of a typical electronic balance.

(b) Schematic diagram of electronic balance; adding a sample moves the balance pan down, allowing more light to reach the detector. The control circuitry directs the electromagnetic servomotor to generate an opposing force, raising the sample up until the original intensity of light at the detector is restored.

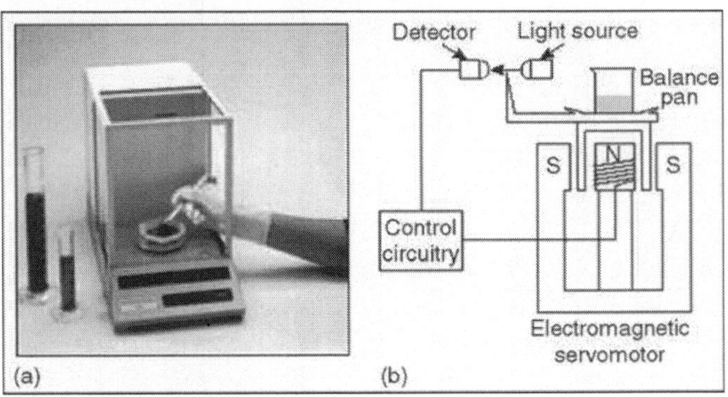

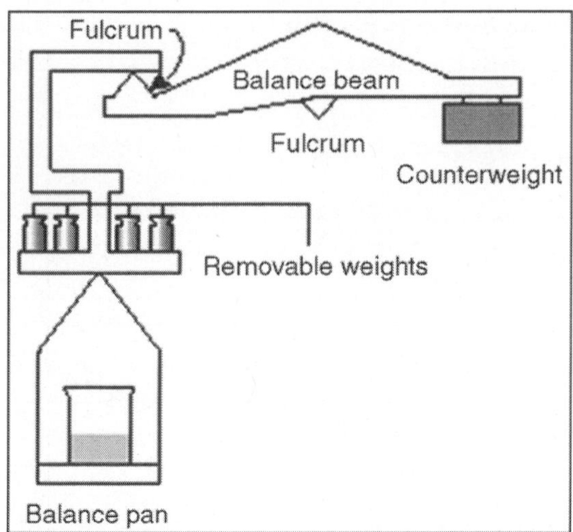

Fig. Schematic Diagram of Single-arm Mechanical Balance.

The mass of a sample is determined by difference. If the material being weighed is not moisture-sensitive, a clean and dry container is placed on the balance.

The mass of this container is called the tare. Most balances allow the tare to be automatically adjusted to read a mass of zero. The sample is then transferred to the container, the new mass is measured and the sample's mass determined by subtracting the tare. Samples that absorb moisture from the air are weighed differently.

The sample is placed in a covered weighing bottle and their combined mass is determined. A portion of the sample is removed, and the weighing bottle and remaining sample are reweighed. The difference between the two masses gives the mass of the transferred sample.

Several important precautions help to minimize errors in measuring an object's mass. Balances should be placed on heavy surfaces to minimize the effect of vibrations in the surrounding environment and should be maintained in a level position.

Analytical balances are sensitive enough that they can measure the mass of a fingerprint. For this reason, materials placed on a balance should normally be handled using tongs or laboratory tissues. Volatile liquid samples should be weighed in a covered container to avoid the loss of sample by evaporation.

Air currents can significantly affect a sample's mass. To avoid air currents, the balance's glass doors should be closed, or the balance's wind shield should be in place.

A sample that is cooler or warmer than the surrounding air will create convective air currents that adversely affect the measurement of its mass. Finally, samples dried in an oven should be stored in a desiccator to prevent them from reabsorbing moisture from the atmosphere.

Equipment for Measuring Volume

Analytical chemists use a variety of glassware to measure volume, several examples of which are shown in Figure.

The type of glassware used depends on how exact the volume needs to be. Beakers, dropping pipets, and graduated cylinders are used to measure volumes approximately, typically with errors of several percent. Pipets and volumetric flasks provide a more accurate means for measuring volume.

When filled to its calibration mark, a volumetric flask is designed to contain a specified volume of solution at a stated temperature, usually 20°C.

The actual volume contained by the volumetric flask is usually within 0.03–0.2 per cent of the stated value. Volumetric flasks containing less than 100 mL generally measure volumes to the hundredth of a milliliter, whereas larger volumetric flasks measure volumes to the tenth of a milliliter. For example, a 10-mL volumetric flask contains 10.00 mL, but a 250-mL volumetric flask holds 250.0 mL (this is important when keeping track of significant figures).

Because a volumetric flask contains a solution, it is useful in preparing solutions with exact concentrations.

The reagent is transferred to the volumetric flask, and enough solvent is added to dissolve the reagent. After the reagent is dissolved, additional solvent is added in several portions, mixing the solution after each addition. The final adjustment of volume to the flask's calibration mark is made using a dropping pipet.

To complete the mixing process, the volumetric flask should be inverted at least ten times.

A pipet is used to deliver a specified volume of solution. Several different styles of pipets are available. Transfer pipets provide the most accurate means for delivering a known volume of solution; their volume error is similar to that from an equivalent volumetric flask. A 250-mL transfer pipet, for instance, will deliver 250.0 mL.

To fill a transfer pipet, suction from a rubber bulb is used to pull the liquid up past the calibration mark (*never* use your mouth to suck a solution into a pipet).

After replacing the bulb with your finger, the liquid's level is adjusted to the calibration mark, and the outside of the pipet is wiped dry. The pipet's contents are allowed to drain into the receiving container with the tip of the pipet touching the container walls. A small portion of the liquid remains in the pipet's tip and should not be blown out.

Measuring pipets are used to deliver variable volumes, but with less accuracy than transfer pipets.

With some measuring pipets, delivery of the calibrated volume requires that any solution remaining in the tip be blown out. Digital pipets and syringes can be used to deliver volumes as small as a microliter. Three important precautions are needed when working with pipets and volumetric flasks. First, the volume delivered by a pipet or contained by a volumetric flask assumes that the glassware is clean. Dirt and grease on the inner glass surface prevents liquids from draining evenly, leaving droplets of the liquid on the container's walls.

For a pipet this means that the delivered volume is less than the calibrated volume, whereas drops of liquid above the calibration mark mean that a volumetric flask contains more than its calibrated volume. Commercially available cleaning solutions can be used to clean pipets and volumetric flasks.

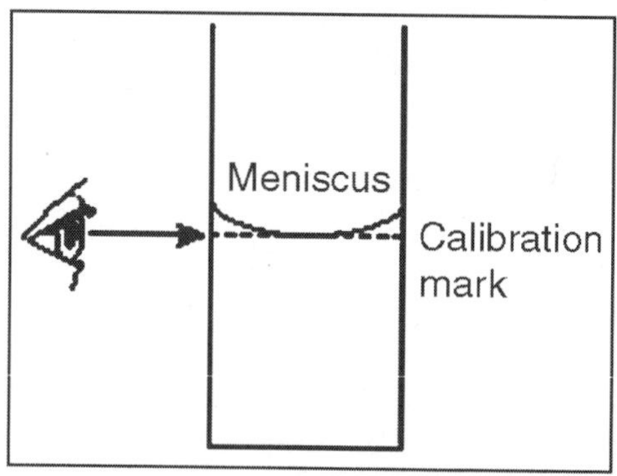

Fig. Means of Reading the Meniscus on a Volumetric Flask or Pipet.

Second, when filling a pipet or volumetric flask, set the liquid's level exactly at the calibration mark. The liquid's top surface is curved into a meniscus, the bottom of which should be exactly even with the glassware's calibration mark. The meniscus should be adjusted with the calibration mark at eye level to avoid parallax errors.

If your eye level is above the calibration mark the pipet or volumetric flask will be overfilled. The pipet or volumetric flask will be underfilled if your eye level is below the calibration mark. Finally, before using a pipet or volumetric flask you should rinse it with several small portions of the solution whose volume is being measured.

This ensures that any residual liquid remaining in the pipet or volumetric flask is removed.

Equipment for Drying Samples

Many materials need to be dried prior to their analysis to remove residual moisture. Depending on the material, heating to a temperature of 110–140 °C is usually sufficient. Other materials need to be heated to much higher temperatures to initiate thermal decomposition. Both processes can be accomplished using a laboratory oven capable of providing the required temperature.

Commercial laboratory ovens are used when the maximum desired temperature is 160–325°C (depending on the model). Some ovens include the ability to circulate heated air, allowing for a more efficient removal of moisture and shorter drying times. Other ovens provide a tight seal for the door, allowing the oven to be evacuated. In some situations a conventional laboratory oven can be replaced with a microwave oven. Higher temperatures, up to 1700°C, can be achieved using a muffle furnace. After drying or decomposing a sample, it should be cooled to room temperature in a desiccator to avoid the readsorption of moisture.

A desiccator is a closed container that isolates the sample from the atmosphere. A drying agent, called a desiccant, is placed in the bottom of the container.

Typical desiccants include calcium chloride and silica gel. A perforated plate sits above the desiccant, providing a shelf for storing samples. Some desiccators are equipped with stopcocks that allow them to be evacuated.

PREPARING SOLUTIONS

Preparing a solution of known concentration is perhaps the most common activity in any analytical lab. The method for measuring out the solute and solvent depend on the desired concentration units, and how exact the solution's concentration needs to be known.

Pipets and volumetric flasks are used when a solution's concentration must be exact; graduated cylinders, beakers, and reagent bottles suffice when concentrations need only be approximate. Two methods for preparing solutions are described in this section.

Preparing Stock Solutions

A stock solution is prepared by weighing out an appropriate portion of a pure solid or by measuring out an appropriate volume of a pure liquid and diluting to a known volume. Exactly how this is done depends on the required concentration units.

For example, to prepare a solution with a desired molarity you would weigh out an appropriate mass of the reagent, dissolve it in a portion of solvent, and bring to the desired volume. To prepare a solution where the solute's concentration is given as a volume percent, you would measure out an appropriate volume of solute and add sufficient solvent to obtain the desired total volume.

Preparing Solutions by Dilution

Solutions with small concentrations are often prepared by diluting a more concentrated stock solution. A known volume of the stock solution is transferred to a new container and brought to a new volume. Since the total amount of solute is the same before and after dilution, we know that

$$Co \,\acute{}\, Vo = Cd \,\acute{}\, Vd \quad 2.4$$

where *Co* is the concentration of the stock solution, *Vo* is the volume of the stock solution being diluted, *Cd* is the concentration of the dilute solution, and *Vd* is the volume of the dilute solution. Again, the type of glassware used to measure *Vo* and *Vd* depends on how exact the solution's concentration must be known.

STOICHIOMETRIC CALCULATIONS

A balanced chemical reaction indicates the quantitative relationships between the moles of reactants and products. These stoichiometric relationships provide the basis for many analytical calculations. Consider, for example, the problem of determining the amount of oxalic acid, $H_2C_2O_4$, in rhubarb. One method for this analysis uses the following reaction in which we oxidize oxalic acid to CO_2.

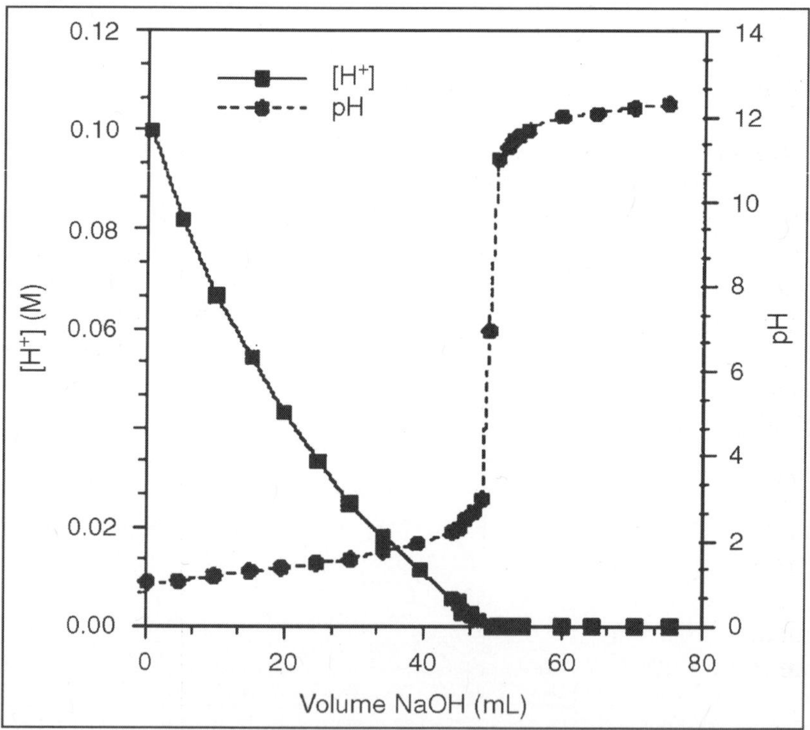

Fig. Graph of [H$^+$] Versus Volume of NaOH and pH Versus Volume of NaOH for the Reaction of 0.10 M HCl with 0.10 M NaOH.

$$2Fe^{3+}(aq) + H_2C_2O_4(aq) + 2H_2O(\ell) \rightarrow 2Fe^{2+}(aq) + 2CO_2(g) + 2H_3O^+(aq)$$

The balanced chemical reaction provides the stoichiometric relationship between the moles of Fe^{3+} used and the moles of oxalic acid in the sample being analyzed— specifically, one mole of oxalic acid reacts with two moles of

Fe_3. As shown in Example, the balanced chemical reaction can be used to determine the amount of oxalic acid in a sample, provided that information about the number of moles of Fe^{3+} is known.

In the analysis described in Example oxalic acid already was present in the desired form.

In many analytical methods the compound to be determined must be converted to another form prior to analysis. For example, one method for the quantitative analysis of tetraethylthiuram disulfide ($C_{10}H_{20}N_2S_4$), the active ingredient in the drug Antabuse (disulfiram), requires oxidizing the S to SO_2, bubbling the SO_2 through H_2O_2 to produce H_2SO_4, followed by an acid–base titration of the H_2SO_4 with NaOH.

Although we can write and balance chemical reactions for each of these steps, it often is easier to apply the principle of the conservation of reaction units. A reaction unit is that part of a chemical species involved in a reaction. Consider, for example, the general unbalanced chemical reaction

$$A + B \rightarrow \text{Products}$$

Conservation of reaction units requires that the number of reaction units associated with the reactant A equal the number of reaction units associated with the reactant B. Translating the previous statement into mathematical form gives

Number of reaction units per A × moles A

Number of reaction units per B × moles B

If we know the moles of A and the number of reaction units associated with A and B, then we can calculate the moles of B. Note that a conservation of reaction units, as defined by equation, can only be applied between two species. There are five important principles involving a conservation of reaction units: mass, charge, protons, electron pairs, and electrons.

Conservation of Mass

The easiest principle to appreciate is conservation of mass. Except for nuclear reactions, an element's total mass at the end of a reaction must be the same as that present at the beginning of the reaction; thus, an element serves as the most fundamental reaction unit. Consider, for example, the combustion of butane to produce CO_2 and H_2O, for which the unbalanced reaction is

$$C_4H_{10}(g) + O_2(g) \rightarrow CO_2(g) + H_2O(g)$$

All the carbon in CO_2 comes from the butane, thus we can select carbon as a reaction unit. Since there are four carbon atoms in butane, and one carbon atom in CO_2, we write

$$4 \times \text{moles } C_4H_{10} = 1 \times \text{moles } CO_2$$

Hydrogen also can be selected as a reaction unit since all the hydrogen in butane ends up in the H_2O produced during combustion. Thus, we can write

$$10 \times \text{moles } C_4H_{10} = 2 \times \text{moles } H_2O$$

Although the mass of oxygen is conserved during the reaction, we cannot apply equation because the O_2 used during combustion does not end up in a single product. Conservation of mass also can, with care, be applied to groups of atoms.

For example, the ammonium ion, NH_4, can be precipitated as $Fe(NH_4)2(SO_4)2.6H_2O$. Selecting NH_4^+ as the reaction unit gives

$$2 \times \text{moles } Fe(NH_4)2(SO_4)^2 \cdot 6H_2O = 1 \times \text{moles } NH_4^+$$

Conservation of Charge

The stoichiometry between two reactants in a precipitation reaction is governed by a conservation of charge, requiring that the total cation charge and the total anion charge in the precipitate be equal.

The reaction units in a precipitation reaction, therefore, are the absolute values of the charges on the cation and anion that make up the precipitate. Applying equation to a precipitate of $Ca_3(PO_4)2$ formed from the reaction of Ca_2 and PO_4^{3-}, we write

$$2 \times \text{moles}\,Ca^{2+} = 3 \times \text{moles}PO_4^{3-}$$

Conservation of Protons

In an acid–base reaction, the reaction unit is the proton. For an acid, the number of reaction units is given by the number of protons that can be donated to the base; and for a base, the number of reaction units is the number of protons that the base can accept from the acid. In the reaction between H_3PO_4 and NaOH, for example, the weak acid H_3PO4 can donate all three of its protons to NaOH, whereas the strong base NaOH can accept one proton. Thus, we write

$$3 \times \text{moles } H_3PO_4 = 1 \times \text{moles NaOH}$$

Care must be exercised in determining the number of reaction units associated with the acid and base. The number of reaction units for an acid, for instance, depends not on how many acidic protons are present, but on how many of the protons are capable of reacting with the chosen base. In the reaction between H_3PO_4 and NH_3

$$H_3PO_4(aq) + 2NH_3(aq) \rightleftharpoons HPO_4^-(aq) + 2nH_4^+(aq)$$

a conservation of protons requires that

$$2 \times \text{moles } H_3PO_4 = \text{moles of } NH_3$$

Conservation of Electron Pairs

In a complexation reaction, the reaction unit is an electron pair. For the metal, the number of reaction units is the number of coordination sites available for binding ligands. For the ligand, the number of reaction units is equivalent to the number of electron pairs that can be donated to the metal. One of the most important analytical complexation reactions is that between the ligand ethylenediaminetetracetic acid (EDTA), which can donate 6 electron pairs and 6 coordinate metal ions, such as Cu^{2+}; thus

$$6 \times \text{mole } Cu^{2+} = 6 \times \text{moles EDTA}$$

Conservation of Electrons

In a redox reaction, the reaction unit is an electron transferred from a reducing agent to an oxidizing agent. The number of reaction units for a reducing agent is equal to the number of electrons released during its oxidation. For an oxidizing agent, the number of reaction units is given by the number of electrons needed to cause its reduction. In the reaction between $Fe^{3+}$and oxalic acid, for example, Fe^{3+}undergoes a 1-electron reduction. Each carbon atom in oxalic acid is initially present in a 3 oxidation state, whereas the carbon atom in CO_2 is in a +4 oxidation state. Thus, we can write

$$1 \times \text{moles } Fe^{3+} = 2 \times \text{moles of } H_2C_2O_4$$

Note that the moles of oxalic acid are multiplied by 2 since there are two carbon atoms, each of which undergoes a 1-electron oxidation.

7

The Importance of Analytical Methodology

INTRODUCTION

The importance of analytical methodology is evident when examining the results of environmental monitoring programmes. The purpose of a monitoring programme is to determine the present status of an environmental system and to assess longterm trends in the quality of the system. These are broad and poorly defined goals. In many cases, such studies are initiated with little thought to the questions the data will be used to answer. This is not surprising since it can be hard to formulate questions in the absence of initial information about the system. Without careful planning, however, a poor experimental design may result in data that has little value.

These concerns are illustrated by the Chesapeake Bay monitoring programme. This research programme, designed to study nutrients and toxic pollutants in the Chesapeake Bay, was initiated in 1984 as a cooperative venture between the federal government, the state governments of Maryland, Virginia, and Pennsylvania, and the District of Columbia. A 1989 review of some of the problems with this programme highlights the difficulties common to many monitoring programmes.

At the beginning of the Chesapeake Bay monitoring programme, little attention was given to the proper choice of analytical methods, in large part because the intended uses of the monitoring data were not specified. The analytical methods initially chosen were those standard methods already approved by the EPA.

In many cases these methods proved to be of little value for this monitoring project. Most of the EPA-approved methods were designed to detect pollutants at their legally mandated maximum allowed concentrations.

The concentrations of these contaminants in natural waters, however, are often well below the detection limit of the EPA methods. For example, the EPA-approved standard method for phosphate had a detection limit of 7.5 ppb. Since actual phosphate concentrations in Chesapeake Bay usually were below the EPA detection limit, the EPA method provided no useful information.

On the other hand, a non-approved variant of the EPA method commonly used in chemical oceanography had a detection limit of 0.06 ppb. In other cases, such as the elemental analysis for particulate forms of carbon, nitrogen, and phosphorus, EPA-approved procedures provided poorer reproducibility than non-approved methods.

CHARACTERIZING EXPERIMENTAL ERRORS

Realizing that our data for the mass of a penny can be characterized by a measure of central tendency and a measure of spread suggests two questions. First, does our measure of central tendency agree with the true, or expected value? Second, why are our data scattered around the central value? Errors associated with central tendency reflect the accuracy of the analysis, but the precision of the analysis is determined by those errors associated with the spread.

Accuracy

Accuracy is a measure of how close a measure of central tendency is to the true, or expected value. Accuracy is usually expressed as either an absolute error

$$E = \overline{X} - \mu$$

or a percent relative error, E_r.

$$E_r = \frac{\overline{X} - \mu}{\mu} \times 100$$

Although the mean is used as the measure of central tendency in equations, the median could also be used. Errors affecting the accuracy of an analysis are called determinate and are characterized by a systematic deviation from the true value; that is, all the individual measurements are either too large or too small. A positive determinate error results in a central value that is larger than the true value, and a negative determinate error leads to a central value that is smaller than the true value. Both positive and negative determinate errors may affect the result of an analysis, with their cumulative effect leading to a net positive or negative determinate error. It is possible, although not likely, that positive and negative determinate errors may be equal, resulting in a central value with no net determinate error.

Determinate errors may be divided into four categories: sampling errors, method errors, measurement errors, and personal errors.

Sampling Errors We introduce determinate sampling errors when our sampling strategy fails to provide a representative sample. This is especially important when sampling heterogeneous materials. For example, determining the environmental quality of a lake by sampling a single location near a point source of pollution, such as an outlet for industrial effluent, gives misleading results. In determining the mass of a U.S. penny, the strategy for selecting pennies must ensure that pennies from other countries are not inadvertently included in the sample. Determinate errors associated with selecting a sample can be minimized with a proper sampling strategy.

Method Errors Determinate method errors are introduced when assumptions about the relationship between the signal and the analyte are invalid. In terms of the general relationships between the measured signal and the amount of analyte

$$S_{meas} = kn_A + S_{reag} \text{(total analysis method)}$$

$$S_{meas} = kC_A + S_{reag} \text{(concentration method)}$$

Method errors exist when the sensitivity, *k,* and the signal due to the reagent blank, *S*reag, are incorrectly determined. For example, methods in which *S*meas is the mass of a precipitate containing the analyte (gravimetric method) assume that the sensitivity is defined by a pure precipitate of known stoichiometry.

When this assumption fails, a determinate error will exist. Method errors involving sensitivity are minimized by standardizing the method, whereas method errors due to interferents present in reagents are minimized by using a proper reagent blank. Method errors due to interferents in the sample cannot be minimized by a reagent blank. Instead, such interferents must be separated from the analyte or their concentrations determined independently.

Measurement Errors Analytical instruments and equipment, such as glassware and balances, are usually supplied by the manufacturer with a statement of the item's maximum measurement error, or tolerance. For example, a 25-mL volumetric flask might have a maximum error of ±0.03 mL, meaning that the actual volume contained by the flask lies within the range of 24.97–25.03 mL. Although expressed as a range, the error is determinate; thus, the flask's true volume is a fixed value within the stated range.

Table. Measurement Errors for Selected Glassware[3]

Glassware	Measurement Errors for Volume (mL)	Class A Glassware (±mL)	Class B Glassware (±mL)
Transfer Pipets	1	0.006	0.012
	2	0.006	0.012
	5	0.01	0.02
	10	0.02	0.04
	20	0.03	0.06
	25	0.03	0.06
	50	0.05	0.10
Vaolumetric Flasks	5	0.02	0.04
	10	0.02	0.04
	25	0.03	0.06
	50	0.05	0.10
	100	0.08	0.16
	250	0.12	0.24
	500	0.20	0.40

	1000	0.30	0.60
	2000	0.50	1.0
Burets	10	0.02	0.04
	25	0.03	0.06
	50	0.05	0.10

Specifications for class A and class B glassware are taken from American Society for Testing and Materials E288, E542 and E694 standards.

Table. Measurement Errors for Selected Balances

Balance	Capacity (g)	Measurement Error
Precisa 160M	160	±1mg
A & DER 120 M	120	±0.1mg
Metler H54	160	±0.01mg

Table. Measurement Errors for Selected Digital Pipets

Pipet Range	Volume (mL or μL)[a]	Measurement Error (±%)
10–100μL[b]	10	1.0
	50	0.6
	100	0.6
200–1000μL[c]	200	1.5
	1000	0.8
1–10mL[d]	1	0.6
	5	0.4
	10	0.3

Unit for volume same as for pitet range.

Volumetric glassware is categorized by class. Class A glassware is manufactured to comply with tolerances specified by agencies such as the National Institute of Standards and Technology.

Tolerance levels for class A glassware are small enough that such glassware normally can be used without calibration. The tolerance levels for class B glassware are usually twice those for class A glassware. Other types of volumetric glassware, such as beakers and graduated cylinders, are unsuitable for accurately measuring volumes.

Determinate measurement errors can be minimized by calibration. A pipet can be calibrated, for example, by determining the mass of water that it delivers and using the density of water to calculate the actual volume delivered by the pipet. Although glassware and instrumentation can be calibrated, it is never safe to assume that the calibration will remain unchanged during an analysis. Many instruments, in particular, drift out of calibration over time. This complication can be minimized by frequent recalibration.

Personal Errors Finally, analytical work is always subject to a variety of personal errors, which can include the ability to see a change in the colour of an indicator used to signal the end point of a titration; biases, such as consistently overestimating or underestimating the value on an instrument's readout scale; failing to calibrate glassware and instrumentation; and misinterpreting procedural directions. Personal errors can be minimized with proper care. Identifying Determinate Errors Determinate errors can be difficult to detect. Without knowing the true value for an analysis, the usual situation in any analysis with meaning, there is no accepted value with which the experimental result can be compared. Nevertheless, a few strategies can be used to discover the presence of a determinate error.

Some determinate errors can be detected experimentally by analyzing several samples of different size. The magnitude of a constant determinate error is the same for all samples and, therefore, is more significant when analyzing smaller samples.

The presence of a constant determinate error can be detected by running several analyses using different amounts of sample, and looking for a systematic change in the property being measured.

For example, consider a quantitative analysis in which we separate the analyte from its matrix and determine the analyte's mass. Let's assume that the sample is 50.0percent w/w analyte; thus, if we analyse a 0.100-g sample, the analyte's true mass is 0.050 g. The first two columns of Table give the true mass of analyte for several additional samples. If the analysis has a positive constant determinate error of 0.010 g, then the experimentally determined mass for any sample will always be 0.010 g, larger than its true mass (column four of Table).

Table. Effect of Constant Positive Determinate Error on Analysis of Sample Containing 50% Analyte (%w/w)

Mass Sample (g)	True Mass of Analyte (g)	Constant Error (g)	Mass of Analyte Determined (g)	Percent Analyte Reported (%w/w)
0.100	0.050	0.010	0.060	60.0
0.200	0.100	0.010	0.110	55.0
0.400	0.200	0.010	0.410	51.5
0.800	0.400	0.010	0.410	51.2
1.000	0.500	0.010	0.510	51.0

The analyte's reported weight percent, which is shown in the last column of Table, becomes larger when we analyse smaller samples. A graph of percent w/w analyte versus amount of sample shows a distinct upward trend for small amounts of sample. A smaller concentration of analyte is obtained when analyzing smaller samples in the presence of a constant negative determinate error.

A proportional determinate error, in which the error's magnitude depends on the amount of sample, is more difficult to detect since the result of an analysis is independent of the amount of sample. Table outlines an example showing the effect of a positive proportional error of 1.0percent on the analysis of a sample that is 50.0percent w/w in analyte. In terms of equations, the reagent blank, *S*reag, is an example of a constant determinate error, and the sensitivity, *k,* may be affected by proportional errors.

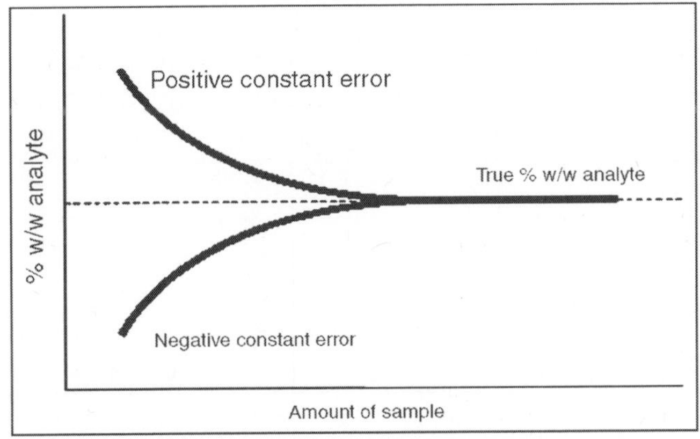

Fig. Effect of a Constant Determinate Error on the Reported Concentration of Analyte.

Potential determinate errors also can be identified by analyzing a standard sample containing a known amount of analyte in a matrix similar to that of the samples being analyzed. Standard samples are available from a variety of sources, such as the National Institute of Standards and Technology (where they are called standard reference materials) or the American Society for Testing and Materials. For example, figure shows an analysis sheet for a typical reference material. Alternatively, the sample can be analyzed by an independent method known to give accurate results, and the results of the two methods can be compared. Once identified, the source of a determinate error can be corrected. The best prevention against errors affecting accuracy, however, is a well-designed procedure that identifies likely sources of determinate errors, coupled with careful laboratory work.

The data in Table were obtained using a calibrated balance, certified by the manufacturer to have a tolerance of less than ±0.002 g. Suppose the Treasury Department reports that the mass of a 1998 U.S. penny is approximately 2.5 g. Since the mass of every penny in Table exceeds the reported mass by an amount significantly greater than the balance's tolerance, we can safely conclude that the error in this analysis is not due to equipment error. The actual source of the error is revealed later in this chapter.

Table. Effect of Constant Positive Determinate Error on Analysis of Sample Containing 50% Analyte (%w/w)

Mass Sample (g)	True Mass of Analyte (g)	Constant Error (%)	Mass of Analyte Determined (g)	Percent Analyte Reported (%w/w)
0.200	0.100	1.00	0.101	50.5
0.400	0.200	1.00	0.202	50.5
0.600	0.300	1.00	0.303	50.5
0.800	0.400	1.00	0.404	50.5
1.000	0.500	1.00	0.505	50.5

Simulated Rainwater (liquid form)

This SRM was developed to aid in the analysis of acidic rainwater by providing a stable, homogeneous material at two levels of acidity.

SRM	Type	Unit of Issue	
2694a	Simulated rainwater	Set of 4:2 of 50 mL at each of 2 levels	
	Constituent Element Parameter	**2694a-I**	**2694a-II**
	pH, 25°C	4.30	3.60
	Electrolytic Conductivity	25.4 (S/cm, 25°C)	129.3
	Acidity, meg/L	0.0544	0.283
	Fluoride, mg/L	0.057	0.108
	Chloride, mg/L	(0.23)*	(0.94)*
	Nitrate, mg/L	(0.53)*	7.19
	Sulfate, mg/L	(2.69)	10.6
	Sodium, mg/L	0.208	0.423
	Potassium, mg/L	0.056	0.108
	Ammonium, mg/L	(0.12)*	(1.06)*]
	Calcium, mg/L	0.0126	0.0364
	Magnesium, mg/L	0.0242	0.0484

Values in parentheses are not certified and given for information only.

Precision

Precision is a measure of the spread of data about a central value and may be expressed as the range, the standard deviation, or the variance. Precision is commonly divided into two categories: repeatability and reproducibility. Repeatability is the precision obtained when all measurements are made by the same analyst during a single period of laboratory work, using the same solutions and equipment.

Reproducibility, on the other hand, is the precision obtained under any other set of conditions, including that between analysts, or between laboratory sessions for a single analyst. Since reproducibility includes additional sources of variability, the reproducibility of an analysis can be no better than its repeatability.

Errors affecting the distribution of measurements around a central value are called indeterminate and are characterized by a random variation in both magnitude and direction. Indeterminate errors need not affect the accuracy of an analysis.

Since indeterminate errors are randomly scattered around a central value, positive and negative errors tend to cancel, provided that enough measurements are made. In such situations the mean or median is largely unaffected by the precision of the analysis.

Sources of Indeterminate Error Indeterminate errors can be traced to several sources, including the collection of samples, the manipulation of samples during the analysis, and the making of measurements. When collecting a sample, for instance, only a small portion of the available material is taken, increasing the likelihood that small-scale inhomogeneities in the sample will affect the repeatability of the analysis. Individual pennies, for example, are expected to show variation from several sources, including the manufacturing process, and the loss of small amounts of metal or the addition of dirt during circulation. These variations are sources of indeterminate error associated with the sampling process. During the analysis numerous opportunities arise for random variations in the way individual samples are treated. In determining the mass of a penny, for example, each penny should be handled in the same manner. Cleaning some pennies but not cleaning others introduces an indeterminate error.

Finally, any measuring device is subject to an indeterminate error in reading its scale, with the last digit always being an estimate subject to random fluctuations, or background noise.

For example, a buret with scale divisions every 0.1 mL has an inherent indeterminate error of ±0.01 – 0.03 mL when estimating the volume to the hundredth of a millilitre. Background noise in an electrical meter can be evaluated by recording the signal without analyte and observing the fluctuations in the signal over time.

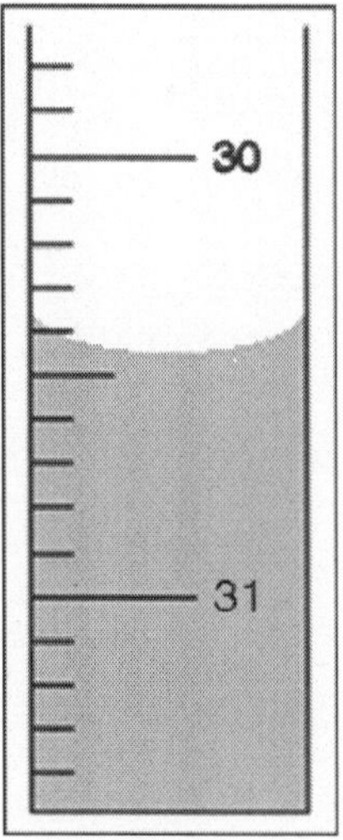

Fig. Close-up of Buret, Showing Difficulty in Estimating Volume. With Scale Divisions Every 0.1 mL it is Difficult to Read the Actual Volume to Better than ±0.01 – 0.03 mL.

Evaluating Indeterminate Error Although it is impossible to eliminate indeterminate error, its effect can be minimized if the sources and relative magnitudes of the indeterminate error are known. Indeterminate errors may be estimated by an appropriate measure of spread. Typically, a standard deviation is used, although in some cases estimated values are used. The contribution from analytical instruments and equipment are easily measured or estimated. Indeterminate errors introduced by the analyst, such as inconsistencies in the treatment of individual samples, are more difficult to estimate.

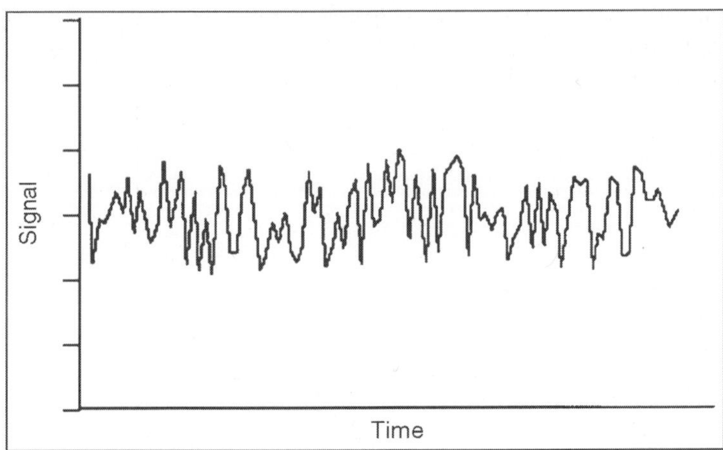

Fig. Background Noise in a Meter Obtained by Measuring Signal Over Time in the Absence of Analyte.

To evaluate the effect of indeterminate error on the data in Table, ten replicate determinations of the mass of a single penny were made, with results shown in Table. The standard deviation for the data in Table is 0.051, and it is 0.0024 for the data in Table. The significantly better precision when determining the mass of a single penny suggests that the precision of this analysis is not limited by the balance used to measure mass, but is due to a significant variability in the masses of individual pennies.

Table. Replicate Determinations of the Mass of a Single United States Penny in Circulation.

Replicate Number	Mass (g)
1	3.025
2	3.024
3	3.028
4	3.027
5	3.028
6	3.023
7	3.022
8	3.021
9	3.026
10	3.024

Error and Uncertainty

Analytical chemists make a distinction between error and uncertainty.3 Error is the difference between a single measurement or result and its true value. In other words, error is a measure of bias. As discussed earlier, error can be divided into determinate and indeterminate sources.

Although we can correct for determinate error, the indeterminate portion of the error remains. Statistical significance testing, which is discussed later in this chapter, provides a way to determine whether a bias resulting from determinate error might be present. Uncertainty expresses the range of possible values that a measurement or result might reasonably be expected to have.

Note that this definition of uncertainty is not the same as that for precision. The precision of an analysis, whether reported as a range or a standard deviation, is calculated from experimental data and provides an estimation of indeterminate error affecting measurements.

Uncertainty accounts for all errors, both determinate and indeterminate, that might affect our result. Although we always try to correct determinate errors, the correction itself is subject to random effects or indeterminate errors.

Table. Experimentally Determined Volumes Delivered by a 10-mL Class A Pipet.

	Volume Delivered Trial (mL)		Volum Delivered Trial(mL)
1	10.002	6	9.983
2	9.993	7	9.991
3	9.984	8	9.990
4	9.996	9	9.988
5	9.989	10	9.999

To illustrate the difference between precision and uncertainty, consider the use of a class A 10-mL pipet for delivering solutions. A pipet's uncertainty is the range of volumes in which its true volume is expected to lie. Suppose you purchase a 10-mL class A pipet from a laboratory supply company and use it without calibration. The pipet's tolerance value of ±0.02 mL represents your uncertainty since your best estimate of its volume is 10.00 mL ±0.02 mL.

Precision is determined experimentally by using the pipet several times, measuring the volume of solution delivered each time. Table shows results for ten such trials that have a mean of 9.992 mL and a standard deviation of 0.006.

This standard deviation represents the precision with which we expect to be able to deliver a given solution using any class A 10-mL pipet. In this case the uncertainty in using a pipet is worse than its precision.

Interestingly, the data in Table allow us to calibrate this specific pipet's delivery volume as 9.992 mL. If we use this volume as a better estimate of this pipet's true volume, then the uncertainty is ±0.006. As expected, calibrating the pipet allows us to lower its uncertainty.

SAMPLING AND SAMPLE HANDLING

REPRESENTATIVE SAMPLE

The importance of obtaining a representative sample for analysis cannot be overemphasised. Without it, results may be meaningless or even grossly misleading. Sampling is particularly crucial where a heterogeneous material is to be analysed. It is vital that the aims of the analysis are understood and an appropriate sampling procedure adopted. In some situations, a sampling plan or strategy may need to be devised so as to optimise the value of the analytical information collected. This is necessary particularly where environmental samples of soil, water or the atmosphere are to be collected or a complex industrial process is to be monitored.

Legal requirements may also determine a sampling strategy, particularly in the food and drug industries. A small sample taken for analysis is described as a laboratory sample. Where duplicate analyses or several different analyses are required, the laboratory sample will be divided into sub-samples which should have identical compositions.

Homogeneous materials (*e.g.*, single or mixed solvents or solutions and most gases) generally present no particular sampling problem as the composition of any small laboratory sample taken from a larger volume will be representative of the bulk solution.

Heterogeneous materials have to be homogenised prior to obtaining a laboratory sample if an average or bulk composition is required. Conversely, where analyte levels in different parts of the material are to be measured, they may need to be physically separated before laboratory samples are taken.

This is known as selective sampling. Typical examples of heterogeneous materials where selective sampling may be necessary include:

- Surface waters such as streams, rivers, reservoirs and seawater, where the concentrations of trace metals or organic compounds in solution and in sediments or suspended particulate matter may each be of importance.
- Materials stored in bulk, such as grain, edible oils, or industrial organic chemicals, where physical segregation (stratification) or other effects may lead to variations in chemical composition throughout the bulk.
- Ores, minerals and alloys, where information about the distribution of a particular metal or compound is sought.
- Laboratory, industrial or urban atmospheres where the concentrations of toxic vapors and fumes may be localised or vary with time.

Obtaining a laboratory sample to establish an average analyte level in a highly heterogeneous material can be a lengthy procedure. For example, sampling a large shipment of an ore or mineral, where the economic cost needs to be determined by a very accurate assay, is typically approached in the following manner.

- Relatively large pieces are randomly selected from different parts of the shipment.
- The pieces are crushed, ground to coarse granules and thoroughly mixed.
- A repeated coning and quartering process, with additional grinding to reduce particle size, is used until a laboratory-sized sample is obtained. This involves creating a conical heap of the material, dividing

it into four equal portions, discarding two diagonally opposite portions and forming a new conical heap from the remaining two quarters. The process is then repeated as necessary.

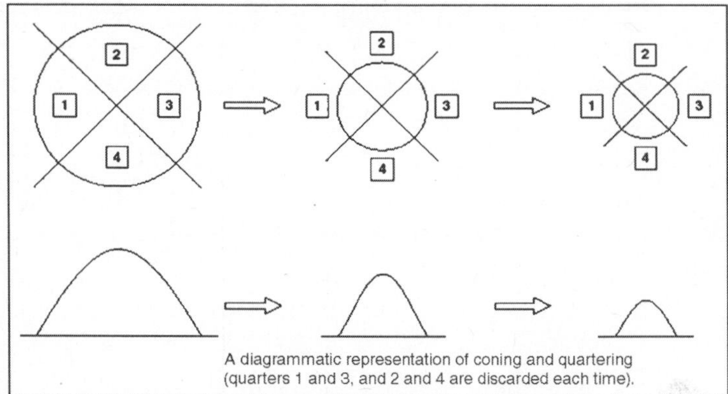

A diagrammatic representation of coning and quartering (quarters 1 and 3, and 2 and 4 are discarded each time).

The distribution of toxic heavy metals or organic compounds in a land redevelopment site presents a different problem. Here, to economise on the number of analyses, a grid is superimposed on the site dividing it up into approximately one- to five-metre squares. From each of these, samples of soil will be taken at several specified depths. A three-dimensional representation of the distribution of each analyte over the whole site can then be produced, and any localised high concentrations, or hot spots, can be investigated by taking further, more closelyspaced, samples. Individual samples may need to be ground, coned and quartered as part of the sampling strategy.

Repeated sampling over a period of time is a common requirement. Examples include the continuous monitoring of a process stream in a manufacturing plant and the frequent sampling of patients' body fluids for changes in the levels of drugs, metabolites, sugars or enzymes, etc., during hospital treatment. Studies of seasonal variations in the levels of pesticide, herbicide and fertilizer residues in soils and surface waters, or the continuous monitoring of drinking water supplies are two further examples.

Having obtained a representative sample, it must be labeled and stored under appropriate conditions. Sample identification through proper labeling, increasingly done by using bar codes and optical readers under computer control, is an essential feature of sample handling.

SAMPLE STORAGE

Samples often have to be collected from places remote from the analytical laboratory and several days or weeks may elapse before they are received by the laboratory and analyzed. Furthermore, the workload of many laboratories is such that incoming samples are stored for a period of time prior to analysis. In both instances, sample containers and storage conditions (*e.g.*, temperature, humidity, light levels and exposure to the atmosphere) must be controlled such that no significant changes occur that could affect the validity of the analytical data. The following effects during storage should be considered:

- Increases in temperature leading to the loss of volatile analytes, thermal or biological degradation, or increased chemical reactivity.
- Decreases in temperature that lead to the formation of deposits or the precipitation of analytes with low solubilities.
- Changes in humidity that affect the moisture content of hygroscopic solids and liquids or induce hydrolysis reactions.
- UV radiation, particularly from direct sunlight, that induces photochemical reactions, photodecomposition or polymerization;
- Air-induced oxidation.
- Physical separation of the sample into layers of different density or changes in crystallinity.

In addition, containers may leak or allow contaminants to enter. A particular problem associated with samples having very low (trace and ultra-trace) levels of analytes in solution is the possibility of losses by adsorption onto the walls of the container or contamination by substances being leached from the container by the sample solvent. Trace metals may be depleted by adsorption or ion-exchange processes if stored in glass containers, whilst sodium, potassium, boron and silicates can be leached from the glass into the sample solution. Plastic containers should always be used for such samples. Conversely, sample solutions containing organic solvents and other organic liquids should be stored in glass containers because the base plastic or additives such as plasticizers and antioxidants may be leached from the walls of plastic containers.

SAMPLE PRETREATMENT

Samples arriving in an analytical laboratory come in a very wide assortment of sizes, conditions and physical forms and can contain analytes from major constituents down to ultra-trace levels. They can have a variable moisture content and the matrix components of samples submitted for determinations of the same analyte(s) may also vary widely. A preliminary, or pre-treatment, is often used to condition them in readiness for the application of a specific method of analysis or to pre-concentrate (enrich) analytes present at very low levels. Examples of pretreatments are:

- Drying at 100°C to 120°C to eliminate the effect of a variable moisture content.
- Weighing before and after drying enables the water content to be calculated or it can be established by thermogravimetric analysis.
- Separating the analytes into groups with common characteristics by distillation, filtration, centrifugation, solvent or solid phase extraction.
- Removing or reducing the level of matrix components that are known to cause interference with measurements of the analytes;

- Concentrating the analytes if they are below the concentration range of the analytical method to be used by evaporation, distillation, co-precipitation, ion exchange, solvent or solid phase extraction or electrolysis.

Sample clean-up in relation to matrix interference and to protect specialized analytical equipment such as chromatographic columns and detection systems from high levels of matrix components is widely practised using solid phase extraction (SPE) cartridges. Substances such as lipids, fats, proteins, pigments, polymeric and tarry substances are particularly detrimental.

SAMPLE PREPARATION

A laboratory sample generally needs to be prepared for analytical measurement by treatment with reagents that convert the analyte(s) into an appropriate chemical form for the selected technique and method, although in some instances it is examined directly as received or mounted in a sample holder for surface analysis. If the material is readily soluble in aqueous or organic solvents, a simple dissolution step may suffice. However, many samples need first to be decomposed to release the analyte(s) and facilitate specific reactions in solution. Sample solutions may need to be diluted or concentrated by enrichment so that analytes are in an optimum concentration range for the method.

The stabilization of solutions with respect to pH, ionic strength and solvent composition, and the removal or masking of interfering matrix components not accounted for in any pre-treatment may also be necessary. An internal standard for reference purposes in quantitative analysis is sometimes added before adjustment to the final prescribed volume. Some common methods of decomposition and dissolution are given in *Table*.

Some methods for sample decomposition and dissolution

Method of attack	Type of sample
Heated with concentrated mineral acids (HCl, HNO_3, aqua regia) or strong alkali, including microwave digestion	Geological, metallurgical
Fusion with flux (Na_2O_2, Na_2CO_3, $LiBO_2$, $KHSO_4$, KOH)	Geological, refractory materials
Heated with HF and H_2SO_4 or $HClO_4$	Silicates where SiO_2 is not the analyte
Acid leaching with HNO_3	Soils and sediments
Dry oxidation by heating in a furnace or wet oxidation by boiling with concentrated H_2SO_4 and HNO_3 or $HClO_4$	Organic materials with inorganic analytes

8

Statistical Analysis of Data in Analytical Chemistry

STATISTICAL METHODS FOR NORMAL DISTRIBUTIONS

The most commonly encountered probability distribution is the normal, or Gaussian, distribution. A normal distribution is characterized by a true mean, μ, and variance, σ^2, which are estimated using $\overline{X}$ and *s*2. Since the area between any two limits of a normal distribution is well defined, the construction and evaluation of significance tests are straightforward.

Comparing $\overline{X}$ to μ

One approach for validating a new analytical method is to analyse a standard sample containing a known amount of analyte, μ. The method's accuracy is judged by determining the average amount of analyte in several samples, $\overline{X}$, and using a significance test to compare it with μ. The null hypothesis is that $\overline{X}$ and μ are the same and that any difference between the two values can be explained by indeterminate errors affecting the determination of $\overline{X}$. The alternative hypothesis is that the difference between $\overline{X}$ and μ is too large to be explained by indeterminate error.

The equation for the test (experimental) statistic, *t*exp, is derived from the confidence interval for μ

$$\mu = \overline{X} \pm \frac{t_{exp} s}{\sqrt{n}}$$

Rearranging equation

$$t_{exp} = \frac{|\mu - \overline{X}| \times \sqrt{n}}{s}$$

gives the value of *t*exp when μ is at either the right or left edge of the sample's apparent confidence interval.

The value of *t*exp is compared with a critical value, $t(\alpha, \upsilon)$, which is determined by the chosen significance level, α, the degrees of freedom for the sample, υ, and whether the significance test is onetailed or two-tailed.

The critical value $t(\alpha, \upsilon)$ defines the confidence interval that can be explained by indeterminate errors. If *t*exp is greater than $t(\alpha, \upsilon)$, then the confidence interval for the data is wider than that expected from indeterminate errors. In this case, the null hypothesis is rejected and the alternative hypothesis is accepted. If *t*exp is less than or equal to $t(\alpha,v)$, then the confidence interval for the data could be attributed to indeterminate error, and the null hypothesis is retained at the stated significance level.

A typical application of this significance test, which is known as a *t*-test of $\overline{X}$ to μ, is outlined in the following example.

(a) The shaded area under the normal distribution curves shows the apparent confidence intervals for the sample based on *t*exp. The solid bars in (b) and (c) show the actual confidence intervals that can be explained by indeterminate error using the critical value of (α,v). In part (b) the null hypothesis is rejected and the alternative hypothesis is accepted. In part (c) the null hypothesis is retained.

Its source should be identified and corrected before analyzing additional samples. Failing to reject the null hypothesis, however, does not imply that the method is accurate, but only indicates that there is insufficient evidence to prove the method inaccurate at the stated confidence level.

The utility of the *t*-test for $\overline{X}$ and μis improved by optimizing the conditions used in determining $\overline{X}$. Examining equation shows that increasing the number of replicate determinations, *n*, or improving the precision of the analysis enhances the utility of this significance test.

A *t*-test can only give useful results, however, if the standard deviation for the analysis is reasonable. If the standard deviation is substantially larger than the expected standard deviation, μ, the confidence interval around – *X* will be so large that a significant difference between – *X* and μ may be difficult to prove. On the other hand, if the standard deviation is significantly smaller than expected, the confidence interval around – *X* will be too small, and a significant difference between – *X* and μ may be found when none exists. A significance test that can be used to evaluate the standard deviation is the subject of the next section.

Comparing S^2 to σ^2

When a particular type of sample is analyzed on a regular basis, it may be possible to determine the expected, or true variance, ?2, for the analysis. This often is the case in clinical labs where hundreds of blood samples are analyzed each day. Replicate analyses of any single sample, however, results in a sample variance, $s2$. A statistical comparison of $s2$ to ?2 provides useful information about whether the analysis is in a state of "statistical control." The null hypothesis is that $s2$ and σ^2 are identical, and the alternative hypothesis is that they are not identical.

The test statistic for evaluating the null hypothesis is called an F-test, and is given as either

$$F_{exp} = \frac{s^2}{\sigma^2} \text{ or } F_{exp} = \frac{\sigma^2}{s^2}$$

$$(s^2 > \sigma^2) \quad (\sigma^2 > s^2)$$

depending on whether $s2$ is larger or smaller than σ^2. Note that Fexp is defined such that its value is always greater than or equal to 1.

If the null hypothesis is true, then Fexp should equal 1. Due to indeterminate errors, however, the value for Fexp usually is greater than 1. A critical value, $F(\alpha, v_{num}, v_{den})$, gives the largest value of F that can be explained by indeterminate error. It is chosen for a specified significance level, α, and the degrees of freedom for the variances in the numerator, v_{num}, and denominator, v_{den}. The degrees of freedom for $s2$ is $n - 1$, where n is the number of replicates used in determining the sample's variance.

Comparing Two Sample Variances

The F-test can be extended to the comparison of variances for two samples, A and B, by rewriting equation as

$$F_{exp} = \frac{s_A^2}{s_B^2}$$

where A and B are defined such that s^2_A is greater than or equal to s^2_B. An example of this application of the F-test is shown in the following example.

Comparing Two Sample Means

The result of an analysis is influenced by three factors: the method, the sample, and the analyst. The influence of these factors can be studied by conducting a pair of experiments in which only one factor is changed. For example, two methods can be compared by having the same analyst apply both methods to the same sample and examining the resulting means. In a similar fashion, it is possible to compare two analysts or two samples.

Significance testing for comparing two mean values is divided into two categories depending on the source of the data. Data are said to be unpaired when each mean is derived from the analysis of several samples drawn from the same source. Paired data are encountered when analyzing a series of samples drawn from different sources. Unpaired Data Consider two samples, A and B, for which mean values, $\overline{X}_A$ and $\overline{X}_B$, and standard deviations, sA and *s*B, have been measured. Confidence intervals for μ_A and μ_B can be written for both samples

$$\mu_A = \overline{X}_A \pm \frac{ts_A}{\sqrt{n_A}}$$

$$\mu_B = \overline{X}_B \pm \frac{ts_B}{\sqrt{n_B}}$$

where *n*A and *n*B are the number of replicate trials conducted on samples A and B. A comparison of the mean values is based on the null hypothesis that $\overline{X}_A$ and $\overline{X}_B$ are identical, and an alternative hypothesis that the means are significantly different. A test statistic is derived by letting μ_A equal μ_B, and combining equations to give

$$\overline{X}_A \pm \frac{ts_A}{\sqrt{n_A}} = \overline{X}_B \pm \frac{ts_B}{\sqrt{n_B}}$$

Solving for $|\overline{X}_A - \overline{X}_B|$ and using a propagation of uncertainty, gives

$$|\overline{X}_A - \overline{X}_B| = t \times \sqrt{\frac{s_A^2}{n_A} + \frac{s_B^2}{n_B}}$$

Finally, solving for *t,* which we replace with *t*exp, leaves us with

$$t_{exp} = \frac{|\overline{X}_A - \overline{X}_B|}{\sqrt{(s_A^2/n_A) + (s_B^2/n_B)}}$$

The value of *t*exp is compared with a critical value, $t(\alpha, v)$, as determined by the chosen significance level, α, the degrees of freedom for the sample,v, and whether the significance test is one-tailed or two-tailed.

It is unclear, however, how many degrees of freedom are associated with $t(\alpha, v)$ since there are two sets of independent measurements. If the variances *s*2Aand *s*2B estimate the same σ^2, then the two standard deviations can be factored out of equation and replaced by a pooled standard deviation, *s*pool, which provides a better estimate for the precision of the analysis. Thus, equation becomes

$$t_{exp} = \frac{|\overline{X}_A - \overline{X}_B|}{s_{pool}\sqrt{(1/n_A) + (1/n_B)}}$$

with the pooled standard deviation given as

$$s_{pool} = \sqrt{\frac{(n_A - 1)s_A^2 + (n_B - 1)s_B^2}{n_A + n_B - 2}}$$

As indicated by the denominator of equation, the degrees of freedom for the pooled standard deviation is $n_A + n_B - 2$.

If *s*A and *s*B are significantly different, however, then *t*exp must be calculated using equation. In this case, the degrees of freedom is calculated using the following imposing equation

$$v = \frac{[(s_A^2 / n_A) + (s_B^2 / n_B)]^2}{[(s_A^2 / n_A)^2 / (n_A + 1)] + [s_B^2 / n_B)^2 / (n_B + 1)]} - 2$$

Since the degrees of freedom must be an integer, the value of ??obtained using equation is rounded to the nearest integer.

Regardless of whether equation is used to calculate *t*exp, the null hypothesis is rejected if *t*exp is greater than *t*(α, v), and retained if *t*exp is less than or equal to *t*(α, v). Paired Data In some situations the variation within the data sets being compared is more significant than the difference between the means of the two data sets. This is commonly encountered in clinical and environmental studies, where the data being compared usually consist of a set of samples drawn from several populations.

For example, a study designed to investigate two procedures for monitoring the concentration of glucose in blood might involve blood samples drawn from ten patients. If the variation in the blood glucose levels among the patients is significantly larger than the anticipated variation between the methods, then an analysis in which the data are treated as unpaired will fail to find a significant difference between the methods.

In general, paired data sets are used whenever the variation being investigated is smaller than other potential sources of variation.

In a study involving paired data the difference, *d*i, between the paired values for each sample is calculated. The average difference, – *d*, and standard deviation of the differences, *sd*, are then calculated. The null hypothesis is that – *d* is 0, and that there is no difference in the results for the two data sets. The alternative hypothesis is that the results for the two sets of data are significantly different, and, therefore, – *d* is not equal to 0. The test statistic, *t*exp, is derived from a confidence interval around –

$$0 = \bar{d} \pm \frac{ts_d}{\sqrt{n}}$$

where *n* is the number of paired samples. Replacing *t* with *t*exp and rearranging gives

$$t_{exp} = \frac{|\bar{d}|\sqrt{n}}{s_d}$$

The value of *t*exp is then compared with a critical value, *t*(α, v), which is determined by the chosen significance level, α, the degrees of freedom for the sample, v, and whether the significance test is one-tailed or two-tailed.

For paired data, the degrees of freedom is *n* – 1. If *t*exp is greater than *t*(α, v), then the null hypothesis is rejected and the alternative hypothesis is accepted. If *t*exp is less than or equal to *t*(α, v), then the null hypothesis is retained, and a significant difference has not been demonstrated at the stated significance level. This is known as the paired *t*-test.

A paired *t*-test can only be applied when the individual differences, *d*i, belong to the same population. This will only be true if the determinate and indeterminate errors affecting the results are independent of the concentration of analyte in the samples. If this is not the case, a single sample with a larger error could result in a value of *d*i that is substantially larger than that for the remaining samples.

Including this sample in the calculation of – *d* and *sd* leads to a biased estimate of the true mean and standard deviation. For samples that span a limited range of analyte concentrations, such as that in Example, this is rarely a problem. When paired data span a wide range of concentrations, however, the magnitude of the determinate and indeterminate sources of error may not be independent of the analyte's concentration. In such cases the paired *t*-test may give misleading results since the paired data with the largest absolute determinate and indeterminate errors will dominate – *d*. In this situation a comparison is best made using a linear regression, details of which are discussed in the next chapter.

Outliers

On occasion, a data set appears to be skewed by the presence of one or more data points that are not consistent with the remaining data points. Such values are called outliers. The most commonly used significance test for identifying outliers is Dixon's *Q*-test. The null hypothesis is that the apparent outlier is taken from the same population as the remaining data. The alternative hypothesis is that the outlier comes from a different population, and, therefore, should be excluded from consideration.

The *Q*-test compares the difference between the suspected outlier and its nearest numerical neighbour to the range of the entire data set. Data are ranked from smallest to largest so that the suspected outlier is either the first or the last data point. The test statistic, Q_{exp}, is calculated using equation if the suspected outlier is the smallest value ($X1$)

$$Q_{exp} = \frac{X_2 - X_1}{X_n - X_1}$$

or using equation if the suspected outlier is the largest value (X_n)

$$Q_{exp} = \frac{X_n - X_{n-1}}{X_n - X_1}$$

where *n* is the number of members in the data set, including the suspected outlier. It is important to note that equations are valid only for the detection of a single outlier. Other forms of Dixon's *Q*-test allow its extension to the detection of multiple outliers. The value of Q_{exp} is compared with a critical value, $Q(\alpha, n)$, at a significance level of α. The *Q*-test is usually applied as the more conservative twotailed test, even though the outlier is the smallest or largest value in the data set. If Q_{exp} is greater than $Q(\alpha, n)$, then the null hypothesis is rejected and the outlier may be rejected. When Qexp is less than or equal to $Q(\alpha, n)$ the suspected outlier must be retained.

The Q-test should be applied with caution since there is a probability, equal to α, that an outlier identified by the Q-test actually is not an outlier. In addition, the Q-test should be avoided when rejecting an outlier leads to a precision that is unreasonably better than the expected precision determined by a propagation of uncertainty. Given these two concerns it is not surprising that some statisticians caution against the removal of outliers.

On the other hand, testing for outliers can provide useful information if we try to understand the source of the suspected outlier. For example, the outlier identified in Example represents a significant change in the mass of a penny (an approximately 17percent decrease in mass), due to a change in the composition of the U.S. penny. In 1982, the composition of a U.S. penny was changed from a brass alloy consisting of 95percent w/w Cu and 5percent w/w Zn, to a zinc core covered with copper.

Detection Limits

The focus of this chapter has been the evaluation of analytical data, including the use of statistics. In this final section we consider how statistics may be used to characterize a method's ability to detect trace amounts of an analyte.

A method's detection limit is the smallest amount or concentration of analyte that can be detected with statistical confidence. The International Union of Pure and Applied Chemistry (IUPAC) defines the detection limit as the smallest concentration or absolute amount of analyte that has a signal significantly larger than the signal arising from a reagent blank. Mathematically, the analyte's signal at the detection limit, (SA)DL, is

$$(S_A)_{DL} = S_{reag} + zQ_{reag}$$

where Sreag is the signal for a reagent blank, σ_{reag} is the known standard deviation for the reagent blank's signal, and z is a factor accounting for the desired confidence level. The concentration, (CA)DL, or absolute amount of analyte, (nA)DL, at the detection limit can be determined from the signal at the detection limit.

$$(C_A)_{DL} = \frac{(S_A)_{DL}}{k}$$

$$(n_A)_{DL} = \frac{(S_A)_{DL}}{k}$$

The value for z depends on the desired significance level for reporting the detection limit. Typically, z is set to 3, which, from Appendix, corresponds to a significance level of $\alpha = 0.00135$. Consequently, only 0.135percent of measurements made on the blank will yield signals that fall outside this range.

When $Z\alpha_{reag}$ is unknown, the term $z\alpha_{reag}$ may be replaced with *ts*reag, where *t* is the appropriate value from a *t*-table for a one-tailed analysis.

In analyzing a sample to determine whether an analyte is present, the signal for the sample is compared with the signal for the blank. The null hypothesis is that the sample does not contain any analyte, in which case (*S*A)DL and *S*reag are identical. The alternative hypothesis is that the analyte is present, and (*S*A)DL is greater than *S*reag. If (*S*A)DL exceeds *S*reag by *z*α(or *ts*), then the null hypothesis is rejected and there is evidence for the analyte's presence in the sample. The probability that the null hypothesis will be falsely rejected, a type 1 error, is the same as the significance level. Selecting *z* to be 3 minimizes the probability of a type 1 error to 0.135percent.

Significance tests, however, also are subject to type 2 errors in which the null hypothesis is falsely retained. Consider, for example, the situation shown in figure, where *S*A is exactly equal to (SA)DL. In this case the probability of a type 2 error is 50percent since half of the signals arising from the sample's population fall below the detection limit. Thus, there is only a 50:50 probability that an analyte at the IUPAC detection limit will be detected. As defined, the IUPAC definition for the detection limit only indicates the smallest signal for which we can say, at a significance level of ?, that an analyte is present in the sample. Failing to detect the analyte, however, does not imply that it is not present. An alternative expression for the detection limit, which minimizes both type 1 and type 2 errors, is the limit of identification, (*S*A)LOI, which is defined as

$$(S_A)_{LOI} = S_{reag} + zQ_{reag} + zQ_{samp}$$

As shown in Figure, the limit of identification is selected such that there is an equal probability of type 1 and type 2 errors. The American Chemical Society's Committee on Environmental Analytical Chemistry recommends the limit of quantitation, (*S*A)LOQ.

$$(S_A)_{LOQ} = S_{reag} + 10Q_{reag}$$

Other approaches for defining the detection limit have also been developed.16 The detection limit is often represented, particularly when used in debates over public policy issues, as a distinct line separating analytes that can be detected from those that cannot be detected.17 This use of a detection limit is incorrect.

Defining the detection limit in terms of statistical confidence levels implies that there may be a gray area where the analyte is sometimes detected and sometimes not detected. This is shown in Figure where the upper and lower confidence limits are defined by the acceptable probabilities for type 1 and type 2 errors. Analytes producing signals greater than that defined by the upper confidence limit are always detected, and analytes giving signals smaller than the lower confidence limit are never detected. Signals falling between the upper and lower confidence limits, however, are ambiguous because they could belong to populations representing either the reagent blank or the analyte. Figure represents the smallest value of *SA* for which no such ambiguity exists.

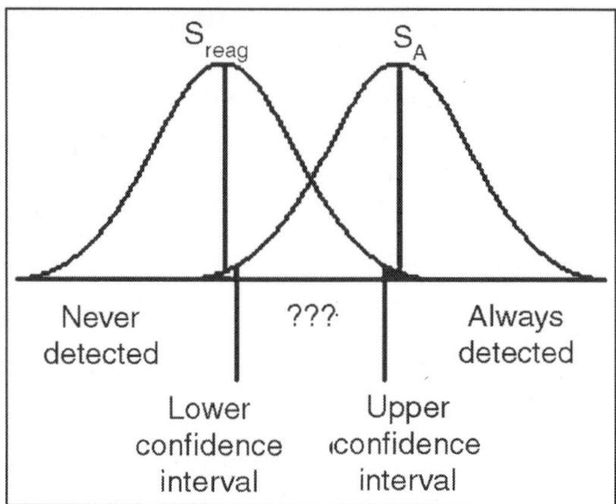

Fig. Establishment of Areas where the Signal is Never Detected, Always Detected, and where Results are Ambiguous. The Upper and Lower Confidence Limits are Defined by the Probability of a Type 1 Error (*dark shading*), and the Probability of a Type 2 Error (*light shading*).

STATISTICAL ANALYSIS OF DATA

In the previous section we noted that the result of an analysis is best expressed as a confidence interval. For example, a 95percent confidence interval for the mean of five results gives the range in which we expect to find the mean for 95percent of all samples of equal size, drawn from the same population. Alternatively, and in the absence of determinate errors, the 95percent confidence interval indicates the range of values in which we expect to find the population's true mean. The probabilistic nature of a confidence interval provides an opportunity to ask and answer questions comparing a sample's mean or variance to either the accepted values for its population or similar values obtained for other samples.

For example, confidence intervals can be used to answer questions such as "Does a newly developed method for the analysis of cholesterol in blood give results that are significantly different from those obtained when using a standard method?" or "Is there a significant variation in the chemical composition of rainwater collected at different sites downwind from a coalburning utility plant?" In this section we introduce a general approach to the statistical analysis of data. Specific statistical methods of analysis are covered in Section.

Significance Testing

Let's consider the following problem. Two sets of blood samples have been collected from a patient receiving medication to lower her concentration of blood glucose. One set of samples was drawn immediately before the medication was administered; the second set was taken several hours later.

The samples are analyzed and their respective means and variances reported. How do we decide if the medication was successful in lowering the patient's concentration of blood glucose?

One way to answer this question is to construct probability distribution curves for each sample and to compare the curves with each other. Three possible outcomes are shown in Figure. In Figure, the probability distribution curves are completely separated, strongly suggesting that the samples are significantly different. In Figure, the probability distributions for the two samples are highly overlapped, suggesting that any difference between the samples is insignificant.

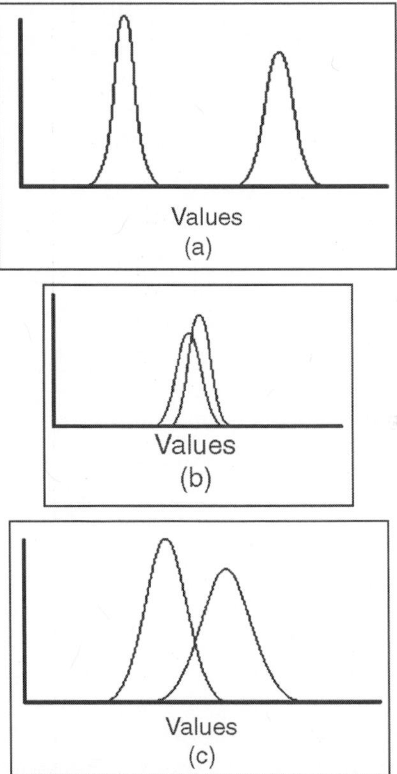

Fig. Three Example of Possible Relationships between the Probability Distributins for two Populations. (a) Completely Separate Distributions; (b) Distributions with a Great deal of Overlap; (c) Distributions with some Overlap.

Figure, however, presents a dilemma. Although the means for the two samples appear to be different, the probability distributions overlap to an extent that a significant number of possible outcomes could belong to either distribution. In this case we can, at best, only make a statement about the probability that the samples are significantly different. The process by which we determine the probability that there is a significant difference between two samples is called significance testing or hypothesis testing. Before turning to a discussion of specific examples, however, we will first establish a general approach to conducting and interpreting significance tests.

Constructing a Significance Test

A significance test is designed to determine whether the difference between two or more values is too large to be explained by indeterminate error. The first step in constructing a significance test is to state the experimental problem as a yes or- no question, two examples of which were given at the beginning of this section. A null hypothesis and an alternative hypothesis provide answers to the question.

The null hypothesis, H_0, is that indeterminate error is sufficient to explain any difference in the values being compared. The alternative hypothesis, HA, is that the difference between the values is too great to be explained by random error and, therefore, must be real. A significance test is conducted on the null hypothesis, which is either retained or rejected. If the null hypothesis is rejected, then the alternative hypothesis must be accepted. When a null hypothesis is not rejected, it is said to be retained rather than accepted. A null hypothesis is retained whenever the evidence is insufficient to prove it is incorrect. Because of the way in which significance tests are conducted, it is impossible to prove that a null hypothesis is true.

The difference between retaining a null hypothesis and proving the null hypothesis is important. To appreciate this point, let us return to our example on determining the mass of a penny. After looking at the data in Table, you might pose the following null and alternative hypotheses

H_0: Any U.S. penny in circulation has a mass that falls in the range of 2.900–3.200 g

H_A: Some U.S. pennies in circulation have masses that are less than 2.900 g or more than 3.200 g.

To test the null hypothesis, you reach into your pocket, retrieve a penny, and determine its mass. If the mass of this penny is 2.512 g, then you have proved that the null hypothesis is incorrect. Finding that the mass of your penny is 3.162 g, however, does not prove that the null hypothesis is correct because the mass of the next penny you sample might fall outside the limits set by the null hypothesis. After stating the null and alternative hypotheses, a significance level for the analysis is chosen. The significance level is the confidence level for retaining the null hypothesis or, in other words, the probability that the null hypothesis will be incorrectly rejected. In the former case the significance level is given as a percentage (*e.g.*, 95percent), whereas in the latter case, it is given as

$$\alpha = 1 - \frac{\text{confidence level}}{100}$$

Thus, for a 95percent confidence level, α is 0.05.

Next, an equation for a test statistic is written, and the test statistic's critical value is found from an appropriate table. This critical value defines the breakpoint between values of the test statistic for which the null hypothesis will be retained or rejected. The test statistic is calculated from the data, compared with the critical value, and the null hypothesis is either rejected or retained. Finally, the result of the significance test is used to answer the original question.

One-Tailed and Two-Tailed Significance Tests

Consider the situation when the accuracy of a new analytical method is evaluated by analyzing a standard reference material with a known μ. A sample of the standard is analyzed, and the sample's mean is determined. The null hypothesis is that the sample's mean is equal to μ

$$H_0 : \overline{X} = \mu$$

If the significance test is conducted at the 95percent confidence level (α= 0.05), then the null hypothesis will be retained if a 95percent confidence interval around $\overline{X}$

contains μ. If the alternative hypothesis is

$$H_A : \overline{X} \neq \mu$$

then the null hypothesis will be rejected, and the alternative hypothesis accepted if μ lies in either of the shaded areas at the tails of the sample's probability distribution. Each of the shaded areas accounts for 2.5percent of the area under the probability distribution curve. This is called a two-tailed significance test because the null hypothesis is rejected for values of μ at either extreme of the sample's probability distribution.

The alternative hypothesis also can be stated in one of two additional ways

$$H_A : \overline{X} > \mu$$

$$H_A : \overline{X} < \mu$$

for which the null hypothesis is rejected if μ falls within the shaded areas shown in Figures, respectively. In each case the shaded area represents 5percent of the area under the probability distribution curve. These are examples of one-tailed significance tests. For a fixed confidence level, a two-tailed test is always the more conservative test because it requires a larger difference between – X and μ to reject the null hypothesis.

Most significance tests are applied when there is no a priori expectation about the relative magnitudes of the parameters being compared. A two-tailed significance test, therefore, is usually the appropriate choice. One-tailed significance tests are reserved for situations when we have reason to expect one parameter to be larger or smaller than the other. For example, a one-tailed significance test would be appropriate for our earlier example regarding a medication's effect on blood glucose levels since we believe that the medication will lower the concentration of glucose.

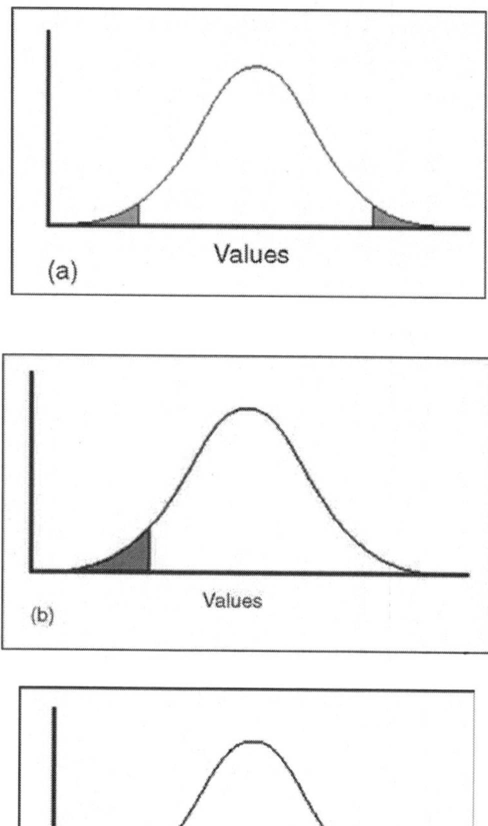

Fig. Examples of (a) two-tailed, (b) and (c) one-tailed, Significance Tests. The Shaded Areas in each Curve Represent the Values for which the Null Hypothesis is Rejected.

Errors in Significance Testing

Since significance tests are based on probabilities, their interpretation is naturally subject to error. As we have already seen, significance tests are carried out at a significance level, α, that defines the probability of rejecting a null hypothesis that is true. For example, when a significance test is conducted at $\alpha = 0.05$, there is a 5percent probability that the null hypothesis will be incorrectly rejected. This is known as a type 1 error, and its risk is always equivalent to α. Type 1 errors in two-tailed and one-tailed significance tests are represented by the shaded areas under the probability distribution curves in Figure

The second type of error occurs when the null hypothesis is retained even though it is false and should be rejected. This is known as a type 2 error, and its probability of occurrence is β. Unfortunately, in most cases β cannot be easily calculated or estimated.

The probability of a type 1 error is inversely related to the probability of a type 2 error. Minimizing a type 1 error by decreasing α, for example, increases the likelihood of a type 2 error. The value of αchosen for a particular significance test, therefore, represents a compromise between these two types of error. Most of the examples in this text use a 95percent confidence level, or $\alpha = 0.05$, since this is the most frequently used confidence level for the majority of analytical work. It is not unusual, however, for more stringent (*e.g.* $\alpha = 0.01$) or for more lenient (*e.g.* $\alpha = 0.10$) confidence levels to be used.

STATISTICAL EVALUATION

DETERMINATE ERRORS

There are three basic sources of determinate or systematic errors that lead to a bias in measured values or results:

- The analyst or operator.
- The equipment (apparatus and instrumentation) and the laboratory environment.
- The method or procedure.

It should be possible to eliminate errors of this type by careful observation and record keeping, equipment maintenance and training of laboratory personnel. Operator errors can arise through carelessness, insufficient training, illness or disability. Equipment errors include substandard volumetric glassware, faulty or worn mechanical components, incorrect electrical signals and a poor or insufficiently controlled laboratory environment.

Method or procedural errors are caused by inadequate method validation, the application of a method to samples or concentration levels for which it is not suitable or unexpected variations in sample characteristics that affect measurements.

Determinate errors that lead to a higher value or result than a true or accepted one are said to show a positive bias; those leading to a lower value or result are said to show a negative bias. Particularly large errors are described as gross errors; these should be easily apparent and readily eliminated. Determinate errors can be proportional to the size of sample taken for analysis. If so, they will have the same effect on the magnitude of a result regardless of the size of the sample, and their presence can thus be difficult to detect. For example, copper(II) can be determined by titration after reaction with potassium iodide to release iodine according to the equation

$$2Cu_2+ + 4I \rightarrow 2CuI + I_2$$

However, the reaction is not specific to copper(II), and any iron(III) present in the sample will react in the same way. Results for the determination of copper in an alloy containing 20percent, but which also contained 0.2percent of iron are shown in *Figure* for a range of sample sizes.

The same absolute error of +0.2percent or relative error of 1percent (*i.e.* a positive bias) occurs regardless of sample size, due to the presence of the iron. This type of error may go undetected unless the constituents of the sample and the chemistry of the method are known.

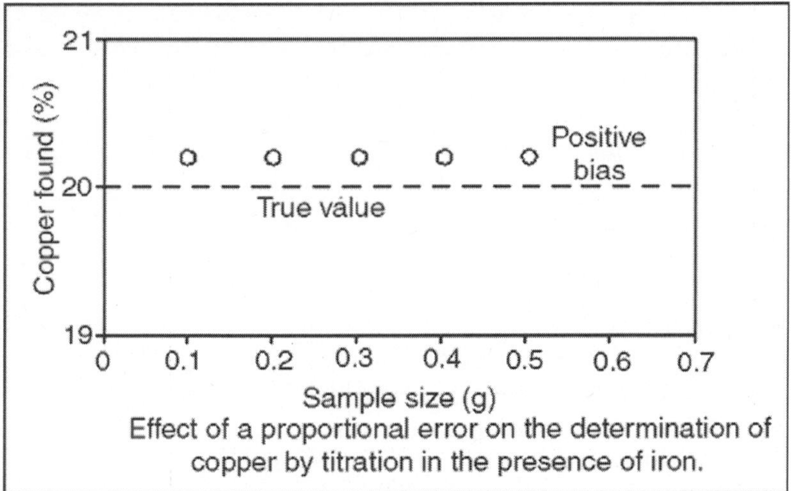

Effect of a proportional error on the determination of copper by titration in the presence of iron.

Constant determinate errors are independent of sample size, and therefore become less significant as the sample size is increased. For example, where a visual indicator is employed in a volumetric procedure, a small amount of titrant is required to change the colour at the end-point, even in a blank solution (*i.e.* when the solution contains none of the species to be determined).

This indicator blank is the same regardless of the size of the titer when the species being determined is present.

The relative error, therefore, decreases with the magnitude of the titer, as shown graphically in *Figure*. Thus, for an indicator blank of 0.02 cm^3, the relative error for a 1 cm^3 titer is 2percent, but this falls to only 0.08percent for a 25 cm^3 titer.

INDETERMINATE ERRORS

Known also as random errors, these arise from random fluctuations in measured quantities, which always occur even under closely controlled conditions.

It is impossible to eliminate them entirely, but they can be minimized by careful experimental design and control. Environmental factors such as temperature, pressure and humidity, and electrical properties such as current, voltage and resistance are all susceptible to small continuous and random variations described as noise.

These contribute to the overall indeterminate error in any physical or physico-chemical measurement, but no one specific source can be identified.

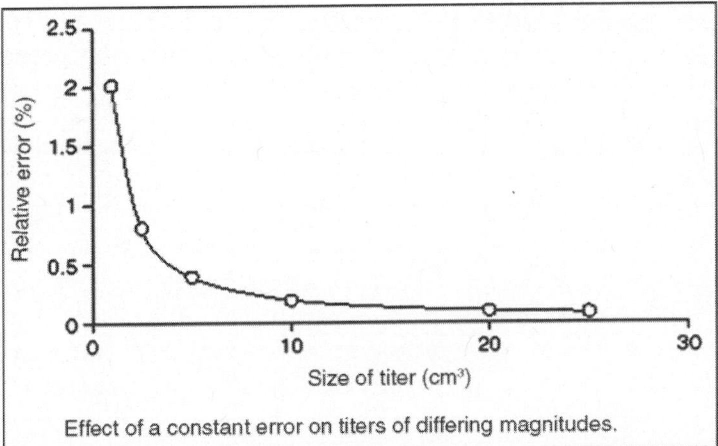

Effect of a constant error on titers of differing magnitudes.

A series of measurements made under the same prescribed conditions and represented graphically is known as a frequency distribution.

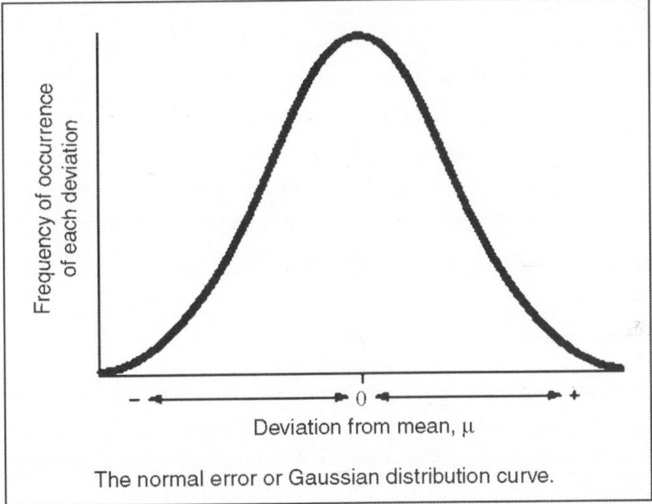

The normal error or Gaussian distribution curve.

The frequency of occurrence of each experimental value is plotted as a function of the magnitude of the error or deviation from the average or mean value.

For analytical data, the values are often distributed symmetrically about the mean value, the most common being the normal error or Gaussian distribution curve. The curve shows that

- Small errors are more probable than large ones.
- Positive and negative errors are equally probable.
- The maximum of the curve corresponds to the mean value.

The normal error curve is the basis of a number of statistical tests that can be applied to analytical data to assess the effects of indeterminate errors, to compare values and to establish levels of confidence in results.

ACCUMULATED ERRORS

Errors are associated with every measurement made in an analytical procedure, and these will be aggregated in the final calculated result. The accumulation or propagation of errors is treated similarly for both determinate (systematic) and indeterminate (random) errors.

Determinate (systematic) errors can be either positive or negative, hence some cancellation of errors is likely in computing an overall determinate error, and in some instances this may be zero. The overall error is calculated using one of two alternative expressions, that is

- Where only a linear combination of individual measurements is required to compute the result, the overall absolute determinate error, *ET*, is given by

$$ET = E_1 + E_2 + E_3 + \ldots\ldots$$

E_1 and E_2 etc., being the absolute determinate errors in the individual measurements taking sign into account

- Where a multiplicative expression is required to compute the result, the overall relative determinate error, *ETR*, is given by

$$ETR = E_1R + E_2R + E_3R + \ldots\ldots$$

E_1R and E_2R etc., being the relative determinate errors in the individual measurements taking sign into account.

The accumulated effect of indeterminate (random) errors is computed by combining statistical parameters for each measurement.

ASSESSMENT OF ACCURACY AND PRECISION

Accuracy and Precision

These two characteristics of numerical data are the most important and the most frequently confused.

It is vital to understand the difference between them, and this is best illustrated diagrammatically as in *Figure*. Four analysts have each performed a set of five titrations for which the correct titer is known to be 20.00 cm^3. The titers have been plotted on a linear scale, and inspection reveals the following:

- The average titers for analysts B and D are very close to 20.00 cm^3 - these two sets are therefore said to have good accuracy.
- The average titers for analysts A and C are well above and below 20.00 cm^3 respectively - these are therefore said to have poor accuracy.
- The five titers for analyst A and the five for analyst D are very close to one another within each set – these two sets therefore both show good precision.
- The five titers for analyst B and the five for analyst C are spread widely within each set - these two sets therefore both show poor precision.

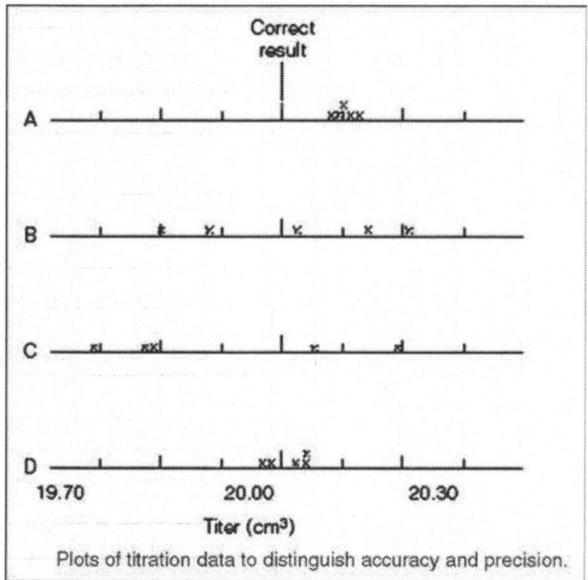

Plots of titration data to distinguish accuracy and precision.

It should be noted that good precision does not necessarily produce good accuracy (analyst A) and poor precision does not necessarily produce poor accuracy (analyst B). However, confidence in the analytical procedure and the results is greater when good precision can be demonstrated (analyst D).

Accuracy is generally the more important characteristic of quantitative data to be assessed, although consistency, as measured by precision, is of particular concern in some circumstances. Trueness is a term associated with accuracy, which describes the closeness of agreement between the average of a large number of results and a true or accepted reference value.

The degree of accuracy required depends on the context of the analytical problem; results must be shown to be fit for the purpose for which they are intended. For example, one result may be satisfactory if it is within 10percent of a true or accepted value whilst it may be necessary for another to be within 0.5percent.

By repeating an analysis a number of times and computing an average value for the result, the level of accuracy will be improved, provided that no systematic error (bias) has occurred.

Accuracy cannot be established with certainty where a true or accepted value is not known, as is often the case. However, statistical tests indicating the accuracy of a result with a given probability are widely used (*vide infra*).

Precision, which is a measure of the variability or dispersion within a set of replicated values or results obtained under the same prescribed conditions, can be assessed in several ways.

The spread or range (*i.e.* the difference between the highest and lowest value) is sometimes used, but the most popular method is to estimate the standard deviation of the data (*vide infra*).

The precision of results obtained within one working session is known as repeatability or within-run precision. The precision of results obtained over a series of working sessions is known as reproducibility or between-runs precision. It is sometimes necessary to separate the contributions made to the overall precision by within-run and between-runs variability. It may also be important to establish the precision of individual steps in an analysis.

STANDARD DEVIATION

This is the most widely used measure of precision and is a parameter of the normal error or Gaussian curve. *Figure* shows two curves for the frequency distribution of two theoretical sets of data, each having an infinite number of values and known as a statistical population.

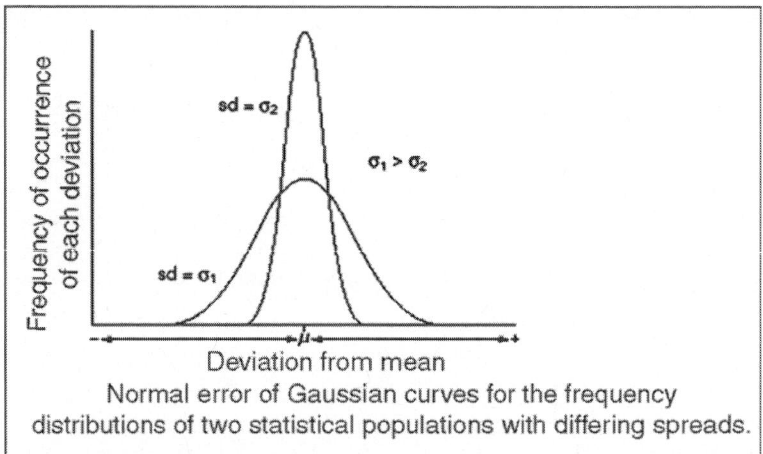

Normal error of Gaussian curves for the frequency distributions of two statistical populations with differing spreads.

The maximum in each curve corresponds to the population mean, which for these examples has the same value, *m*. However, the spread of values for the two sets is quite different, and this is reflected in the half-widths of the two curves at the points of inflection, which, by definition, is the population standard deviation, *s*. As *s2* is much less than *s1*, the precision of the second set is much better than that of the first. The abscissa scale can be calibrated in absolute units or, more commonly, as positive and negative deviations from the mean, *m*.

In general, the smaller the spread of values or deviations, the smaller the value of *s* and hence the better the precision. In practice, the true values of *m* and *s* can never be known because they relate to a population of infinite size. However, an assumption is made that a small number of experimental values or a statistical sample drawn from a statistical population is also distributed normally or approximately so. The experimental mean, $\overline{x}$, of a set of values x_1, x_2, x_3, …, x_n is therefore considered to be an estimate of the true or population mean, *m*, and the experimental standard deviation, *s*, is an estimate of the true or population standard deviation, *s*.

A useful property of the normal error curve is that, regardless of the magnitude of *m* and *s*, the area under the curve within defined limits on either side of *m* (usually expressed in multiples of ±*s*) is a constant proportion of the total area.

Expressed as a percentage of the total area, this indicates that a particular percentage of the population will be found between those limits. Thus, approximately 68percent of the area, and therefore of the population, will be found within ±1*s* of the mean, approximately 95percent will be found within ±2*s* and approximately 99.7percent within ±3*s*.

More practically convenient levels, as shown in *Figure*, are those corresponding to 90percent, 95percent and 99percent of the population, which are defined by ±1.64*s*, ±1.96*s* and ±2.58*s* respectively. Many statistical tests are based on these probability levels.

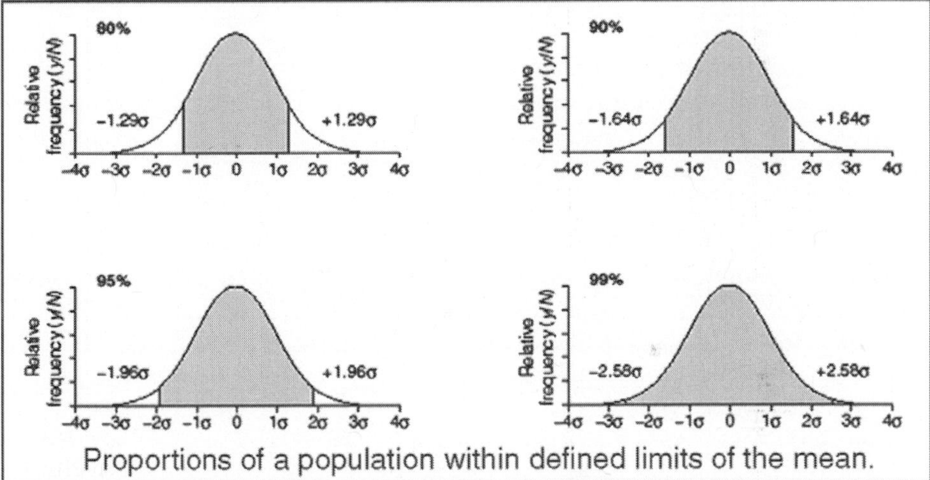

Proportions of a population within defined limits of the mean.

The value of the population standard deviation, σ, is given by the formula

$$\sigma = \sqrt{\frac{\sum_{i=1}^{i=N} (x_i - \mu)^2}{N}}$$

where x_i represents any individual value in the population and *N* is the total number of values, strictly infinite.

The summation symbol, S, is used to show that the numerator of the equation is the sum for $i = 1$ *to* $i = N$ of the squares of the deviations of the individual *x* values from the population mean, *m*. For very large sets of data (*e.g.*, when *N* > 50), it may be justifiable to use this formula as the difference between *s* and *s* will then be negligible. However, most analytical data consists of sets of values of less than ten and often as small as three. Therefore, a modified formula is used to calculate an estimated standard deviation, *s*, to replace *s*, and using an experimental mean, $\overline{x}$, to replace the population mean, *m*:

$$s = \sqrt{\frac{\sum_{i=1}^{i=N}(x_i - \overline{x})^2}{N-1}}$$

Note that N in the denominator is replaced by $N - 1$, which is known as the number of degrees of freedom and is defined as the number of independent deviations $(x_i - \overline{x})$ used to calculate s.

For single sets of data, this is always one less than the number in the set because when $N - 1$ deviations are known the last one can be deduced as, taking sign into account, $t, \sum_{i=1}^{i=N}(x_i - \overline{x})$) must be zero. In summary, the calculation of an estimated standard deviation, s, for a small number of values involves the following steps:

- Calculation of an experimental mean.
- Calculation of the deviations of individual xi values from the mean.
- Squaring the deviations and summing them.
- Dividing by the number of degrees of freedom, $N - 1$.
- Taking the square root of the result.

Note that if N were used in the denominator, the calculated value of s would be an underestimate of s. Estimated standard deviations are easily obtained using a calculator that incorporates statistical function keys or with one of the many computer software packages. It is, however, useful to be able to perform a stepwise arithmetic calculation, and an example using the set of five replicate titers by analyst A is shown below.

Example.

x_i/cm^3	$(x_i - x)$	$(x_i - x)^2$
20.16	– 0.04	1.6×10^{-3}
20.22	+ 0.02	4×10^{-4}
20.18	–0.02	4×10^{-4}
20.20	0.00	0
20.24	+ 0.04	1.6×10^{-3}
Σ101.00		4×10^{-3}

$\overline{x}$ 20.20 $s = \sqrt{\frac{4\times10^{-3}}{4}}$

$= 0.032\ cm^3$

RELATIVE STANDARD DEVIATION

The relative standard deviation, RSD or sr, is also known as the coefficient of variation, CV.

It is a measure of relative precision and is normally expressed as a percentage of the mean value or result

$$s_r = (s/\overline{x}) \times 100$$

It is an example of a relative error and is particularly useful for comparisons between sets of data of differing magnitude or units, and in calculating accumulated (propagated) errors. The RSD for the data in *Example 1* is given below.

Example. 2.

$$s_r = \frac{0.032}{20.20} \times 100 = 0.16\%$$

POOLED STANDARD DEVIATION

Where replicate samples are analyzed on a number of occasions under the same prescribed conditions, an improved estimate of the standard deviation can be obtained by pooling the data from the individual sets. A general formula for the pooled standard deviation, *spooled*, is given by the expression

$$s_{period} = \sqrt{\frac{\sum_{i=1}^{i=N_1}(x_i - \overline{x}_1)^2 + \sum_{i=1}^{i=N_2}(x_i - \overline{x}_2)^2 + \sum_{i=1}^{i=N_3}(x_i - \overline{x}_3)^2 + ... + \sum_{i=1}^{i=N_1}(x_i - \overline{x}_k)^2}{\sum_{i=1}^{i=k} N_i = k}}$$

where N_1, N_2, N_3 ... N_k are the numbers of results in each of the *k* sets, and $\overline{x}_1$, $\overline{x}_2$, $\overline{x}_3$,... $\overline{x}_k$, are the means for each of the *k* sets.

VARIANCE

The square of the standard deviation, s_2, or estimated standard deviation, s_2, is used in a number of statistical computations and tests, such as for calculating accumulated (propagated) errors or when comparing the precisions of two sets of data.

OVERALL PRECISION

Random errors accumulated within an analytical procedure contribute to the overall precision. Where the calculated result is derived by the addition or subtraction of the individual values, the overall precision can be found by summing the variances of all the measurements so as to provide an estimate of the overall standard deviation, *i.e.*

$$s_{overall} = \sqrt{\left(s_1^2 + s_2^3 + s_3^2 + ...\right)}$$

Example. In a titrimetric procedure, the buret must be read twice, and the error associated with each reading must be taken into account in estimating the overall precision. If the reading error has an estimated standard deviation of 0.02 cm^3, then the overall estimated standard deviation of the titration is given by $s_{overall} = \sqrt{(0.02^2 + 0.02^2)} = 0.028$ cm^3

Note that this is less than twice the estimated standard deviation of a single reading. The overall standard deviation of weighing by difference is estimated in the same way.

If the calculated result is derived from a multiplicative expression, the overall *relative* precision is found by summing the squares of the *relative* standard deviations of all the measurements, *i.e.*

$$s_{overall} = \sqrt{\left(s_{r_1}^2 + s_{r_2}^2 + r_{r_3}^2 + ...\right)}$$

CONFIDENCE INTERVAL

The true or accepted mean of a set of experimental results is generally unknown except where a certified reference material is being checked or analyzed for calibration purposes. In all other cases, an estimate of the accuracy of the experimental mean, $\overline{x}$, must be made. This can be done by defining a range of values on either side of $\overline{x}$ within which the true mean, *m*, is expected to lie with a defined level of probability. This range, which ideally should be as narrow as possible, is based on the standard deviation and is known as the confidence interval, CI, and the upper and lower limits of the range as confidence limits, CL. Confidence limits can be calculated using the standard deviation, *s*, if it is known, or the estimated standard deviation, *s*, for the data. In either case, a probability level must be defined, otherwise the test is of no value. When the standard deviation is already known from past history, the confidence limits are given by the equation

$$CL(\mu) = \overline{x} \pm \frac{z\sigma}{\sqrt{N}}$$

where *z* is a statistical factor related to the probability level required, usually 90percent, 95percent or 99percent. The values of *z* for these levels are 1.64, 1.96 and 2.58, respectively, and correspond to the multiples of the standard deviation shown in *Figure*. Where an estimated standard deviation is to be used, *s* is replaced by *s,* which must first be calculated from the current data. The confidence limits are then given by the equation

$$CL(\mu) = \overline{x} \pm \frac{ts}{\sqrt{N}}$$

where *z* is replaced by an alternative statistical factor, *t*, also related to the probability level but in addition determined by the number of degrees of freedom for the set of data, *i.e.* one less than the number of results. It should be noted that (i) the confidence interval is inversely proportional to $\overline{N}$, and (ii) the higher the selected probability level, the greater the confidence interval becomes as both z and t increase.

The calculation of confidence limits using each of the two formulae.

Example 3. The chloride content of water samples has been determined a very large number of times using a particular method, and the standard deviation found to be 7 ppm.

Further analysis of a particular sample gave experimental values of 350 ppm for a single determination, for the mean of two replicates and for the mean of four replicates. Using equation, and at the 95percent probability level, z = 1.96 and the confidence limits are:

1. Determination $CL(\mu) = 350 \pm \frac{1.96 \times 7}{\sqrt{1}} = 350 \pm 14 ppm$

2. Determinations $CL(\mu) = 350 \pm \frac{1.96 \times 7}{\sqrt{2}} = 350 \pm 10 ppm$

3. Determinations $CL(\mu) = 350 \pm \frac{1.96 \times 7}{\sqrt{4}} = 350 \pm 7 ppm$

Example 4. The same chloride analysis as in *Example 3*, but using a new method for which the standard deviation was not known, gave the following replicate results, mean and estimated standard deviation:

Chloride/ppm	*Mean*	*Estimated Standard deviation*
346	351.67 ppm	6.66 ppm
359		
350		

Using equation, and at the 95percent probability level, t = 4.3 for two degrees of freedom, and the confidence limits are:

$$\text{3 determinations } CL(\mu) = 352 \pm \frac{4.3 \times 6.66}{\sqrt{3}} = 352 \pm 17 ppm$$

The wider limits given by equation when the standard deviation is estimated with only three results reflects the much greater uncertainty associated with this value, which in turn affects the confidence in the degree of accuracy. To demonstrate good accuracy, the confidence interval, CI, should be as small as possible and increasing the number of replicates will clearly achieve this. However, due to the $\overline{N}$ term in the denominator, to reduce the interval by, say, a factor of two requires an increase in the number of replicates by a factor of four.

Unfortunately, the law of diminishing returns applies here, so if the CI is to be halved again, the number of replicates must be increased from *four* to *sixteen*. Similarly, the number of replicates would have to be increased from three to twelve to halve the CI, which would represent an unacceptable amount of time and money for most analytical laboratories.

INSTRUMENT CONTROL AND DATA PROCESSING

These aspects of HPLC closely parallel those described for GC. Additional instrument parameters to be set and monitored include:

- Solvent composition, flow rate and pressure limit;
- Gradient elution programmes;
- Wavelength(s) and wavelength-switching for UV-visible absorbance detectors.
- Wavelength range, sampling frequency and mode of display for a DAD.

MODES OF HPLC

Almost any type of solute mixture can be separated by HPLC because of the wide range of stationary phases available, and the additional selectivity provided by varying the mobile phase composition. Both normal- and reversedphase separations are possible, depending on the relative polarities of the two phases.

Although these are sometimes referred to as modes of HPLC, the nature of the stationary phase and/or the solute sorption mechanism provide a more specific means of classification, and modes based on these and the types of solutes to which they are best suited are summarized below.

- *Adsorption chromatography:* Separations are usually normal-phase with a silica gel stationary phase and a mobile phase of a non-polar solvent blended with additions of a more polar solvent to adjust the overall polarity or eluting power, *e.g.* *n*-hexane + dichloromethane or di-ethyl ether. The choice of solvent is limited if a UV absorbance detector is to be used. Traces of water in the solvents must be controlled, otherwise solute retention will not be reproducible. Solutes are retained by surface adsorption; they compete with solvent molecules for active silanol sites (Si-OH), and are eluted in order of increasing polarity. This mode is not used extensively, but is suitable for mixtures of structural isomers and solutes with differing functional groups. Members of a homologous series can not be separated by adsorption chromatography because the non-polar parts of a solute do not interact with the polar adsorbent surface.
- Modified partition or bonded-phase chromatography (BPC). Most HPLC stationary phases are chemically-modified silicas, or bonded phases, by far the most widely used being those modified with non-polar hydrocarbons. The solute sorption mechanism is described as modified partition, because, although the bonded hydrocarbons are not true liquids, organic solvent molecules from the mobile phase form a liquid layer on the surface. The most popular phase is octadecyl (C_{18} or ODS), and most separations are reversed-phase, the mobile phase being a blend of methanol or acetonitrile with water or an

aqueous buffer. For weakly acidic or basic solutes, the role of pH is crucial because the ionized or protonated forms have a much lower affinity for the ODS than the corresponding neutral species, and are therefore eluted more quickly. The dissociation of weak acids and the protonation of weak bases are shown by the following equations

$$RCOOH \rightarrow RCOO- + H^+$$

$$RNH_2 + H^+ \rightarrow RNH_3^+$$

Thus, at low pH, bases are eluted more quickly than at high pH, whilst the opposite holds for weak acids.

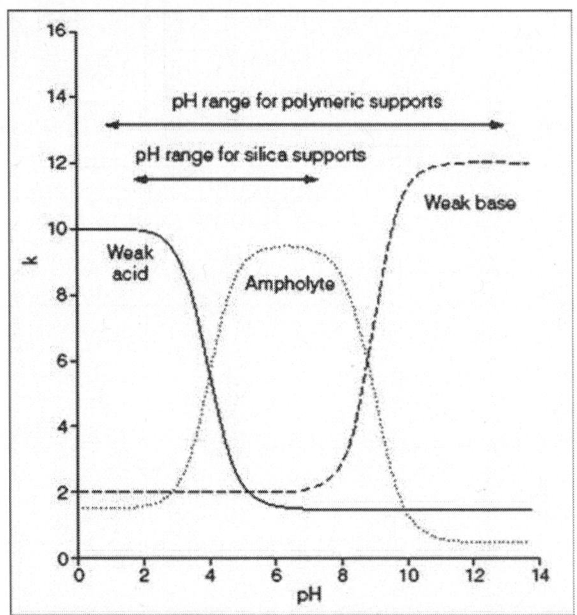

Fig. Relation between Retention Factor, k, and the pH of the Mobile Phase for Weak Acids, Bases and Ampholytes in Reversed-phase Separations.

ODS and other hydrocarbon stationary phases will separate many mixtures, and are invariably a first choice in developing new HPLC methods. They are particularly suited to the separation of moderately polar to polar solutes.

Aminoalkyl and cyanoalkyl (nitrile) bonded phases (the alkyl group is usually propyl) are moderately polar. The former is particularly useful in separating mixtures of sugars and other carbohydrates, whilst the latter is used as a substitute for unmodified silica, giving more reproducible retention factors and less tailing, especially with basic solutes. Both normalphase and reversed-phase chromatography is possible by appropriate choice of eluents.

- *Ion-exchange chromatography (IEC):* Stationary phases for the separation of mixtures of ionic solutes, such as inorganic cations and anions, amino acids and proteins, are based either on microparticulate ion-exchange resins, which are crosslinked co-polymers of styrene and divinyl benzene, or on bonded phase silicas. Both types have either sulfonic acid cation-exchange sites ($–SO_3^- H^+$) or quaternary ammonium anion-exchange sites ($–N + R_3OH^-$) incorporated into their structures.

Ion-exchange chromatography is not that widely used. Inorganic ions and some cations are better separated by a related mode known as ion chromatography (*vide infra*), whilst for organic ions, ion-pair chromatography is generally preferred because of its superior efficiency, resolution and selectivity.

- *Ion chromatography (IC):* This is a form of ion-exchange chromatography for the separation of inorganic and some organic cations and anions with conductometric detection after suppressing (removing) the mobile phase electrolyte.

The stationary phase is a pellicular material (porous-layer beads), the particles consisting of an impervious central core surrounded by a thin porous outer layer (~2 *mm* thick) incorporating cation- or anion-exchange sites. The thin layer results in much faster rates of exchange (mass transfer) than is normally the case with ion-exchange and therefore higher efficiencies.

Mobile phases are electrolytes such as NaOH, $NaCO_3$ or $NaHCO_3$ for the separation of anions, and HCl or CH_3SO_3H for the separation of cations. The detection of low levels of ionic solutes in the presence of high levels of an eluting electrolyte is not feasible unless the latter can be removed. This is accomplished by a suppressor cartridge that essentially converts the electrolyte into water, leaving the solute ions as the only ionic species in the mobile phase. The following equations summarize the reactions for the separation of inorganic anions on an anion-exchange column in the HCO_3 form using a sodium hydrogen carbonate mobile phase:

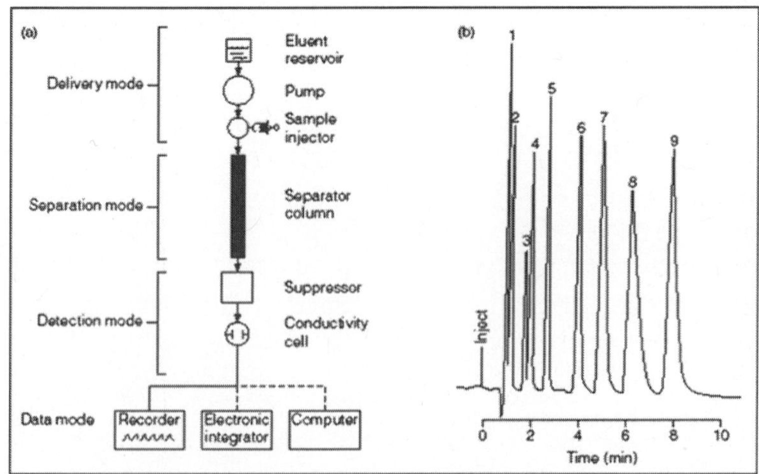

Fig. Ion Chromatography. (a) Schematic Diagram of an Ion Chromatograph. (b) Anions in Water separated on an Anion-exchange Column.

Column Reaction:

$$n(\text{Resin}-N^+R_3HCO_3^-)+X^{n-}\rightarrow(\text{Resin}-N^+R_3)_nX^{n-}+nHCO_3^-$$

where $X^{n-}=F^-,Cl^-,NO_3^-,SO_4^{2-},PO_4^{3-}$ etc.

Suppressor Reactions:

$$Na^+HCO_3^- + \underset{\text{introduced via a membrane}}{H^+} \rightarrow H_2O+CO_2+\underset{\text{removed via a membrane}}{Na^+}$$

$$\underset{\text{separated by the column}}{Na_n{}^+X^{n-}} + \underset{\text{introduced via a membrane}}{nH^+} \rightarrow \underset{\text{Detected by conductance}}{H_n{}^+X^{n-}} + \underset{\text{removed via a membrane}}{nNa^+}$$

Similar reactions form the basis of the separation of cations. An example of the separation of inorganic anions at the ppm level is shown in *Figure*.

- *Size exclusion chromatography (SEC):* This is suitable for mixtures of solutes with relative molecular masses (RMM) in the range 102–108 Da. Stationary phases are either microparticulate cross-linked co-polymers of styrene and divinyl benzene with a narrow distribution of pore sizes, or controlled-porosity silica gels, usually end-capped with a short alkyl chain reagent to prevent adsorptive interactions with solutes. Exclusion is not a true sorption mechanism because solutes do not interact with the stationary phase. They can be divided into three groups:
  - Those larger than the largest pores are excluded completely, and are eluted in the same volume as the interstitial space in the column, V_o.
  - Those smaller than the smallest pores, can diffuse throughout the entire network and are eluted in a total volume, *Vtot*.
  - Those of an intermediate size separate according to the extent to which they diffuse through the network of pores, of volume V_p and are eluted in volumes between V_o and *Vtot*.

Only those solutes in the third group will be separated from one another, and their retention volumes are directly proportional to the logarithm of their relative molecular mass (RMM; molecular weight). Columns can be calibrated with standards of known RMM before analyzing unknowns.

Figure shows a typical plot of elution volume against log (RMM) and a chromatogram of a mixture with a range of solutes of differing molecular mass. SEC is of particular value in characterizing polymer mixtures and in separating biological macromolecules such as peptides and proteins. It is also used for preliminary separations prior to further analysis by other more efficient modes of HPLC.

- *Chiral Chromatography:* Chiral stationary phases (CSP) enable enantiomers (mirror image forms) of a solute to be separated. Several types of these stereoselective materials have been investigated and marketed commercially, some of the most useful being cyclodextrins bonded to silica. The cyclodextrins are cyclic chiral carbohydrates with barrel-shaped cavities into which solutes can fit and be bound by H-bonding, *p-p* and dipolar interactions. Where the total adsorptive binding energies of two enantiomers differ, they will have different retention factors and can be resolved. Steric repulsion and the pH, ionic strength and temperature of the mobile phase all affect the resolution. Although of great interest to the pharmaceutical industry for the separation of enantiomeric forms of drugs having different pharmacological activities, chiral columns are expensive and most have very limited working lives. Capillary electrophoresis provides a cheaper alternative.

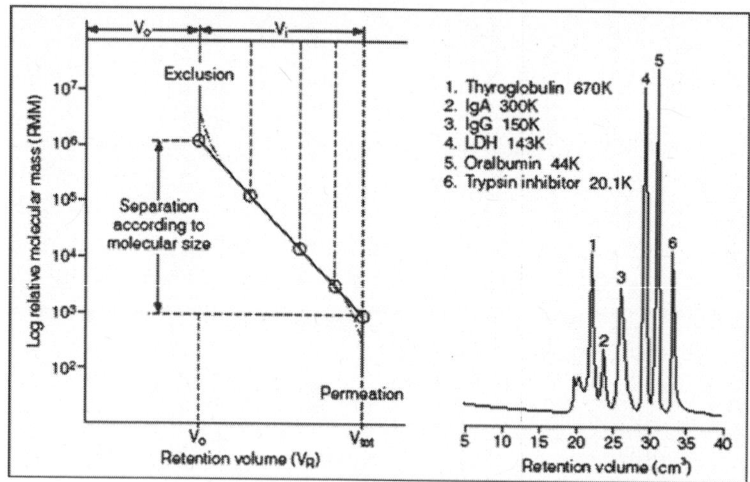

Fig. A Typical Size Excusion Calibration Curve and Chromatogram of the Separation of a Protein Mixture. Column: BIOSEP-SEC-S3000. Mobile Phase: pH 6.8 Phosphate Buffer. Detector: UV abs, at 280 nm.

OPTIMIZATION OF SEPARATIONS

The optimum conditions for an HPLC separation are those which give the required resolution in the minimum time. A stepwise approach based on the characteristics of the solutes to be separated, and trial chromatograms with different mobile phase compositions is usually adopted. The following is an outline of the procedure for a reversed-phase separation where a hydrocarbon stationary phase, usually C18 (ODS), is the first choice.

- The mode of HPLC most suited to the structures and properties of the solutes to be separated is selected, having regard to their relative molecular mass, polarity, ionic or ionizable character, and solubility in organic and aqueous solvents.
- The stationary phase and column are selected. The shortest column and the smallest particle size of stationary phase consistent with adequate resolution should be used.

- The detector, subject to availability, should match the solute characteristics. UV-visible absorbance detectors are suitable for many solutes except those that are fully saturated. Fluorescence and electrochemical detectors should be considered where high sensitivity is required.

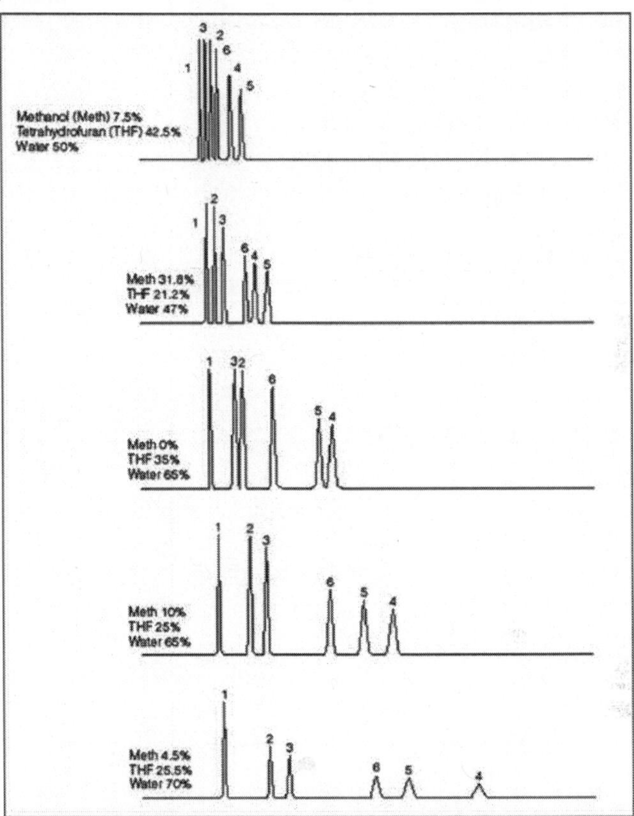

Fig. Optimizing an HPLC Separation using Ternay Mobile Praser. Solute: 1 Benzyl Alcohol; 2. Phenol; 3. 3-phenyl-propanot; 4. 2, 4-dimethylphenol; 5. Benzene; 6. Diethyl o-phtnalate.

- Mobile phase composition is optimized by obtaining and evaluating a number of trial chromatograms, often with the aid of computer optimization software packages. A typical series of chromatograms for a reversed-phase separation on a hydrocarbon bonded phase column is shown in *Figure*.

Note how the elution order of the six components in the mixture alters with the mobile phase composition.

QUALITATIVE ANALYSIS

Methods of identifying unknown solutes separated by chromatographic techniques are described in Topic. In the case of HPLC, there are four alternatives:

- Comparisons of retention factors (k) or retention times (tR) with those of known solutes under identical conditions, preferably on two columns of differing selectivity to reduce the chances of ambiguous identifications.
- Comparisons of chromatograms of samples spiked with known solutes with the chromatogram of the unspiked sample.
- Comparisons of UV-visible spectra recorded by a diode-array detector with those of known solutes. This is of limited value because most spectra have only two or three broad peaks so many solutes have very similar spectral features.
- Interfacing a high-performance liquid chromatograph with a mass spectrometer. This enables spectral information for an unknown solute to be recorded and interpreted.

QUANTITATIVE ANALYSIS

Methods used in quantitative chromatography are decribed in Topic, and alternative calibration procedures are described in Topic. Peak areas are more reliable than peak heights as they are directly proportional to the quantity of a solute injected when working within the linear range of the detector. Most HPLC detectors have a wide linear dynamic range, but response factors must be established for each analyte as these can vary considerably. Calibration is normally with external standards chromatographed separately from the samples, or by standard addition.

Constant volume loops for sample injection give very good reproducibility (about 0.5percent relative precision), making internal standards unnecessary, *cf* gas chromatography, and auto-injectors are frequently employed for routine work. An overall relative precision of between 1 and 3percent can be expected.

9

Classifying Separation Techniques

An analyte and an interferent can be separated if there is a significant difference in at least one of their chemical or physical properties. Table provides a partial list of several separation techniques, classified by the chemical or physical property that is exploited.

EXTRACTION TECHNIQUES

The purpose of an extraction technique is physically to separate components of a mixture (solutes) by exploiting differences in their relative solubilities in two immiscible liquids or between their affinities for a solid sorbent. Substances reach an equilibrium distribution through intimate contact between the two phases, which are then physically separated to enable the species in either phase to be recovered for completion of the analysis.

An equilibrium distribution of the solutes between the two phases is established by dissolving the sample in a suitable solvent, then shaking the solution with a second, immiscible, solvent or by passing it through a sorbent bed or disk. Where the equilibrium distributions of two solutes differ, a separation is possible. The principal factors that determine how a solute will distribute between two phases are its polarity and the polarities of each phase. Degree of ionization, hydrogen bonding and other electrostatic interactions also play a part.

Most solvent-extraction procedures involve the extraction of solutes from an aqueous solution into a non-polar or slightly polar organic solvent, such as hexane, methylbenzene or trichloromethane, although the reverse is also possible. Readily extractable solutes, therefore, include covalent neutral molecules with few polar and no ionized substituents; polar, ionized or ionic species will remain in the aqueous phase. A wide range of solid sorbents with different polarities are used in solid phase extraction making this a very versatile technique, but the factors affecting the distribution of solutes are essentially the same as in solvent extraction. Two examples illustrate these points.

Example 1. If an aqueous solution containing iodine and sodium chloride is shaken with a hydrocarbon or chlorinated hydrocarbon solvent, the iodine, being a covalent molecule, will be extracted largely into the organic phase, whilst the completely ionized and hydrated sodium chloride will remain in the aqueous phase.

Example 2. If an aqueous solution containing a mixture of weakly polar vitamins or drugs and highly polar sugars is passed through a hydrocarbon-modified silica sorbent which has a non-polar surface, the vitamins or drugs will be retained on the surface of the sorbent whilst the sugars pass through. Solvent extraction is governed by the Nernst Distribution or Partition Law.

This states that at constant temperature and pressure, a solute, *S*, will always be distributed in the same proportions between two particular immiscible solvents. The ratio of the equilibrium concentrations (strictly their activities in the two phases defines a distribution or partition coefficient, *KD*, given by the expression

$$K_D = \frac{[S]_{org}}{[S]_{ag}}$$

where [] denotes concentrations (activities) of the distributing solute species, *S*, and the value of *KD* is independent of the total solute concentration. In practice, the solute often exists in different chemical forms due to dissociation (ionization), protonation, complexation or polymerization, so a more practically useful expression which defines a distribution or partition ratio, *D*, is

$$D = \frac{(C_S)_{org}}{(C_S)_{ag}}$$

where (C_S) represents the total concentration of all forms of the distributing solute *S* in each phase. If no interactions involving *S* occur in either phase, then *D* and K_D would be identical. However, the value of *D* is determined by the experimental conditions and it can be adjusted over a wide range to suit the requirements of the analytical procedure.

A distribution ratio can also be defined for solid phase extraction, i.e

$$D = \frac{(C_S)_{sorg}}{(C_S)_{Max}}$$

where the numerator represents the solute concentration in the solid sorbent and the denominator the solute concentration in the liquid phase.

Solutes with large values of *D* (*e.g.* 104 or more) will be essentially quantitatively extracted into the organic phase or retained by the sorbent, although the nature of an equilibrium process means that 100percent extraction or retention can never be achieved.

EXTRACTION EFFICIENCY AND SELECTIVITY

The efficiency of an extraction depends on the value of the distribution ratio, *D*. For solvent extraction, it also depends on the relative volumes of the two liquid phases and for solid-phase extraction on the surface area of the sorbent. With solvent extraction, the percentage of a solute extracted, *E*, is given by the expression

$$E = \frac{100D}{[D + \left[\left(V_{aq} / V_{org}\right)\right]}$$

where V_{aq} and V_{org} are the volumes of the aqueous and organic phases, respectively. For solutes with small values of *D*, multiple extractions will improve the overall efficiency, and an alternative expression enables this to calculated

$$(C_{aq})_n = C_{aq}\left[V_{aq} / \left(DV_{org} + V_{aq}\right)\right]^n$$

where C_{aq} and $(C_{aq})n$ are the amounts of solute in the aqueous phase initially and remaining after *n* extractions, respectively.

Selectivity in extraction procedures is the degree to which solutes in a mixture can be separated by virtue of having different distribution ratios. For two solutes with distribution ratios D_1 and D_2, a separation or selectivity factor, *b*, is defined as

$$b = D_1/D_2 \text{ where } D_1 > D_2$$

Selectivity factors exceeding 104 or 105 (log *b* values exceeding four or five) are necessary to achieve a quantitative separation of two solutes, as for most practical purposes, a separation would be considered complete if one solute could be extracted with greater than 99 percent efficiency whilst extracting less than 1 per cent of another. The extraction of many solutes can be enhanced or supressed by adjusting solution conditions, *e.g.* pH or the addition of complexing agents (*vide infra*).

SOLVENT EXTRACTION

Solvent extraction (SE) is used as a means of sample pre-treatment or clean-up to separate analytes from matrix components that would interfere with their detection or quantitation.

It is also used to pre-concentrate analytes present in samples at very low levels and which might otherwise be difficult or impossible to detect or quantify.

Most extractions are carried out batchwise in a few minutes using separating funnels. However, the efficient extraction of solutes with very small distribution ratios (< 1) can be achieved only by continuously exposing the sample solution to fresh solvent that is recycled by refluxing in a specially designed apparatus.

Broad classes of organic compounds, such as acids and bases, can be separated by pH control, and trace metal ions complexed with organic reagents can be separated or concentrated prior to spectrometric analysis.

Extraction of Organic Acids and Bases

Organic compounds with acidic or basic functionalities dissociate or protonate in aqueous solutions according to the pH of the solution. Their extraction can, therefore, be optimized by pH adjustments. The relation between pH and the distribution ratio, *D*, of a weak acid can be derived in the following way: A weak acid, HA, dissociates in water according to the equation

$$HA = H^+ + A^-$$

The acid dissociation constant, K_g, is defined as

$$K_g = \frac{\left[H^+\right]_{aq}\left[A^-\right]_{aq}}{[HA]_{aq}}$$

Only the undissociated form, [*HA*], can be extracted into a non-polar or slightly polar solvent such as diethyl ether, the distribution or partition coefficient, *KD*, being given by

$$K_D = \frac{[HA]_{other}}{[HA]_{aq}}$$

However, the distribution ratio takes account of both the dissociated and undissociated forms of the acid in the aqueous phase and is given by

$$D = \frac{[HA]_{ether}}{\left([HA]_{aq} + \left[A^-\right]_{aq}\right)}$$

Re-arrangement of equation and substitution for $[A^-]_{aq}$ in equation gives

$$D = \frac{[HA]_{ether}}{[HA]_{aq}\left(1 + K_a / \left[H^+\right]_{aq}\right)}$$

and substituting K_D for $[HA]_{ether}/[HA]_{aq}$ from equation gives

$$D = \frac{K_D}{\left(1 + K_a[H^+]\right)}$$

Equation shows that at low pH, when the acid is undissociated, it is extracted with the greatest efficiency as $D \cong KD$, whereas as the pH is increased the value of *D* decreases until at high pH the acid is completely dissociated into the anion A^-, and none will be extracted. This is shown graphically in *Figure* as pH versus *E* plots for two weak acids (curves 1 and 2) with different acid dissociation constants (*Ka* or *pKa* values), curve 1 being for the stronger of the two acids. For a weak organic base, such as an amine, protonation occurs at low pH according to the equation

$$RNH_2 + H^+ = RNH_3^+$$

The relation between pH and the distribution ratio for a weak base (curve 3) is therefore the opposite of a weak acid so that it is possible to separate acids from bases in a mixture either by extracting the acids at low pH or the bases at high pH. Separating mixtures of acids or mixtures of bases is possible only if their dissociation or protonation constants differ by several *pK* units.

Extraction of Metals

Metal ions in aqueous solutions are not themselves extractable into organic solvents, but many can be complexed with a variety of organic reagents to form extractable species. Some inorganic complex ions can be extracted as neutral ionic aggregates by association with suitable ions of opposite charge (counter ions). There are two principal metal extraction systems:

- Uncharged metal chelate complexes (ring structures that satisfy the coordination requirements of the metal) with organic reagents.

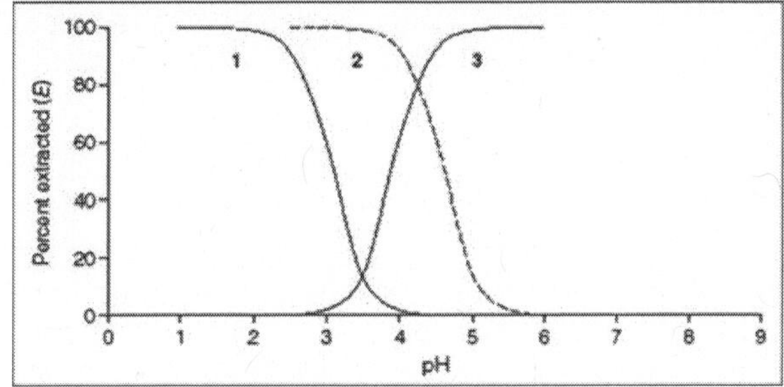

Fig. Solvent extraction Curves for Two Organic Acids having different pK_2 Values (Curves 1 and 2) and a Base (Curve 3). Curve 1 Represents the Acid with the Larger K_2 (Smaller pK_2).

- Electrically neutral ion-association complexes.

The Nernst distribution law applies to metal complexes, but their distribution ratios are determined by several interrelated equilibria. As in the case of organic acids and bases, the efficiency of extraction of metal chelates is pH dependent, and for some ion-association complexes, notably oxonium systems (hydrogen ions solvated with ethers, esters or ketones), inorganic complex ions can be extracted from concentrated solutions of mineral acids.

Reagents that form neutral metal chelate complexes (5- or 6-membered ring structures) are weakly acidic and contain one or more additional co-ordinating sites (O, N or S atoms). Protons are displaced according to the general equation

$$\underset{\text{not extractable}}{M^{n+}(H_2O)_x} + nHR = \underset{\text{extractable}}{MR_n} + nH^- + xH_2O$$

where HR is a weakly acidic and co-ordinating reagent (ligand), and the metal ion Mn^+ has a formal valency n. The removal of hydrogen ions is necessary to drive the reaction to completion, and pH control, which is essential, is achieved by buffering the aqueous solution. It is sometimes possible to improve the selectivity of the procedure by adding an additional reagent, known as a masking agent, that reacts preferentially with one of the metals to form a non-extractable water-soluble complex. Typical masking agents include EDTA (ethylenediaminetetraacetic acid), citrate, tartrate, fluoride, cyanide and thiourea. Solvents commonly used to extract metal chelate complexes include trichloromethane, 4- methyl-pentan-2-one and methylbenzene.

Electrically neutral ion-association complexes consist of cationic (positively charged) and anionic (negatively charged) species that form an overall neutral aggregate extractable by an organic solvent. Either the cations or the anions should be bulky organic hydrophobic groups to provide high solubility in non-polar solvents.

The metal ion can be a cationic or an anionic complex, and can be an inorganic species such as $FeCl^{4-}$, MnO^{4-} or a chelated organic complex such as Fe(1,10-phenanthroline)$_3^{2+}$ or UO_2(oxine)$^{3-}$. Suitable counter-ions of opposite charge to the metal-complex ions include $(C_4H_9)_4N^+$, ClO^{4-} $(C_6H_5CH_2)_3NH^+$, $[(C_4H_9O)_3P = O]H^+$ and an oxonium ion such as $[(C_2H_5)^2O]_3H^+$.

A list of metal chelate complexing agents and ion-association complexes is given in *Table*. Most complexing agents react with a large number of metals (up to 50 or more), but pH control, the use of masking agents and a variety of ion-association systems can enable the selective extraction and separation of just one or two metals to be accomplished.

Reagent	Type of metal complex
8-Hydroxyquinoline (oxine) Di-alkyldithiocarbamates e.g. sodium diethyldithiocarbamate (NaDDTC)	Neutral metal chelate complexes, extractable into organic solvents. Intense color of many facilitates colorimetric determinations.
1,10-Phenanthroline (o-phen) 2,9-Dimethyl-1,10-phenanthroline (neocuproine) Ethylenediaminetetraacetic acid (EDTA)	Ion-association complexes. Metals as cationic or anionic chelated complexes extracted with suitable counter ion.
Oxonium systems: i.e. protons solvated with alkyl ethers, ketones, esters or alcohols	Ion-pairs with anionic metal halide or thiocyanate complexes. Chloride complexes extractable from strong HCl solutions.

SOLID-PHASE SORBENTS

These are generally either silica or chemically-modified silica similar to the bonded phases used in high-performance liquid chromatography but of larger particle size, typically 40–60 *m*m diameter. Solutes interact with the surface of the sorbent through van der Waals forces, dipolar interactions, H-bonding, ion-exchange and exclusion. The four chromatographic sorption mechanisms described in Topic can be exploited depending on the sorbent selected and the nature of the sample.

Sorbents can be classified according to the polarity of the surface. Hydrocarbon-modified silicas are non-polar, and therefore hydrophobic, but are capable of extracting a very wide range of organic compounds from aqueous solutions.

However, they do not extract very polar compounds well, if at all, and these are best extracted by unmodified silica, alumina or Florisil, all of which have a polar surface. Ionic and ionizable solutes are readily retained by an ion-exchange mechanism using cationic or anionic sorbents.

Weak acids can be extracted from aqueous solutions of high pH when they are ionized, and weak bases from aqueous solutions of low pH when they are protonated. It should be noted that this is the opposite way around compared to solvent extraction into non-polar solvents.

However, by suppressing ionization through pH control, extraction by hydrocarbonmodified silica sorbents is possible. Sorbents of intermediate polarity, such as cyanopropyl and aminopropyl modified silicas may have different selectivities to nonpolar and polar sorbents. Some SPE sorbents are listed in *Table* along with the predominant interaction mechanism for each one.

SOLID-PHASE EXTRACTION

Compared to solvent extraction, solid-phase extraction (SPE), is a relatively new technique, but it has rapidly become established as the prime means of sample pre-treatment or the clean-up of dirty samples, *i.e.* those containing high levels of matrix components such as salts, proteins, polymers, resins, tars etc. In addition to being potential sources of interference with the detection and quantitation of analytes, their presence can be detrimental to the stability and performance of columns and detectors when a chromatographic analysis is required.

The removal of interfering matrix components in general and the pre-concentration of trace and ultra-trace level analytes are other important uses of SPE which is versatile, rapid and, unlike solvent extraction, requires only small volumes of solvents, or none at all in the case of solid phase microextraction (*vide infra*).

Sorbent		Polarity	Interaction mechanisms
Silica	SiO_2	Polar	Adsorption: H-bonding
Florisil,	$MgSiO_3$	Polar	H-bonding
alumina	Al_2O_3	Polar	H-bonding
Bonded phases (modified silica)			
$-C_{18}H_{37}$	(C18 or ODS)	Nonpolar	Van der Waals interactions
$-C_8H_{17}$	(C8 or octyl)	Nonpolar	Van der Waals interactions
$-C_6H_5$	(phenyl)	Nonpolar	Van der Waals interactions and $\pi-\pi$ interactions
$-(CH_2)_3CN$	(cyanopropyl)	Polar	Polar interactions: H-bonding
$-(CH_2)_3NH_2$	(aminopropyl)	Polar	H-bonding
$-(CH_2)_3C_6H_4SO_3H$		Ionic	Cation exchange
$-(CH_2)_3N(CH_3)_3Cl$		Ionic	Anion exchange
Chiral	(cyclodextrin)	Polar	Adsorption: H-bonding dipolar interactions steric effects
Styrene/divinyl benzene co-polymer		Nonpolar	Size exclusion

Furthermore, SPE sorbents are cheap enough to be discarded after use thus obviating the need for regeneration. The analysis of environmental, clinical, biological and pharmaceutical samples have all benefited from the rapid growth in the use of SPE where it has largely replaced solvent extraction. Specific examples include the determination of pesticides and herbicides in polluted surface waters and soils, polycyclic aromatic hydrocarbons (PAHs) in drinking water, polluted industrial and urban atmospheres, and drugs in biological fluids.

Sorbents are either packed into disposable cartridges the size of a syringe barrel, fabricated into disks or incorporated into plastic pipette tips or well plates. Most SPE is carried out using a small packed bed of sorbent (25–500 mg) contained in a cartridge made from a polypropylene syringe barrel, the sorbent being retained in position by polyethylene fritted disks. The sorbent generally occupies only the lower half of the cartridge, leaving space above to accommodate several millilitres of the sample solution or washing and eluting solvents. A typical cartridge procedure is illustrated in *Figure* and consists of four distinct steps:

- *Sorbent conditioning:* The cartridge is flushed through with the sample solvent to wet the surface of the sorbent and to create the same pH and solvent composition as those of the sample, thus avoiding undesirable chemical changes when the sample is applied.
- *Sample loading or retention:* The sample solution is passed through the cartridge with the object of *either* retaining the analytes of interest whilst the matrix components pass through or retaining the matrix components whilst the analytes pass through. In some procedures, the analyte(s) and one or more of the matrix components are retained whilst the remainder of the matrix components pass through.
- *Rinsing:* This is necessary to remove all those components not retained by the sorbent during the retention step and which may remain trapped in the interstitial solvent.
- *Elution:* This final step is to recover retained analytes, otherwise the matrixfree solution and rinsings from the second and third steps are combined for quantitative recovery of the analytes before completion of the analysis.

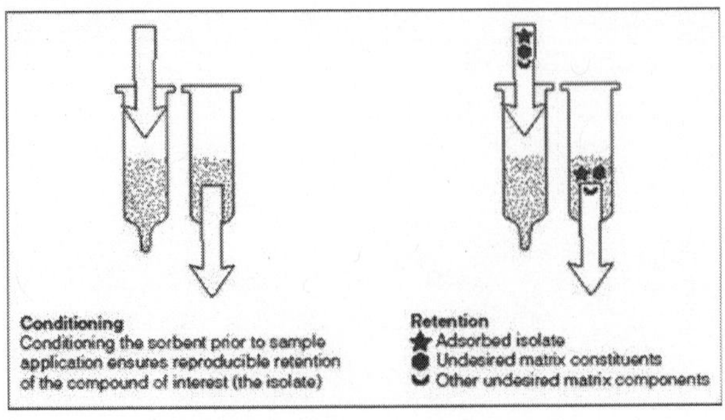

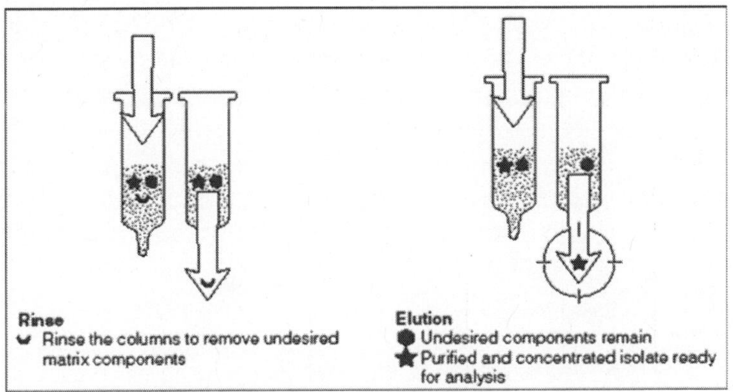

Fig. Diagrammatic Representation of a Cartridge-based Solid Phase Extraction Procedure.

SPE can be semi- or fully-automated to increase sample throughput and to improve both precision and accuracy. The degree of automation ranges from the parallel off-line processing of batches of up to about 10 samples using a vacuum manifold to provide suction, to on-line autosamplers, xyz liquid handlers and robotic workstations. Several alternative formats for SPE are available. These include

- Disks, which have relatively large cross-sectional areas compared to packedbed cartridges and thin sorbent layers (0.5–1 mm thick) containing about 15 mg of material. This reduced bed-mass results in low void volumes (as little as 15 *m*l) thus minimizing solvent consumption in the rinsing and elution steps, improving selectivity and facilitating high solvent flow rates when large volumes of sample are to be processed.
- Plastic pipette-tips incorporating small sorbent beds designed for processing very small volumes of sample and solvents rapidly, and having the advantage of allowing flow in both directions if required.
- Well plates containing upwards of 96 individual miniature samplecontainers in a rectangular array and fitted with miniature SPE packed beds or disks. Well plates are used in xyz liquid handlers for processing large numbers of samples prior to the transfer of aliquots to analytical instruments, particularly gas and liquid chromatographs and mass spectrometers.
- Solid-phase microextraction (SPME) is an important variation of SPE that allows trace and ultra-trace levels of analytes in liquid or gaseous samples to be concentrated. The sorbent is a thin layer of a polymeric substance such as polydimethylsiloxane (PDMS) coated onto a fused-silica optical fibre about 1 cm long and attached to a modified microsyringe. The fibre is exposed to the sample and then inserted directly into the injection port of a gas or liquid chromatograph to

complete the analysis. An advantage of SPME over SPE is the avoidance of solvents, but good precision for quantitative determinations is more difficult to achieve, and automated systems are only just being developed. SPME is finding particular use in water analysis, the analysis of fragrances and volatiles in foodstuffs by headspace sampling, the detection of drugs and their metabolites in urine, blood and breath samples, and the monitoring of air quality in working environments.

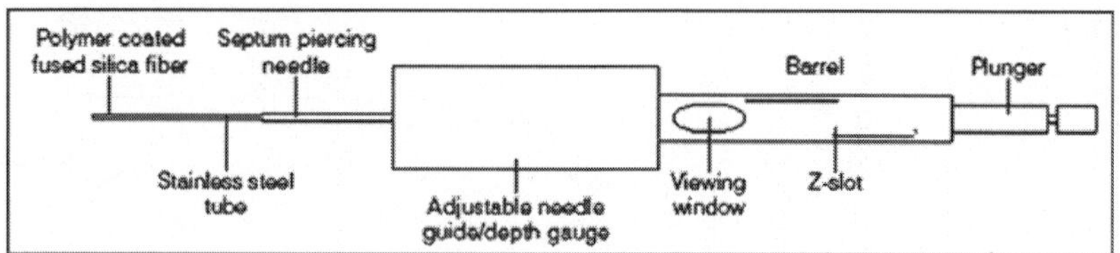

Fig. Diagram of a Solid Phase Microextraction Device.

SEPARATIONS BASED ON SIZE

The simplest physical property that can be exploited in a separation is size. The separation is accomplished using a porous medium through which only the analyte or interferent can pass. Filtration, in which gravity, suction, or pressure is used to pass a sample through a porous filter is the most commonly encountered separation technique based on size.

Table. Classification of Separation Techniques

Basis of Separation	Separation Technique
Size	filtration
dialysis	
size-exclusion chromatography	
Mass and density	centrifugation
Complex formation	masking
Change in Physical state	distillation
sublimation recrystallization	
Change in chemical state	precipitation
	ion exchange
	electrodeposition
volatilization	extraction
Partitioning between phases	chromatography

Particulate interferents can be separated from dissolved analytes by filtration, using a filter whose pore size retains the interferent. This separation technique is important in the analysis of many natural waters, for which the presence of suspended solids may interfere in the analysis.

Filtration also can be used to isolate analytes present as solid particulates from dissolved ions in the sample matrix. For example, this is a necessary step in gravimetry, in which the analyte is isolated as a precipitate.

Another example of a separation technique based on size is dialysis, in which a semipermeable membrane is used to separate the analyte and interferent.

Dialysis membranes are usually constructed from cellulose, with pore sizes of 1–5 nm. The sample is placed inside a bag or tube constructed from the membrane. The dialysis membrane and sample are then placed in a container filled with a solution whose composition differs from that of the sample.

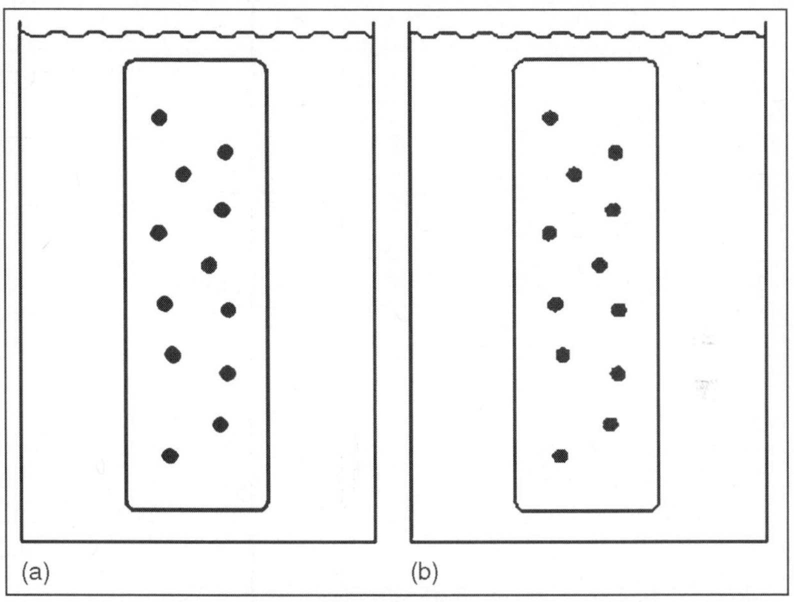

Fig. Illustration of a Dialysis Membrane in Action. In (a) the Sample Solution is Placed in the Dialysis Tube and Submerged in the Solvent. (b) Smaller Partides pass Through the Membrane, but Larger Particles Remain within the Dialysis Tube.

If the concentration of a particular species is not the same on the two sides of the membrane, the resulting concentration gradient provides a driving force for its diffusion across the membrane. Although small particles may freely pass through the membrane, larger particles are unable to pass. Dialysis is frequently used to purify proteins, hormones, and enzymes. During kidney dialysis, metabolic waste products, such as urea, uric acid, and creatinine, are removed from blood by passing it over a dialysis membrane.

Size-exclusion chromatography, which also is called gel permeation or molecularexclusion chromatography, is a third example of a separation technique based on size.

In this technique a column is packed with small, approximately 10 – m, porous particles of cross-linked dextrin or polyacrylamide. The pore size of the particles is controlled by the degree of cross-linking, with greater cross-linking resulting in smaller pore sizes. The sample to be separated is placed into a stream of solvent that is pumped through the column at a fixed flow rate. Particles too large to enter the pores are not retained and pass through the column at the same rate as the solvent.

Those particles capable of entering into the pore structure take longer to pass through the column. Smaller particles, which penetrate more deeply into the pore structure, take the longest time to pass through the column. Size-exclusion chromatography is widely used in the analysis of polymers and in biochemistry, where it is used for the separation of proteins.

SEPARATION VERSUS PRECONCENTRATION

Two frequently encountered analytical problems are: (1) the presence of matrix components interfering with the analysis of the analyte; and (2) the presence of analytes at concentrations too small to analyse accurately. We have seen how a separation can be used to solve the former problem. Interestingly, separation techniques can often be used to solve the second problem as well. For separations in which a complete recovery of the analyte is desired, it may be possible to transfer the analyte in a manner that increases its concentration. This step in an analytical procedure is known as a preconcentration.

Two examples from the analysis of water samples illustrate how a separation and preconcentration can be accomplished simultaneously. In the gas chromatographic analysis for organophosphorous pesticides in environmental waters, the analytes in a 1000-mL sample may be separated from their aqueous matrix by a solidphase extraction using 15 mL of ethyl acetate.

After the extraction, the analytes are present in the ethyl acetate at a concentration that is 67 times greater than that in the original sample (if the extraction is 100 per cent efficient). The preconcentration of metal ions is accomplished by a liquid–liquid extraction with a metal chelator.

For example, before their analysis by atomic absorption spectrophotometry, metal ions in aqueous samples can be concentrated by extraction into methyl isobutyl ketone (MIBK) using ammonium pyrrolidine dithiocarbamate (APDC) as a chelating agent.

Typically, a 100-mL sample is treated with 1 mL of APDC, and extracted with ten mL of MIBK. The result is a ten-fold increase in the concentration of the metal ions. This procedure can be adjusted to increase the concentrations of the metal ions by as much as a factor of 40.

SEPARATIONS BASED ON MASS OR DENSITY

If there is a difference in the mass or density of the analyte and interferent, then a separation using centrifugation may be possible. The sample, as a suspension, is placed in a centrifuge tube and spun at a high angular velocity (high numbers of revolutions per minute, rpm). Particles experiencing a greater centrifugal force have faster sedimentation rates and are preferentially pulled towards the bottom of the centrifuge tube. For particles of equal density the separation is based on mass, with heavier particles having greater sedimentation rates.

When the particles are of equal mass, those with the highest density have the greatest sedimentation rate.

Table. Conditions for the Separation of Selected Cellular Components by Centrifugation

Components	Centrifugal Force (×g)	Time (min)
Eukaryotic cell	1000	5
Cell membranes, nuclei	4000	10
Mitochondria, bacterial cells	15,000	20
Lysosomes, bacterial membranes	30,000	30
Ribosomes	100,000	180

Centrifugation is of particular importance as a separation technique in biochemistry. As shown in Table, cellular components can be separated by centrifugation. For example, lysosomes can be separated from other cellular components by repeated differential centrifugation, in which the sample is divided into a solid residue and a solution called the supernatant.

After destroying the cell membranes, the solution is centrifuged at 15,000 x*g* (a centrifugal field strength that is 15,000 times that of the Earth's gravitational field) for 20 min, leaving a residue of cell membranes and mitochondria. The supernatant is isolated by decanting from the residue and is centrifuged at 30,000 x*g* for 30 min, leaving a residue of lysosomes.

An alternative approach to differential centrifugation is equilibrium–density–gradient centrifugation. The sample is either placed in a solution with a preformed density gradient or in a solution that, when centrifuged, forms a density gradient. For example, density gradients can be established with solutions of sucrose or CsCl.

SEPARATIONS BASED ON A PARTITIONING BETWEEN PHASES

The most important class of separation techniques is based on the selective partitioning of the analyte or interferent between two immiscible phases. When a phase containing a solute, *S*, is brought into contact with a second phase, the solute partitions itself between the two phases.

$$S_{\text{phase 1}} \rightleftharpoons S_{\text{phase 2}}$$

The equilibrium constant for reaction 7.18

$$K_D = \frac{[S_{\text{phase 2}}]}{[S_{\text{phase 1}}]}$$

is called the distribution constant, or partition coefficient. If K_D is sufficiently large, then the solute will move from phase 1 to phase 2. The solute will remain in phase 1, however, if the partition coefficient is sufficiently small. If a phase containing two solutes is brought into contact with a second phase, and K_D is favorable for only one of the solutes, then a separation of the solutes may be possible. The physical states of the two phases are identified when describing the separation process, with the phase containing the sample listed first. For example, when the sample is in a liquid phase and the second phase is a solid, the separation involves liquid–solid partitioning.

Extraction Between Two Phases: When the sample is initially present in one of the phases, the separation is known as an extraction. In a simple extraction the sample is extracted one or more times with portions of the second phase.

Simple extractions are particularly useful for separations in which only one component has a favorable distribution ratio. Several important separation techniques are based on simple extractions, including liquid–liquid, liquid–solid, solid–liquid, and gas–solid extractions.

Liquid Liquid Extractions: Liquid–liquid extractions are usually accomplished with a separatory funnel. The two liquids are placed in the separatory funnel and shaken to increase the surface area between the phases. When the extraction is complete, the liquids are allowed to separate, with the denser phase settling to the bottom of the separatory funnel. Liquid–liquid extractions also may be carried out in the sample container by adding the extracting solvent when the sample is collected.

Pesticides in water, for example, may be preserved for longer periods by extracting into a small volume of hexane added to the sample in the field. Liquid–liquid microextractions, in which the extracting phase is a 1-?L drop suspended from a microsyringe also have been described.16 Because of its importance, a more thorough discussion of liquid–liquid extraction is given in Section 7G.

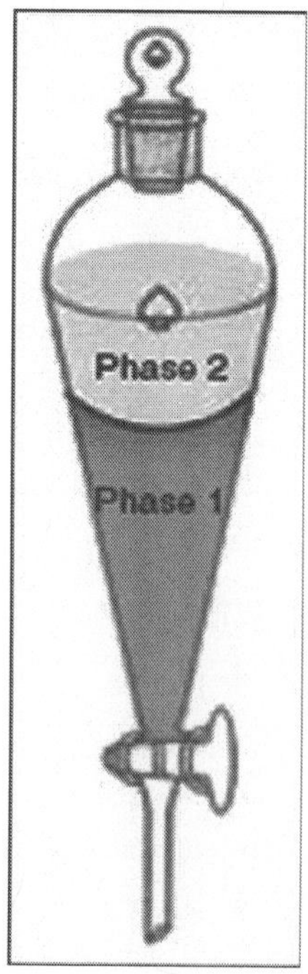

Fig. Separatory Funnel for Use in a Liquid-liquid Extraction.

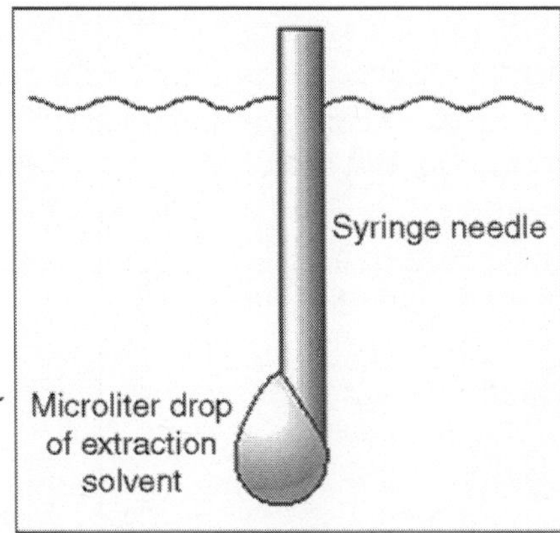

Fig. Schematic of a Liquid-liquid Microextraction Showing Syringe needle with attached 1-μL Droplet..

Solid-Phase Extractions: In a solid-phase extraction the sample is passed through a cartridge containing solid particulates that serve as the adsorbent material. For liquid samples the solid adsorbent is isolated in either a disk cartridge or a column. The choice of adsorbent is determined by the properties of the species being retained and the matrix in which it is found. Representative solid adsorbents

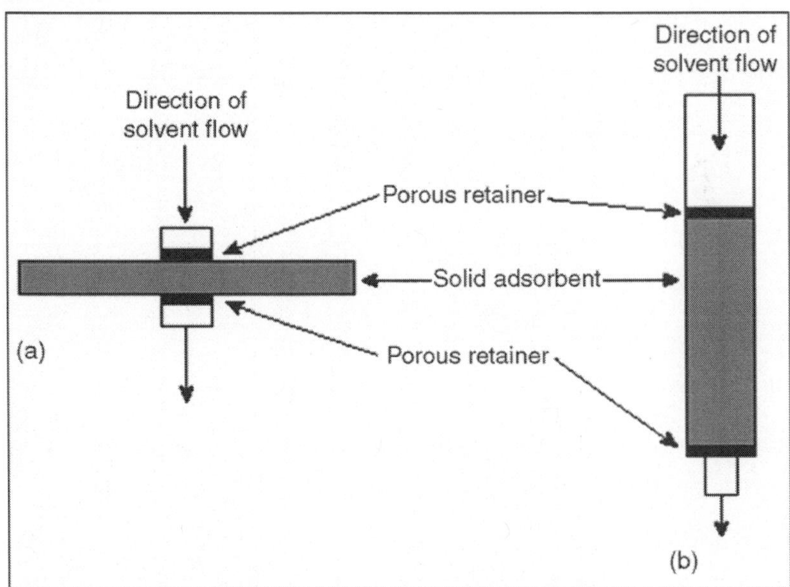

Fig. Solid-phase Extraction Cartridges: (a) Disk Cartridge; (b) Column Cartridge.

are listed in Table. For example, sedatives, such as secobarbital and phenobarbital, can be isolated from serum by a solid-phase extraction using a C-18 solid adsorbent. Typically a 500-?L sample of serum is passed through the cartridge, with the sedatives being retained by a liquid–solid extraction.

The cartridge is then washed with distilled water to remove any residual traces of the serum's matrix. Finally, the retained sedatives are eluted from the cartridge by a solid–liquid extraction using 500 ?L of acetone. For many analyses, solid–phase extractions are replacing liquid–liquid extractions due to their ease of use, faster extraction times, decreased volumes of solvent, and their superior ability to concentrate the analytes. The last advantage is discussed in more detail in the final section of this chapter.

Solid-phase microextractions also have been developed. In one approach, a fused silica fibre is placed inside a syringe needle. The fibre, which is coated with a thin organic film, such as poly(dimethyl siloxane), is lowered into the sample by depressing a plunger and exposed to the sample for a predetermined time. The fibre is then withdrawn into the needle and transferred to a gas chromatograph for analysis.

In gas–solid extractions the sample is passed through a container packed with a solid adsorbent. One example of the application of gas–solid extraction is in the analysis of organic compounds for carbon and hydrogen. The sample is combusted in a flowing stream of O_2, and the gaseous combustion products are passed through a series of solid-phase adsorbents that remove the CO_2 and H_2O.

10

Solvent Extraction

INTRODUCTION TO SOLVENTS

The use of solvents for analytical work is determined by their properties, as shown in *Table.* Solvents with high dielectric constants (*e*r > 10), for example, water and ammonia, are referred to as polar and are ionizing solvents, promoting the formation and separation of ions in their solutions, whereas those where *e*r is about 2, such as diethyl ether, tetrachloromethane and hexane are non-polar and are non-ionizing solvents. There are also many solvents whose behaviour is intermediate between these extremes.

The solution process in a liquid may be represented by a general equation:

$$A\ (1) + B = B(sol)$$

solvent solute solution

The action of solution changes the properties of both solute and solvent. The solute is made more mobile in solution, and its species may solvate by attraction to the solvent. The solvent structure is also disrupted by the presence of species different in size, shape and polarity from the solvent molecules.

Table. Properties of Some Solvents

Properties of some solvents

Solvent	Boiling point (°C)	Density, (g cm^{-3})	Dielectric constant, ε_r
Water	100	1.00	78.6
Ammonia	−34	0.68	22.0
Ethanol	78	0.79	24.3
n-hexane	69	0.66	1.88
Diethyl ether	34	0.71	4.33

Note: density at 25°C or at BP; dielectric constant = relative permittivity

Note: Density at 25°C or at BP; dielectric constant = relative permittivity

Ideally, the behaviour should depend on the concentration *m* (in molarity, mole fraction or other units), but often this must be modified and the activity, *a,* used:

$$a = m\gamma = p/p^{\phi}$$

where g is called the activity coefficient. The vapour pressure of the solution is p, and that in the standard state is p_n. Activities are dimensionless.

Solvents, such as water, with high dielectric constants (or relative permittivities) reduce the force F between ions of charges z_1e and z_2e a distance r apart:

$$F = z_1 z_2 e^2 / \varepsilon_0 \varepsilon_t r^2$$

where e_0 is the permittivity of free space. Also, they will solvate ions more strongly and thus assist ion formation and separation.

Hexane, diethyl ether and tetrachloromethane (CCl_4) all have low dielectric constants and are non-polar. They are very poor at ionizing solutes. However, they are very good solvents for non-polar substances.

SOLUBILITY

The equilibrium amount of solute which will dissolve in a given amount of solvent at a given temperature and pressure is called the solubility. The solubility may be quoted in any units of concentration, for example, mol m^{-3}, molarity, mole fraction, mass per unit volume or parts per million (ppm).

There is a general 'rule of thumb' that 'like dissolves like'. For example, a non-polar hydrocarbon solvent such as hexane would be a very good solvent for solid hydrocarbons such as dodecane or naphthalene. An ester would be a good solvent for esters, and water or other polar solvents are appropriate for polar and ionic compounds.

- Gases dissolve in solvents according to Henry's Law, provided they do not react with the solvent:

$$p_B = x_B K$$

where x_B is the mole fraction of solute gas B which dissolves at a partial pressure p_B of B, and K is a constant at a given temperature. This is analytically important for several reasons. For example, nitrogen is bubbled through solutions to decrease the partial pressure of oxygen in electrochemical experiments. Similarly, air is removed from liquid chromatography solvents by passing helium through them, or by boiling them, since gas solubility decreases as the temperature is increased.

- *Liquids:* When different liquids are mixed, many types of behaviour may occur. If the molecules in the liquids are of similar size, shape, polarity and chemical nature they may mix in all proportions. For example, benzene and methylbenzene (toluene) mix completely. In such ideal solutions, obeying Raoult's law, the activity coefficient is close to 1:

$$a = p/p^{\ominus} = x$$

If the component molecules differ greatly in polarity, size or chemical nature (*e.g.*, water and tetrachloromethane) they may not mix at all. This is an important condition for solvent extraction. The distribution of a solute between a pair of immiscible liquids depends primarily on the solubility of the solute in each liquid.

- Solids generally follow the 'like dissolves like' rule. Nonpolar, covalent materials dissolve best in non-polar solvents. Solid triglycerides such as tristearin are extracted by diethyl ether, but are nearly insoluble in water. Salts, such as sodium chloride are highly soluble in water, but virtually insoluble in ether.

IONS IN SOLUTION

The behaviour of ions in solution may be summarized as follows.

- Solids whose structure consists of ions held together by electrostatic forces (*e.g.* NaCl) must be separated into discrete ions when they dissolve. These ions often gain stability by solvation with molecules of the solvent. Such solutions are described as strong electrolytes.
- Some covalent molecules, such as ethanoic acid, may form ions by the unequal breaking of a covalent bond, followed by stabilization by solvation.

This occurs only partially and these are called weak electrolytes.

$$H:OCOCH_3 \rightleftharpoons H^+ + {}^-OCOCH_3$$

- In some cases ions do not separate completely. In concentrated solutions, oppositely charged ions may exist as ion-pairs, large ions of surfactants may aggregate into micelles, which are used in capillary electrophoresis, and the dissociation of covalent molecules may be only partial.
- At 'infinite dilution', that is, as the concentration approaches zero, ions are truly separate. However, even at quite low concentrations, the attractions between ions of opposite charge will cause each ion to become surrounded by an irregular cloud, or ionic atmosphere. As the solution becomes even more concentrated, the ionic atmosphere becomes more compact around each ion and alters its behaviour greatly.

In very dilute solutions, the effects of the ionic atmosphere may be approximated by using the Debye-Hückel theory, which predicts that the mean ionic activity coefficient, $g\pm$ is, for an electrolyte with positive ions with charge $z+$, negative ions with charge z^-, is given by:

$$\log(\gamma_\pm) = -A(Z_+, Z_-)\sqrt{(1)}$$

where I is the ionic strength = 1D 2 S (ci zi)

- For all the ions in the solution. For more concentrated ionic solutions, above 0.1 M, no general theory exists, but additional terms are often added to the Debye-Hückel equation to compensate for the change in the activity.

THE pX NOTATION

The concentration of species in solution may range from very small to large. For example in a saturated aqueous solution of silver chloride, the concentration of silver ions is about 10-5 M, while for concentrated hydrochloric acid the concentration of hydrogen and chloride ions is about 10 M. For convenience, a logarithmic scale is often used:

$$pX = -\log(X)$$

where X is the concentration of the species, or a related quantity. Thus, for the examples above, pAg = 5 in saturated aqueous silver chloride and pH = –1 in concentrated HCl.

Since equilibrium constants are derived from activities or concentrations as noted below, this notation is also used for them:

$$pK = -\log(K)$$

EQUILIBRIA IN SOLUTION

Most reactions will eventually reach equilibrium. That is, the concentrations of reactants and products change no further, since the rates of the forward and reverse reactions are the same.

From the above arguments concerning solutions, and from the laws of thermodynamics, any equilibrium in solution involving species D, F, U and V:

$$D + F \rightleftharpoons U + V$$

will have an equilibrium constant *KT*, at a particular temperature *T* given by:

$$K_T = (a_U . a_V)/(a_D / a_P)$$

where the activities are the values at equilibrium. It should be noted that K_T changes with temperature. The larger the equilibrium constant, the greater will be the ratio of products to reactants at equilibrium.

There are many types of equilibria that occur in solution, but for the important analytical conditions of ionic equilibria in aqueous solution, four examples are very important.

- *Acid and base dissociation:* In aqueous solution, strong electrolytes (*e.g.*, NaCl, HNO_3, NaOH) exist in their ionic forms all the time. However, weak electrolytes exhibit dissociation equilibria. For ethanoic acid, for example:

$$HOOCCH_3 + H_3O \rightleftharpoons H_3O^+ + CH_3COO^-$$

$$K_a = (a_H.a_A)/(a_{HA}.a_W) = 1.75 \times 10^{-5}$$

where H_A, *W, H* and *A* represent each of the species in the above equilibrium. In dilute solutions the activity of the water a_W is close to 1.

$$NH_3 + H_2O \rightleftharpoons NH_4^+ + OH^-$$

$$K_b = \left(a_{NH_4^+} . a_{OH^-}\right)/\left(a_{NH_3} . a_W\right) = 1.76 \times 10^{-5}$$

Water behaves is an similar way:

$$2H_2O \rightleftharpoons H_3O^+ + OH^-$$

$$K_W = \left(a_{H_3O^+} . a_{OH^-}\right) = 10^{-14}$$

- *Complexation equilibria:* The reaction between an acceptor metal ion M and a ligand L to form a complex ML is characterized by an equilibrium constant. But a simple example will suffice here:

$$M(aq) + L(aq) \rightleftharpoons ML(aq)$$

$$K_t = (a_{ML})/(a_M . a_L)$$

For example, for the copper-EDTA complex at 25°C: K_f = 6.3 ¥ 1018

- *Solubility equilibria:* If a compound is practically insoluble in water, this is useful analytically because it provides a means of separating this compound from others that are soluble. The technique of gravimetric analysis has been developed to give very accurate analyses of materials by weighing pure precipitates of insoluble compounds to give quantitative measurements of their concentration. For the quantitative determination of sulfate ions, SO_4^{2-}, the solution may be treated with a solution of a soluble barium salt such as barium chloride $BaCl_2$, when the following reaction occurs:

$$Ba^{2+} + SO_4^{2-} \rightleftharpoons BaSO_4(s)$$

Conversely, if solid barium sulfate is put into water:

$$BaSO_4(s) = Ba^{2+} + SO_4^{2-}$$

The solubility product, K_{sp}, is an equilibrium constant for this reaction

$$K_{sp} = a(Ba^{2+}).a(SO_4^{2-}) = 1.2 \times 10^{-10}$$

bearing in mind that the pure, solid $BaSO_4$ has a = 1. This means that a solution of barium sulfate in pure water has a concentration of sulfate ions of only 1.1 ¥ 10^{-5} M. The concentration of the barium ions is the same.

- *Redox equilibria:* When a species gains electrons during a reaction, it undergoes reduction and, conversely, when a species loses electrons it undergoes oxidation. In the total reaction, these processes occur simultaneously, for example:

$$Ce^{4+} + Fe^{2+} = Ce^{3+} + Fe^{3+}$$

The cerium is reduced from oxidation state 4 to 3, while the iron is oxidized from 2 to 3. Any general 'redox process' may be written:

$$Ox1 + Red\ 2 = Red\ 1 + Ox2$$

The equilibrium constant of redox reactions is generally expressed in terms of the appropriate electrode potentials, but for the above reaction:

$$K = \left(a\left(Ce^{3+}\right).a\left(Fe^{3+}\right)\right) / \left(a\left(Ce^{4+}\right).a\left(Fe^{2+}\right)\right) = 2.2 \times 10^{12}$$

ALIPHATIC HYDROCARBONS IN CHEMICAL SOLVENTS

Aliphatic hydrocarbons are organic chemical compounds that are used in an industrial setting for chemical solvents, fumigants, and insecticides, or as a chemical intermediate. Chemical solvents that include aliphatic hydrocarbons like isoparaffinic solvents are good for dissolving and removing oil residues from metal surfaces. Companies that manufacture chemical solvents use aliphatic hydrocarbons to create aliphatic compounds like GaroXOL, a popular and highly refined aliphatic solvent with a narrow boiling range and relatively low phytotoxicity. The different properties of various aliphatic hydrocarbons make it popular for use in chemical solvents designed for many different industrial uses.

USES FOR CHEMICAL SOLVENTS WITH ALIPHATIC HYDROCARBONS

Some everyday items that are made with hydrocarbon solvents include paints, adhesive tapes, shoe polish, photocopier fluid, and more. A wide range of solvents can be created out of aliphatic hydrocarbons because of their properties. Industrial companies can create a complete line of solvents for use in different situations and most consumers are unaware of just how many everyday items include aliphatic hydrocarbon solvents.

Aliphatic Hydrocarbons in Everyday Items

Different industries use aliphatic solvents including the dry cleaning, plastics, rubber, textile, chemical, and even pharmaceutical industries. Aliphatic hydrocarbons are using in metal cleaning and for fire extinguishers, as well as for degreasing. They are also used in a wide variety of solvents such as industrial solvents, drug solvents, perfume solvents, detergent, and deodorant solvents. Certain aliphatic compounds can be used in paraffin products and resins.

The Hazards of Aliphatic Solvents

Solvents are derived from different methods depending on the properties of the chemicals being used. The use of aliphatic hydrocarbons when creating chemical solvents can create a hazardous environment and should be approached with caution. Most aliphatic compounds are flammable and are considered toxic because they can have narcotic properties, although the concentration of elements in the compound can make the margin for this very small. The different properties of aliphatic compounds are perfect for deriving a variety of chemical solvents, but precautions should be considered to safeguard the health of anyone working with them.

Safety Measures and Health Precautions

Safety measures must be taken by anyone dealing with aliphatic hydrocarbons and related chemical solutions. Some compounds are created by joining carbon atoms in straight chains, branched chains, or non-aromatic rings. The compounds can be saturated or unsaturated, and can have other elements besides hydrogen bound to the carbon chain. It is important to follow all health precautions set forth when working with aliphatic hydrocarbons because of toxicity levels and the possibility of damage to the nervous system or vital organs like the liver.

Chemical manufacturing companies include aliphatic hydrocarbons in many of their popular solvents. There are manufacturing companies all around the world that use aliphatic hydrocarbons in their chemical processes. Working with aliphatic hydrocarbons to create chemical solvents helps develop everyday items that some people might take for granted.

ALIPHATIC SOLVENTS

Molecules of aliphatic solvents have straight-chain structure. Hexane (C_6H_{14}), gasoline (petrol, benzine), kerosene are aliphatic solvents. Aliphatic solvents are used in oil extraction, degreasing, for manufacturing rubber and paints, as carriers for aerosols and disinfectants.

HEALTH EFFECT OF ALIPHATIC SOLVENTS:

Hexane

Short-term inhalation of hexane may cause dizziness, giddiness, nausea, and headache. Long-term inhalation may cause muscular weakness, blurred vision headache, fatigue and even numbness. No information about carcinogenic effect.

Gasoline

Short-term inhalation of gasoline may result in irritating to the eyes and respiratory system, rapid onset of unconsciousness. Prolonged skin exposure to gasoline may cause dermatitis, neurological disorders. No reliable evidences of carcinogenic effect.

Kerosene

Exposure to kerosene may cause irritating to eyes and skin, lung injury, dermatitis, irritability, restlessness, drowsiness, convulsions, coma and even death. Kerosene is not considered to be carcinogenic to humans.

[HH] White spirits (mineral turpentine spirits)

White spirit is a mixture of aromatic and paraffinic hydrocarbons. White spirits have boiling ranges (temperatures at which they start and finish boiling) 300°F - 430°F (150°C - 220°C). White spirits are used as solvents or diluents in thinners for paints and varnishes, paint driers, colour printing of fabrics, coal benification, for metal cleaning and degreasing, in furniture and rubber industry, for waxes and polishes, in dry cleaning. Health effect of white spirits:

Short-term exposure to white spirits may cause irritation of the respiratory tract, skin and eyes irritation, dizziness and euphoria leading to unconsciousness in severe cases. Long-term inhalation may result in central nervous system complications, blood changes (aplastic anemia, a rare occurrence that is potentially fatal) and dermatitis. No data has been reported concerning the carcinogenic effect of white spirit.

Aromatic solvents

Molecules of pure aromatic solvents have benzene ring structure. Examples of pure aromatic solvents are benzene (C_6H_6), toluene ($C_6H_5CH_3$)and xylene (C_8H_{10}). Pure (high) aromatic solvents are used for degreasing, as thinners, for manufacturing paints, printing inks, insecticides and agricultural chemicals.

HEALTH EFFECT OF AROMATIC SOLVENTS:

Benzene

Short-term inhalation of benzene may cause drowsiness, dizziness, headaches, as well as eye, skin, and respiratory tract irritation, and, at high levels, unconsciousness. Long-term inhalation may cause disorders in the blood, including reduced numbers of red blood cells and aplastic anemia. Benzene is carcinogen (Group A).

Toluene

Short-term inhalation of toluene may cause dysfunction of the central nervous system: fatigue, sleepiness, headaches, and nausea. Short-term inhalation results in irritation of the upper respiratory tract and eyes, sore throat, dizziness, and headache. Not classified as carcinogen.

Xylene

Short-term inhalation of xylene results in irritation of the eyes, nose, and throat, gastrointestinal effects, eye irritation, and neurological effects. Long-term inhalation of xylene may cause dysfunction of the central nervous system (CNS), such as headache, dizziness, fatigue, tremors, and incoordination; respiratory, cardiovascular, and kidney effects have also been reported. Not classified as carcinogen.

TITRATIONS IN NONAQUEOUS SOLVENTS

Thus far we have assumed that the acid and base are in an aqueous solution. Indeed, water is the most common solvent in acid–base titrimetry. When considering the utility of a titration, however, the solvent's influence cannot be ignored.

The dissociation, or autoprotolysis constant for a solvent, SH, relates the concentration of the protonated solvent, SH_2^+, to that of the deprotonated solvent, S^-. For amphoteric solvents, which can act as both proton donors and proton acceptors, the autoprotolysis reaction is

$$2SH \rightleftharpoons SH_2^+ + S^-$$

with an equilibrium constant of

$$K_s = [SH_2^+][S^-]$$

You should recognize that *K*w is just the specific form of *K*s for water. The pH of a solution is now seen to be a general statement about the relative abundance of protonated solvent

$$pH = -\log[SH_2^+]$$

where the pH of a neutral solvent is given as

$$pH_{neut} = \frac{1}{2}pK_s$$

Perhaps the most obvious limitation imposed by *K*s is the change in pH during a titration. To see why this is so, let's consider the titration of a 50 mL solution of 10–4 M strong acid with equimolar strong base. Before the equivalence point, the pH is determined by the untitrated strong acid, whereas after the equivalence point the concentration of excess strong base determines the pH. In an aqueous solution the concentration of H_3O^+ when the titration is 90 per cent complete is

$$[H_3O^+] = \frac{M_aV_a - M_bV_b}{V_a + V_b}$$

$$= \frac{(1 \times 10^{-4} M)(50 \text{ mL}) - (1 \times 10^{4} M)(45 \text{ mL})}{50 + 45} = 5.3 \times 10^{-6} M$$

Corresponding to a pH of 5.3. When the titration is 110 per cent complete, the concentration of OH^- is

$$[OH^-] = \frac{M_bV_b - M_aV_a}{V_a + V_b}$$

$$= \frac{(1 \times 10^{-4} M)(55 \text{ mL}) - (1 \times 10^{-4} M)(50 \text{ mL})}{50 + 55} = 4.8 \times 10^{-6} M$$

or a pOH of 5.3. The pH, therefore, is

$$pH = pKw - pOH = 14.0 - 5.3 = 8.7$$

The Change in pH when the titration passes from 90 per cent to 110 per cent completion is

$$\Delta pH = 8.7 - 5.3 = 3.4$$

If the same titration is carried out in a nonaqueous solvent with a *K*s of 1.0 x 10^{-20}, the pH when the titration is 90 per cent complete is still 5.3. However, the pH when the titration is 110 per cent complete is now

$$pH = pKs - pOH = 20.0 - 5.3 = 14.7$$

In this case the change in pH of

$$\Delta pH = 14.7 - 5.3 = 9.4$$

is significantly greater than that obtained when the titration is carried out in water. Figure shows the titration curves in both the aqueous and non-aqueous solvents. Nonaqueous solvents also may be used to increase the change in pH when titrating weak acids or bases.

Another parameter affecting the feasibility of a titration is the dissociation constant of the acid or base being titrated. Again, the solvent plays an important role. In the Bronsted–Lowry view of acid–base behaviour, the strength of an acid or base is a relative measure of the ease with which a proton is transferred from the acid to the solvent, or from the solvent to the base. For example, the strongest acid that can exist in water is H_3O^+. The acids HCl and HNO_3 are considered strong because they are better proton donors than H_3O^+. Strong acids essentially donate all their protons to H_2O, "leveling" their acid strength to that of H_3O^+. In a different solvent HCl and HNO_3 may not behave as strong acids.

When acetic acid, which is a weak acid, is placed in water, the dissociation reaction

$$CH_3COOH(aq) + H_2O(\ell) \rightleftharpoons H_3O^+(aq) + CH_3COO^-(aq)$$

does not proceed to a significant extent because acetate is a stronger base than water and the hydronium ion is a stronger acid than acetic acid. If acetic acid is placed in a solvent that is a stronger base than water, such as ammonia, then the reaction

$$CH_3COOH + NH_3 \rightleftharpoons NH_4{}^+ + CH_3COO^-$$

proceeds to a greater extent. In fact, HCl and CH3COOH are both strong acids in ammonia.

All other things being equal, the strength of a weak acid increases if it is placed in a solvent that is more basic than water, whereas the strength of a weak base increases if it is placed in a solvent that is more acidic than water. In some cases, however, the opposite effect is observed. For example, the p*K*b for ammonia is 4.76 in water and 6.40 in the more acidic glacial acetic acid. In contradiction to our expectations, ammonia is a weaker base in the more acidic solvent. A full description of the solvent's effect on a weak acid's p*K*a or on the p*K*b of a weak base is beyond the scope of this text. You should be aware, however, that titrations that are not feasible in water may be feasible in a different solvent.

Quantitative Applications

Although many quantitative applications of acid–base titrimetry have been replaced by other analytical methods, there are several important applications that continue to be listed as standard methods. In this section we review the general application of acid–base titrimetry to the analysis of inorganic and organic compounds, with an emphasis on selected applications in environmental and clinical analysis. First, however, we discuss the selection and standardization of acidic and basic titrants.

Selecting and Standardizing a Titrant: Most common acid–base titrants are not readily available as primary standards and must be standardized before they can be used in a quantitative analysis.

Standardization is accomplished by titrating a known amount of an appropriate acidic or basic primary standard. The majority of titrations involving basic analytes, whether conducted in aqueous or non-aqueous solvents, use HCl, $HClO_4$, or H_2SO_4 as the titrant. Solutions of these titrants are usually prepared by diluting a commercially available concentrated stock solution and are stable for extended periods of time.

Since the concentrations of concentrated acids are known only approximately,* the titrant's concentration is determined by standardizing against one of the primary standard weak bases listed in Table.

The most common strong base for titrating acidic analytes in aqueous solutions is NaOH.

Sodium hydroxide is available both as a solid and as an approximately 50 per cent w/v solution. Solutions of NaOH may be standardized against any of the primary weak acid standards listed in Table. The standardization of NaOH, however, is complicated by potential contamination from the following reaction between CO_2 and OH^-.

$$CH_2(g) + 2OH^-(aq) \rightarrow CO_3^{2-}(aq) + H_2O(\ell)$$

When CO_2 is present, the volume of NaOH used in the titration is greater than that needed to neutralize the primary standard because some OH^- reacts with the CO_2.

Table. Selected Primary Standards for the Standardization of Strong Acid and Strong Base Titrants.

Standard	Standardization of Acidic Titrants Titration Reaction	Primary Comment
Na_2CO_3	$Na_2CO_3 + 2H_3O^+ \rightarrow H_2CO_3 + 2Na^+ + 2H_2O$	a
TRIS	$(HOCH_2)_3CNH_2 + H_3O^+ \rightarrow (HOCH_2)_3CNH_3^+ + H_2O)$	b
$Na_2B_4O_7$	$Na_2B_4O_7 + 2H_3O^+ + 3H_2O \rightarrow 2Na^+ + 4H_3BO3$	
Standard	**Standardization of Basic Titrants Titration Reaction**	**Primary Commnet**
$KHC_8H_4O_4$	$KHC_8H_4O_4 + OH– \rightarrow K^+ + C_8H_4O_4^{2-} + H_2O$	c
C_6H_5COOH	$C_6H_5COOH + OH^- \rightarrow C_6H_5COO^- + H_2O$	d
$KH(IO_3)_2$	$KH(IO_3)_2 + OH^- \rightarrow K^+ + 2IO_3^- + H_2O$	

The calculated concentration of OH^-, therefore, is too small. This is not a problem when titrations involving NaOH are restricted to an end point pH less than 6. Below this pH any $CO_3{}^{2-}$ produced in reaction 9.7 reacts with H_3O^+ to form carbonic acid.

$$CO_3^{2-}(aq) + 2H_3O^+(aq) \rightarrow H_2CO_3(aq) + 2H_2O(\ell)$$

Combining reactions 9.7 and 9.8 gives an overall reaction of

$$CO_2(g) + H_2(O)(\ell) \rightarrow H_2CO_3(aq)$$

which does not include OH^-. Under these conditions the presence of CO_2 does not affect the quantity of OH^- used in the titration and, therefore, is not a source of determinate error. For pHs between 6 and 10, however, the neutralization of CO_3^{2-} requires only one proton

$$CO_3^{2-}(aq) + H_3O^+(aq) \rightarrow HCO_3^-(aq) + H_2O(\ell)$$

and the net reaction between CO_2 and OH^- is

$$CO_2(g) + OH^-(aq) \rightarrow HCO_3^-(aq)$$

Under these conditions some OH^- is consumed in neutralizing CO_2. The result is a determinate error in the titrant's concentration. If the titrant is used to analyse an analyte that has the same end point pH as the primary standard used during standardization, the determinate errors in the standardization and the analysis cancel, and accurate results may still be obtained.

Solid NaOH is always contaminated with carbonate due to its contact with the atmosphere and cannot be used to prepare carbonate-free solutions of NaOH. Solutions of carbonate-free NaOH can be prepared from 50 per cent w/v NaOH since Na_2CO_3 is very insoluble in concentrated NaOH.

When CO_2 is absorbed, Na_2CO_3 precipitates and settles to the bottom of the container, allowing access to the carbonate-free NaOH. Dilution must be done with water that is free from dissolved CO_2. Briefly boiling the water expels CO_2 and, after cooling, it may be used to prepare carbonate-free solutions of NaOH. Provided that contact with the atmosphere is minimized, solutions of carbonate-free NaOH are relatively stable when stored in polyethylene bottles. Standard solutions of sodium hydroxide should not be stored in glass bottles because NaOH reacts with glass to form silicate.

Inorganic Analysis: Acid–base titrimetry is a standard method for the quantitative analysis of many inorganic acids and bases. Standard solutions of NaOH can be used in the analysis of inorganic acids such as H_3PO_4 or H_3AsO4, whereas standard solutions of HCl can be used for the analysis of inorganic bases such as Na_2CO_3. Inorganic acids and bases too weak to be analyzed by an aqueous acid–base titration can be analyzed by adjusting the solvent or by an indirect analysis.

For example, the accuracy in titrating boric acid, H_3BO_3, with NaOH is limited by boric acid's small acid dissociation constant of 5.8×10^{-10}. The acid strength of boric acid, however, increases when mannitol is added to the solution because it forms a complex with the borate ion. The increase in Ka to approximately 1.5×10^{-4} results in a sharper end point and a more accurate titration. Similarly, the analysis of ammonium salts is limited by the small acid dissociation constant of 5.7×10^{-10} for NH_4^+. In this case, NH_4^+ can be converted to NH_3 by neutralizing with strong base. The NH_3, for which K_b is 1.8×10^{-5}, is then removed by distillation and titrated with a standard strong acid titrant.

Inorganic analytes that are neutral in aqueous solutions may still be analyzed if they can be converted to an acid or base. For example, NO_3^- can be quantitatively analyzed by reducing it to NH_3 in a strongly alkaline solution using Devarda's alloy, a mixture of 50 per cent w/w Cu, 45 per cent w/w Al, and 5 per cent w/w Zn.

$$3NO_3^-(aq) + 8Al(s) + 5OH^-(aq) + 2H_2O(\ell) \rightarrow 8AlO_2^-(aq) + 3NH_3(aq)$$

The NH3 is removed by distillation and titrated with HCl. Alternatively, NO_3^- can be titrated as a weak base in an acidic nonaqueous solvent such as anhydrous acetic acid, using $HClO_4$ as a titrant.

Acid–base titrimetry continues to be listed as the standard method for the determination of alkalinity, acidity, and free CO_2 in water and wastewater analysis. Alkalinity is a measure of the acid-neutralizing capacity of a water sample and is assumed to arise principally from OH^-, HCO_3^-, and CO_3^{2-}, although other weak bases, such as phosphate, may contribute to the overall alkalinity. Total alkalinity is determined by titrating with a standard solution of HCl or H_2SO_4 to a fixed end point at a pH of 4.5, or to the bromocresol green end point. Alkalinity is reported as milligrams $CaCO_3$ per litre.

When the sources of alkalinity are limited to OH^-, HCO_3^-, and CO_3^{2-}, titrations to both a pH of 4.5 (bromocresol green end point) and a pH of 8.3 (phenolphthalein or metacresol purple end point) can be used to determine which species are present, as well as their respective concentrations.

Titration curves for OH^-, HCO_3^-, and CO_3^{2-} are shown in figure. For a solution containing only OH^- alkalinity, the volumes of strong acid needed to reach the two end points are identical.

If a solution contains only HCO_3^- alkalinity, the volume of strong acid needed to reach the end point at a pH of 8.3 is zero, whereas that for the pH 4.5 end point is greater than zero. When the only source of alkalinity is CO_3^{2-}, the volume of strong acid needed to reach the end point at a pH of 4.5 is exactly twice that needed to reach the end point at a pH of 8.3.

Mixtures of OH^- and CO_3^{2-}, or HCO_3^- and CO_3^{2-} alkalinities also are possible. Consider, for example, a mixture of OH^- and CO_3^{2-}. The volume of strong acid needed to titrate OH^- will be the same whether we titrate to the pH 8.3 or pH 4.5 end point. Titrating CO_3^{2-} to the end point at a pH of 4.5, however, requires twice as much strong acid as when titrating to the pH 8.3 end point. Consequently, when titrating a mixture of these two ions, the volume of strong acid needed to reach the pH 4.5 end point is less than twice that needed to reach the end point at a pH of 8.3. For a mixture of HCO_3^- and CO_3^{2-}, similar reasoning shows that the volume of strong acid needed to reach the end point at a pH of 4.5 is more than twice that need to reach the pH 8.3 end point. Solutions containing OH^- and HCO_3^- alkalinities are unstable with respect to the formation of CO_3^{2-} and do not exist. Table summarizes the relationship between the sources of alkalinity and the volume of titrant needed to reach the two end points.

Acidity is a measure of a water sample's capacity for neutralizing base and is conveniently divided into strong acid and weak acid acidity. Strong acid acidity is due to the presence of inorganic acids, such as HCl, HNO_3, and H_2SO_4, and is commonly found in industrial effluents and acid mine drainage. Weak acid acidity is usually dominated by the formation of H_2CO_3 from dissolved CO_2, but also

Table. Relationship Between End Point Volumes and Sources of Alkalinity

Source of Alkalinity	Relationship Between End Point Volumes
OH^-	$V_{pH\ 4.5} = V_{pH\ 83}$
CO_3^{2-}	$V_{pH\ 4.5} = 2 \times V_{pH\ 8.3}$
HCO_3^-	$V_{pH\ 8.3} = 0$; $V_{pH\ 45} > 0$
OH^- and CO_3^{2-}	$V_{pH\ 4.5} < 2 \times V_{pH\ 8.3}$
CO_3^{2-} and HCO_3^-	$V_{pH\ 45} > 2 \times V_{pH\ 8.3}$

includes contributions from hydrolyzable metal ions such as Fe^{3+}, Al^{3+}, and Mn^{2+}. In addition, weak acid acidity may include a contribution from organic acids.

Acidity is determined by titrating with a standard solution of NaOH to fixed end points at pH 3.7 and pH 8.3. These end points are located potentiometrically, using a pH meter, or by using an appropriate indicator (bromophenol blue for pH 3.7, and metacresol purple or phenolphthalein for pH 8.3). Titrating to a pH of 3.7 provides a measure of strong acid acidity, and titrating to a pH of 8.3 provides a measure of total acidity. Weak acid acidity is given indirectly as the difference between the total and strong acid acidities. Results are expressed as the milligrams of $CaCO_3$ per litre that could be neutralized by the water sample's acidity. An alternative approach for determining strong and weak acidity is to obtain a potentiometric titration curve and use Gran plot methodology to determine the two equivalence points. This approach has been used, for example, in determining the forms of acidity in atmospheric aerosols.

Water in contact with either the atmosphere or carbonate-bearing sediments contains dissolved or free CO_2 that exists in equilibrium with gaseous CO_2 and the aqueous carbonate species H_2CO_3, HCO_3^-, and CO_3^{2-}. The concentration of free CO_2 is determined by titrating with a standard solution of NaOH to the phenolphthalein end point, or to a pH of 8.3, with results reported as milligrams CO_2 per litre. This analysis is essentially the same as that for the determination of total acidity, and can only be applied to water samples that do not contain any strong acid acidity.

Organic Analysis: The use of acid–base titrimetry for the analysis of organic compounds continues to play an important role in pharmaceutical, biochemical, agricultural, and environmental laboratories. Perhaps the most widely employed acid–base titration is the Kjeldahl analysis for organic nitrogen, described earlier in Method.

This method continues to be used in the analysis of caffeine and saccharin in pharmaceutical products, as well as for the analysis of proteins, fertilizers, sludges, and sediments.

Any nitrogen present in the –3 oxidation state is quantitatively oxidized to NH_4^+. Some aromatic heterocyclic compounds, such as pyridine, are difficult to oxidize.

A catalyst, such as HgO, is used to ensure that oxidation is complete. Nitrogen in an oxidation state other than –3, such as nitro- and azonitrogens, is often oxidized to N_2, resulting in a negative determinate error. Adding a reducing agent, such as salicylic acid, reduces the nitrogen to a –3 oxidation state,

Table. Selected Elemental Analyses Based on Acid–Base Titrimetry

Element	Liberated as	Reaction Producing Acid or Base to Be Titrated[a]	Titration
N	$NH_3(g)$	$NH_3(g) + H_3O(aq) \rightarrow NH_4+(aq) + H_2O(\ell)$	excess H_3O^+ with strong base
S	$SO_2(g)$	$SO_2(g) + H_2O_2(aq) \rightarrow H_2SO_4(aq)$	H_2SO_4 with strong base excess
C	$CO_2(g)$	$CO_2(g) + Ba(OH)_2(aq) \rightarrow BaCO_3(s) + H_2O(\ell)$	$Ba(OH)_2$ with strong acid
Cl	$HCl(g)$	$HCl(g) + H_2O(\ell) \rightarrow H_3O+(aq) + Cl–(aq)$	H_3O+ with strong base
F	$SiF_4(g)$	$3SiF_4(g) + 2H_2O(\ell) \rightarrow 2H_2SiF_6(aq) + SiO_2(s)$	H_2SiF_6 with strong base

The acid or base that is eventually titrated is indicated in bold.

Table. Selected Acid–Base Titrimetric Procedures for Organic Functional Groups Based on the Production or Consumption of Acid or Base.

Finctional Group	Reaction Producing Acid or Base to Be Titrated[a]	Titration
Ester	$RCOOR'(aq) + OH–(aq) \rightarrow$ $RCOO–(aq) + HOR'(aq)$	excess OH–with strong acid
Carbonyl	$R_2C = O(aq) + NH_2OH. HCl(aq) \rightarrow$ $R_2C = NOH(aq) + HCl(aq) + H_2O(\ell)$	HCl with strong base
Alcohol[b]	[1] $(CH_3CO)_2O + ROH \rightarrow$ $CH_3COOR + CH_3COOH$ [2] $(CH_3CO)_2O + H_2O \rightarrow$ $2CH_3COOH$	CH_3COOH with strong base; ROH is determined from the difference in the amount of titrant needed to react with a blank consisting only of acetic anhydride, and the amount reacting with the sample.

eliminating this source of error. Other examples of elemental analyses based on the conversion of the element to an acid or base are outlined in Table.

Several organic functional groups have weak acid or weak base properties that allow their direct determination by an acid–base titration. Carboxylic (—COOH), sulfonic (—SO_3H), and phenolic (—C_6H_5OH) functional groups are weak acids that can be successfully titrated in either aqueous or non-aqueous solvents. Sodium hydroxide is the titrant of choice for aqueous solutions. Nonaqueous titrations are often carried out in a basic solvent, such as ethylenediamine, using tetrabutylammonium hydroxide, $(C_4H_9)_4NOH$, as the titrant. Aliphatic and aromatic amines are weak bases that can be titrated using HCl in aqueous solution or $HClO_4$ in glacial acetic acid.

Other functional groups can be analyzed indirectly by use of a functional group reaction that produces or consumes an acid or base. Examples are shown in Table.

Many pharmaceutical compounds are weak acids or bases that can be analyzed by an aqueous or non-aqueous acid–base titration; examples include salicylic acid, phenobarbital, caffeine, and sulfanilamide. Amino acids and proteins can be analyzed in glacial acetic acid, using HClO4 as the titrant. For example, a procedure for determining the amount of nutritionally available protein has been developed that is based on an acid–base titration of lysine residues.

Quantitative Calculations: In acid–base titrimetry the quantitative relationship between the analyte and the titrant is determined by the stoichiometry of the relevant reactions. As outlined in Section 2C, stoichiometric calculations may be simplified by focusing on appropriate conservation principles. In an acid–base reaction the number of protons transferred between the acid and base is conserved; thus

$$\frac{\text{moles of H}^+\text{ donated}}{\text{mole acid}} \times \text{moles acid} = \frac{\text{moles of H}^+\text{ accepted}}{\text{mole base}} \times \text{moles base}$$

In an indirect analysis the analyte participates in one or more preliminary reactions that produce or consume acid or base. Despite the additional complexity, the stoichiometry between the analyte and the amount of acid or base produced or consumed may be established by applying the conservation principles.

Earlier we noted that an acid–base titration may be used to analyse a mixture of acids or bases by titrating to more than one equivalence point. The concentration of each analyte is determined by accounting for its contribution to the volume of titrant needed to reach the equivalence points.

Qualitative Applications

We have already come across one example of the qualitative application of acid–base titrimetry in assigning the forms of alkalinity in waters. This approach is easily extended to other systems. For example, the composition of solutions containing one or two of the following species

$$H_3PO_4 \; H_2PO_4^- \; HPO_4^{2-} \; PO_4^{3-} \; NaOH \; HCl$$

can be determined by titrating with either a strong acid or a strong base to the methyl orange and phenolphthalein end points. As outlined in Table, each species or mixture of species has a unique relationship between the volumes of titrant needed to reach these two end points.

Table. Relationship Between End Point Volumes for Solutions of Phosphate Species with HCl and NaOH

Solution Composition	Relationship Between End Point Volumes with Strong Base Titrant	Relationship Between End Point Volumes with Strong Acid Titrant
H_3PO_4	$V_{PH} = 2 \times V_{MO}$	–[b]
$H_2PO_4^-$	$V_{PH} > 0; V_{MO} = 0$	–
HPO_4^{2-}	–	$V_{MO} > 0; V_{PH} = 0$
PO_4^{3-}	–	$V_{MO} = 2 \times V_{PH}$
HCl	$V_{PH} = V_{MO}$	–
NaOH	–	$V_{MO} = V_{PH}$
HCl and H_3PO_4	$V_{PH} < 2 \times V_{MO}$	–
H_3PO_4 and $H_2PO_4^-$	$V_{PH} > 2 \times V_{MO}$	–
$H_2PO_4^-$ and HPO_4^{2-}	$V_{PH} > 0; V_{MO} = 0$	$V_{MO} > 0; V_{PH} = 0$
HPO_4^{2-} and PO_4^{3-}	–	$V_{MO} > 2 \times V_{PH}$
PO_4^{3-} and NaOH	–	$V_{MO} < 2 \times V_{PH}$

Characterization Applications

Two useful characterization applications involving acid–base titrimetry are the determination of equivalent weight, and the determination of acid–base dissociation constants.

Equivalent Weights: Acid–base titrations can be used to characterize the chemical and physical properties of matter. One simple example is the determination of the equivalent weight* of acids and bases. In this method, an accurately weighed sample of a pure acid or base is titrated to a well-defined equivalence point using a monoprotic strong acid or strong base. If we assume that the titration involves the transfer of n protons, then the moles of titrant needed to reach the equivalence point is given as

$$\text{Moles titrant} = n \times \text{moles analyte}$$

and the formula weight is

$$FW = \frac{\text{g analyte}}{\text{moles analyte}} = n \times \frac{\text{g analyte}}{\text{moles titrant}}$$

Since the actual number of protons transferred between the analyte and titrant is uncertain, we define the analyte's equivalent weight (EW) as the apparent formula weight when $n = 1$. The true formula weight, therefore, is an integer multiple of the calculated equivalent weight.

$$FW = n \times EW$$

Thus, if we titrate a monoprotic weak acid with a strong base, the EW and FW are identical. If the weak acid is diprotic, however, and we titrate to its second equivalence point, the FW will be twice as large as the EW. Equilibrium Constants Another application of acid–base titrimetry is the determination of equilibrium constants. Consider, for example, the titration of a weak acid, HA, with a strong base. The dissociation constant for the weak acid is

$$K_a = \frac{[A^-][H_3O^+]}{[HA]}$$

When the concentrations of HA and A– are equal, equation reduces to *K*a = [H3O+], or pH = p*K*a. Thus, the p*K*a for a weak acid can be determined by measuring the pH for a solution in which half of the weak acid has been neutralized. On a titration curve, the point of half-neutralization is approximated by the volume of titrant that is half of that needed to reach the equivalence point. As shown in Figure, an estimate of the weak acid's p*K*a can be obtained directly from the titration curve.

This method provides a reasonable estimate of the p*K*a, provided that the weak acid is neither too strong nor too weak. These limitations are easily appreciated by considering two limiting cases.

For the first case let's assume that the acid is strong enough that it is more than 50 per cent dissociated before the titration begins. As a result the concentration of HA before the equivalence point is always less than the concentration of A–, and there is no point along the titration curve where [HA] = [A⁻].

At the other extreme, if the acid is too weak, the equilibrium constant for the titration reaction

$$HA(aq) + OH^-(aq) \rightleftharpoons H_2O(\ell)A^-(aq)$$

may be so small that less than 50 per cent of HA will have reacted at the equivalence point. In this case the concentration of HA before the equivalence point is always greater

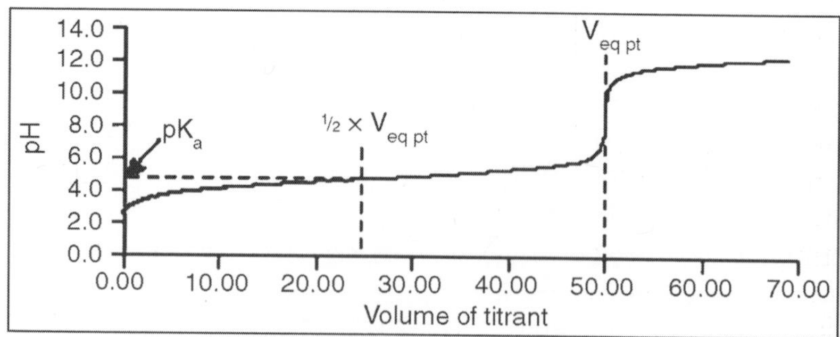

Fig. Estimating the pK_a for a Weak Acid from its Titration Curve with a Strong Base.

than that of A^-. Determining the p*K*a by the half-equivalence point method overestimates its value if the acid is too strong and underestimates its value if the acid is too weak.

A second approach for determining the p*K*a of an acid is to replot the titration curve in a linear form as a Gran plot. For example, earlier we learned that the titration of a weak acid with a strong base can be plotted in a linear form using the following equation

$$V_b \times [H_3O^+] = K_a \times V_{eq} - K_a \times V_b$$

Plotting $V_b \times [H_3O^+]$ versus *V*b, for volumes less than the equivalence point volume yields a straight line with a slope of –*K*a. Other linearizations have been developed that use all the points on a titration curve or require no assumptions.

This approach to determining acidity constants has been used to study the acid–base properties of humic acids, which are naturally occurring, large-molecular-weight organic acids with multiple acidic sites. In one study, a sample of humic acid was found to have six titratable sites, three of which were identified as carboxylic acids, two of which were believed to be secondary or tertiary amines, and one of which was identified as a phenolic group.

PRECIOUS METALS REFINING BY SOLVENT EXTRACTION

GENERAL

If a compound (= solute) is dissolved in a liquid (water) and the solution is brought into contact with a second immiscible liquid (= solvent, maybe a hydrocarbon etc.), then a part of the solute is transferred to the second liquid phase by a force called the chemical potential. The physical-chemical process of transferring the solute between the bulk of the two immiscible liquid phases is calledsolvent extraction.

During intensive mixing of both liquids in the course of time the mass transfer of the solute between the two liquid phases deminishes and finally at very long contact times vanishes; an equilibriumis encountered. The time necessary to reach 90 per cent of the equilibrium is characteristic for a given solute/solvent system and is in the range of seconds up to hours for technically relevant systems. This time is a function of the product k_t a ; (mass transfer coefficient) x (interface area/liquid volume). In technical equipment equilibrium is never reached to a hundred percent; 90 to 99 per cent is considered to be sufficient.

Commonly for practical process layout the equilibrium is characterized by a distribution coefficient which is defined as the ratio D = (all species of solute in organic phase)/(all species of solute in aqueous phase). D is dependent on the initial concentration of the solute and the concentration of other reaction components in question. The distribution coefficient is independent of phase flow conditions and somewhat characteristic for a given solute/solvent/aqueous system. It is one of the key parameters in the design of a solvent extraction process.

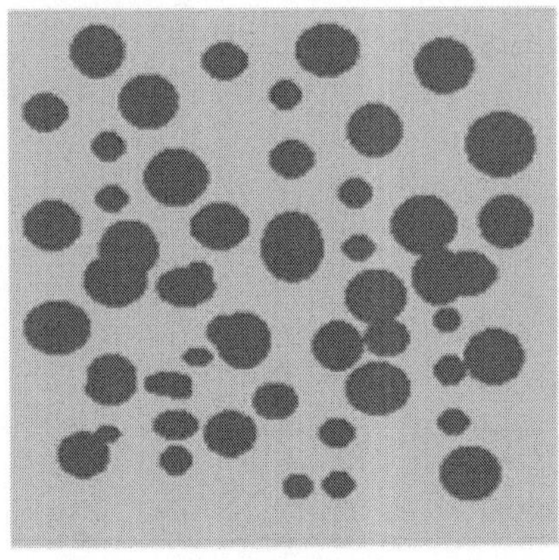

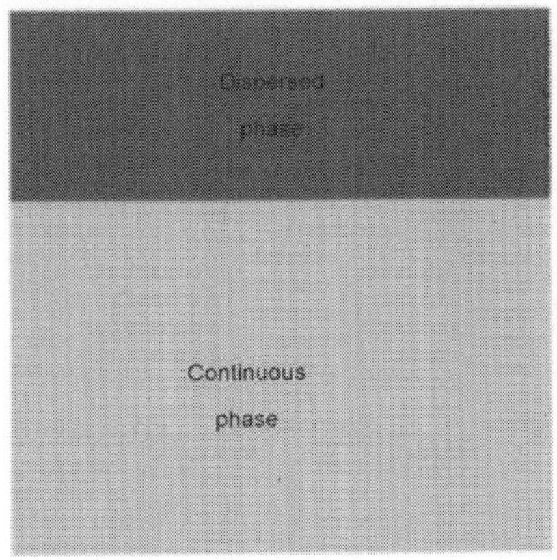

During mixing After setting

Fig. Liquid two-phase system vigorously mixed

After the calculated mixing time in a batch process or residence time in a continuous mixing process has elapsed, mixing of the two liquid phases is stopped and both liquid phases are allowed to coalesce and settle by gravity or centrifugal force. The rate of coalescence is highly depending on the viscosity, density and interfacial tension of the liquids and drop size of the dispersed phase.

The basic equipment to perform a continuous mixing and coalescence process on a technical scale is called a mixer-settler. It is often used in laboratories as an ideal tool for basic system design of continuous solvent extraction processes because it offers reproducable phase flow and contact times.

The device comprises a continuously fed and stirred mixing compartment and a gravity settler compartment where the liquids are allowed to separate. At the end of the settler two individual weirs care for good separation of the liquid bulk phases. The flow capacity of a mixer-settler is reached when the emulsion phase dispersion band overflows the light phase weir or underflows the heavy phase weir.

In technical mixer-settler devices the volume of the settler compartment is often 10-fold the size of the mixer compartment to provide sufficient settling time even for systems with low coalescence rates (in case of high viscosity and/or low density and/or low interfacial tension).

For a good performance of the settler the liquid phase ratio organic/aqueous in a mixer-settler should be kept close to one. To adjust a definite feed ratio of organic/aqueous- independent of optimum internal phase ratio - an external or internal phase recycle can be installed. In the sketch of a conventional mixer-settler below a feed ratio of O/A = 1/10 is shown. Hence such a process would concentrate a solute in the loaded organic by a factor of 10 if the distribution coefficient is high enough.

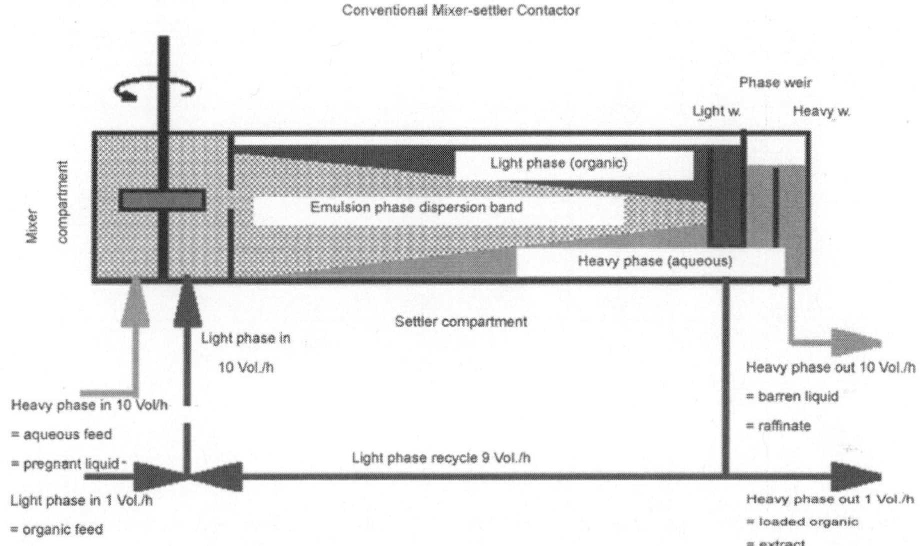

In nearly all practical cases it is necessary to have more than one mixer-settler unit. Repeated mixing and settling can provide a high concentration and purity of the solute especially if the distribution coefficient is low or even close to one. In multistage solvent extraction mixer-settler batteries often are built as contiguous boxes which can save a lot of space.

In applications where a very high number of stages is required theoretically and in large scale industrial application of more than 100 m^3/h solvent extraction is preferably run in continuous countercurrent column equipment (reciprocating plate, rotating disc, stirred column). In column extractors there are no longer discrete compartments where mixing and settling of the liquids occurs but new formation of droplet with "fresh interface" and coalescence of droplets prevails almost in each section of the column in the same extent. On the right the principle scheme of a perforated plate reciprocating column is shown.

Organic feed enters the bottom part of the column through a spray nozzle or ring distributor system and is allowed to coalesce at the top in a separate horizontally mounted settling device to increase the interfacial surface. The control of the hydraulic stability of solvent extraction columns is much more difficult to maintain than in mixer-settlers.

In some cases if solvent extraction is governed by an irreversible chemical reaction of the solute with other components in the solvent the process can be performed in a stirred tank reactor as a batch mixer-settler process.

Please note that the feed for a solvent extraction process containing the solute in question is not necessarily always aqueous. There are lots of examples in industrial organic synthesis where crude organic product solutions have to be treated with aqueous solutions or water for purifying. Moreover the solute is not always the valuable target component in a solution, it might just be an impurity which has to be removed.

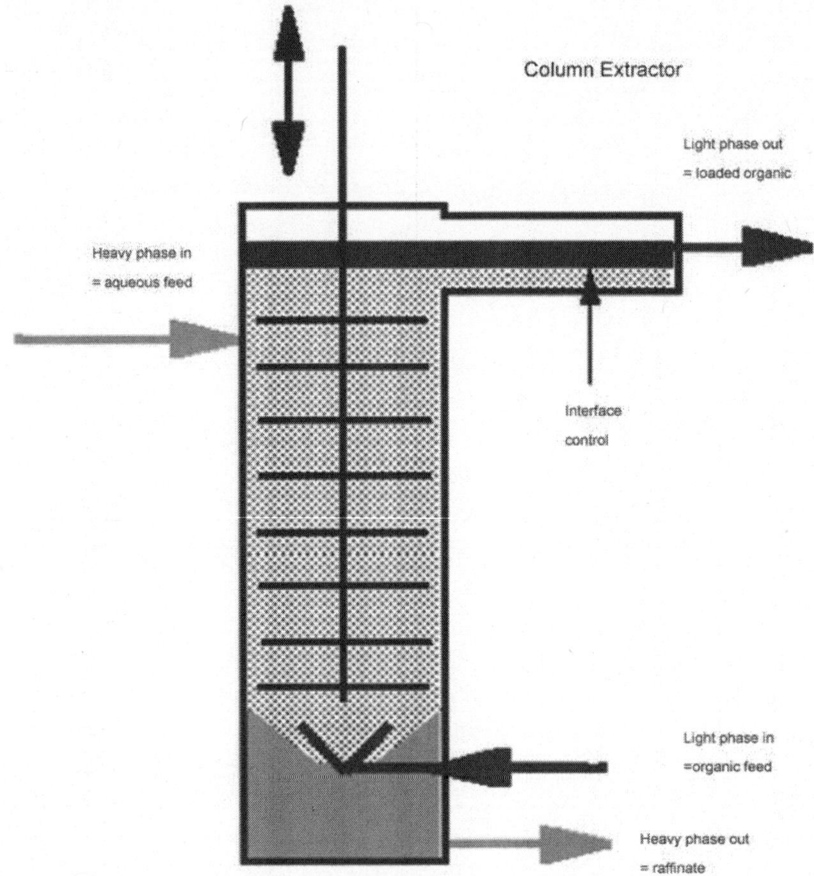

Reactive Extraction

In cases if solvent extraction is enabled by a chemical reaction (at the interphase of aqueous/organic or in one or both of the bulk phases) or if a prevailing chemical reaction is accelerating the basic physical mass transfer, this process is called reactive extraction. The item was first used and introduced into litreature in 1981 by Werner Halwachs in his publication "Reaktivextraktion".

Metal Extraction

As far as solvent extraction of metals or precious metals is concerned an additional chemical, a so-called extractant, is necessary to form a chemical compound (= complex) with the solute and thus make it soluble in the organic phase. Furthermore additional chemicals, so-called modifiers, are used to enhance coalescence of the emulsion and increase the solubility of metal complexes formed in the solvent to suppress formation of solids which may lead to a total collapse of the process by clogging..

Solvent extraction of metals or precious metals is used in refineries for

1. the removal of individual precious metals or group of metals from a pregnant liquor comprising other pgm, base metals and/or salts (= separation).
2. the removal/reduction of impurities from crude solutions of individual precious metals or a group of metals (= refining)

In most cases a complete industrial solvent extraction cycle for metal extraction comprises four consecutive solvent extraction steps in order to win concentrated or purified solute and recover solvent and extractant - otherwise the process would be uneconomic and environmentally hazardous; those subprocesses are:

1. extraction; transfer of the metal into the organic phase by chemical reaction with the extractant
2. scrubbing; removal of coextracted material/metals or excess acid etc. (optional)
3. stripping; transfer of the metal back into a second pure aqueous phase for winning or further processing
4. solvent make-up; treatment of the organic by a third aqueous phase for purification of solvent or extractant; removal of crud or degradation products; topping with fresh organic (optional)

BATCHWISE SINGLE STAGE EXTRACTIONS TECHNIQUES

This is commonly used on the small scale in chemical labs. It is normal to use a separating funnel. For instance, if a chemist were to extract anisole from a mixture of water and 5 per cent acetic acid using ether, then the anisole will enter the organic phase. The two phases would then be separated. The acetic acid can then be scrubbed (removed) from the organic phase by shaking the organic extract with sodium bicarbonate. The acetic acid reacts with the sodium bicarbonate to form sodium acetate, carbon dioxide, and water.

Multistage Countercurrent Continuous Processes

These are commonly used in industry for the processing of metals such as the lanthanides; because the separation factors between the lanthanides are so small many extraction stages are needed. In the multistage processes, the aqueous raffinate from one extraction unit is fed to the next unit as the aqueous feed, while the organic phase is moved in the opposite direction. Hence, in this way, even if the separation between two metals in each stage is small, the overall system can have a higher decontamination factor.

Multistage countercurrent arrays have been used for the separation of lanthanides. For the design of a good process, the distribution ratio should be not too high (>100) or too low (<0.1) in the extraction portion of the process. It is often the case that the process will have a section for scrubbing unwanted metals from the organic phase, and finally a stripping section to obtain the metal back from the organic phase.

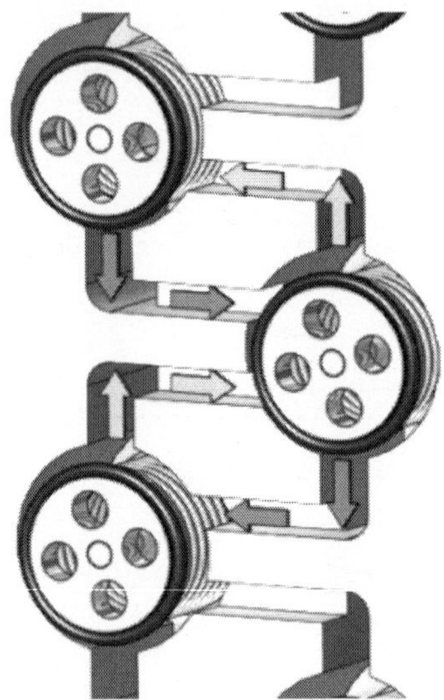

Fig. Coflore continuous countercurrent extractor

Mixer-settlers

Battery of mixer-settlers counter currently interconnected. Each mixer-settler unit provides a single stage of extraction. A mixer settler consists of a first stage that mixes the phases together followed by a quiescent settling stage that allows the phases to separate by gravity.

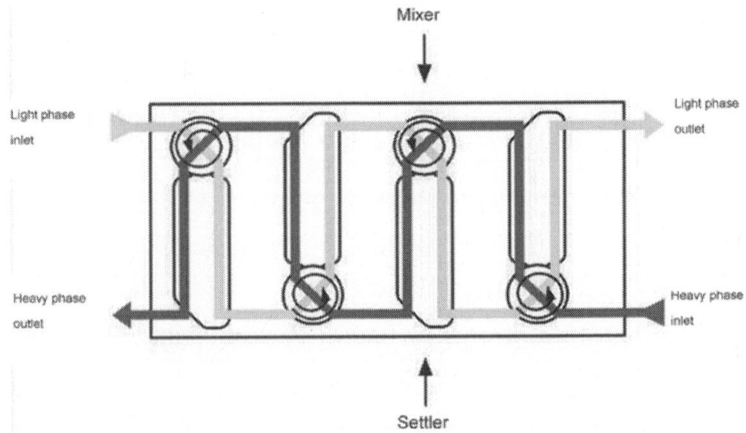

Fig. 4 stage battery of mixer-settlers for counter-current extraction

In the multistage countercurrent process, multiple mixer settlers are installed with mixing and settling chambers located at alternating ends for each stage (since the outlet of the settling sections feed the inlets of the adjacent stage's mixing sections). Mixer-settlers are used when a process requires longer residence times and when the solutions are easily separated by gravity. They require a large facility footprint, but do not require much headspace, and need limited remote maintenance capability for occasional replacement of mixing motors.

Centrifugal Extractors

Centrifugal extractors mix and separate in one unit. Two liquids will be intensively mixed between the spinning rotor and the stationary housing at speeds up to 6000 RPM. This develops great surfaces for an ideal mass transfer from the aqueous phase into the organic phase. At 200 – 2000 g both phases will be separated again. Centrifugal extractors minimize the solvent in the process, optimize the product load in the solvent and extract the aqueous phase completely. Counter current and cross current extractions are easily established.

Extraction without Chemical Change

Some solutes such as noble gases can be extracted from one phase to another without the need for a chemical reaction. This is the simplest type of solvent extraction. When a solvent is extracted, two immiscible liquids are shaken together. The more polar solutes dissolve preferentially in the more polar solvent, and the less polar solutes in the less polar solvent. Some solutes that do not at first sight appear to undergo a reaction during the extraction process do not have distribution ratio that is independent of concentration. A classic example is the extraction of carboxylic acids (HA) into non-polar media such as benzene. Here, it is often the case that the carboxylic acid will form a dimer in the organic layer so the distribution ratio will change as a function of the acid concentration (measured in either phase).

For this case, the extraction constant k is described by $k = [[HA_{organic}]]^2/[[HA_{aqueous}]]$

Solvation Mechanism

Using solvent extraction it is possible to extract uranium, plutonium, or thorium from acid solutions. One solvent used for this purpose is the organophosphate tri-n-butyl phosphate. The PUREX process that is commonly used in nuclear reprocessing uses a mixture of tri-n-butyl phosphate and an inert hydrocarbon (kerosene), the uranium(VI) are extracted from strong nitric acid and are back-extracted (stripped) using weak nitric acid. An organic soluble uranium complex $[UO_2(TBP)_2(NO_3)_2]$ is formed, then the organic layer bearing the uranium is brought into contact with a dilute nitric acid solution; the equilibrium is shifted away from the organic soluble uranium complex and towards the free TBP and uranyl nitrate in dilute nitric acid. The plutonium(IV) forms a similar complex to the uranium(VI), but it is possible to strip the plutonium in more than one way; a reducing agent that converts the plutonium to the trivalent oxidation state can be added. This oxidation state does not form a stable complex with TBP and nitrate unless the nitrate concentration is very high (circa 10 mol/L nitrate is required in the aqueous phase). Another method is to simply use dilute nitric acid as a stripping agent for the plutonium. This PUREX chemistry is a classic example of a solvation extraction.

Here in this case $D_U = k\ TBP^2[[NO_3]]^2$

Ion Exchange Mechanism

Another extraction mechanism is known as the ion exchange mechanism. Here, when an ion is transferred from the aqueous phase to the organic phase, another ion is transferred in the other direction to maintain the charge balance. This additional ion is often a hydrogen ion; for ion exchange mechanisms, the distribution ratio is often a function of pH. An example of an ion exchange extraction would be the extraction of americium by a combination of terpyridine and a carboxylic acid in *tert*-butyl benzene. In this case

$$D_{Am} = k\ \text{terpyridine}^1\text{carboxylic acid}^3\text{H}+^{-3}$$

Another example is the extraction of zinc, cadmium, or lead by a dialkyl phosphinic acid (R_2PO_2H) into a non-polar diluent such as an alkane. A non-polar diluent favours the formation of uncharged non-polar metal complexes.

Some extraction systems are able to extract metals by both the solvation and ion exchange mechanisms; an example of such a system is the americium (and lanthanide) extraction from nitric acid by a combination of 6,6'-*bis*-(5,6-dipentyl-1,2,4-triazin-3-yl)-2,2'-bipyridine and 2-bromohexanoic acid in *tert*-butyl benzene. At both high- and low-nitric acid concentrations, the metal distribution ratio is higher than it is for an intermediate nitric acid concentration.

Ion Pair Extraction

It is possible by careful choice of counterion to extract a metal. For instance, if the nitrate concentration is high, it is possible to extract americium as an anionic nitrate complex if the mixture contains a lipophilic quaternary ammonium salt. An example that is more likely to be encountered by the *'average'* chemist is the use of a phase transfer catalyst. This is a charged species that transfers another ion to the organic phase. The ion reacts and then forms another ion, which is then transferred back to the aqueous phase.

For instance, the 31.1 kJ mol^{-1} is required to transfer an acetate anion into nitrobenzene, while the energy required to transfer a chloride anion from an aqueous phase to nitrobenzene is 43.8 kJ mol^{-1}. Hence, if the aqueous phase in a reaction is a solution of sodium acetate while the organic phase is a nitrobenzene solution of benzyl chloride, then, when a phase transfer catalyst, the acetate anions can be transferred from the aqueous layer where they react with the benzyl chloride to form benzyl acetate and a chloride anion. The chloride anion is then transferred to the aqueous phase. The transfer energies of the anions contribute to that given out by the reaction.

A 43.8 to 31.1 kJ mol^{-1} = 12.7 kJ mol^{-1} of additional energy is given out by the reaction when compared with energy if the reaction had been done in nitrobenzene using one equivalent weight of a tetraalkylammonium acetate.

Aqueous Two-phase Extraction

Aqueous two-phase extraction, also known as *two-phase liquid extraction*, is a unique form of solvent extraction. In an aqueous two-phase extraction, compounds are still separated based on their solubility, but the two immiscible phases are both water-based, an aqueous two phase system. Aqueous two-phase extractions can have a number of advantages over traditional solvent extraction. Solvents are often destructive to proteins, making the traditional extraction impossible for purifying proteins. In addition, organic solvents can be flammable, and their use can cause both environmental and health concerns. Aqueous-two phase extractions do not require solvents, and so avoid these concerns.

Types of Aqueous Two-phase Extractions

Polymer–polymer systems. In a Polymer–polymer system, both phases are generated by a dissolved polymer. The heavy phase will generally be Polyethylene glycol (PEG), and the light phase is generally a polysaccharide. Traditionally, the polymer used is dextran. However, dextran is relatively expensive, and research has been exploring using less expensive polysaccharides to generate the light phase. If the target compound being separated is a protein or enzyme, it is possible to incorporate a ligand to the target into one of the polymer phases. This improves the target's affinity to that phase, and improves its ability to partition from one phase into the other. This, as well as the absence of solvents or other denaturing agents, makes polymer–polymer extractions an attractive option for purifying proteins. The two phases of a polymer–polymer system often have very similar densities, and very low surface tension between them. Because of this, demixing a polymer–polymer system is often much more difficult than demixing a solvent extraction. Methods to improve the demixing include centrifugation, and application of an electric field.

Polymer–salt systems. Aqueous two-phase systems can also be generated by introducing a high concentration of salt to a polymer solution. The polymer phase used is generally still PEG. Generally, a kosmotropic salt, such as Na_3PO_4 is used, however PEG–NaCl systems have been documented when the salt concentration is high enough. Since polymer–salt systems demix readily they are easier to use. However, at high salt concentrations, proteins generally either denature, or precipitate from solution. Thus, polymer–salt systems are not as useful for purifying proteins. Ionic liquids systems. Ionic liquids are ionic compounds with low melting points. While they are not technically aqueous, recent research has experimented with using them in an extraction that does not use organic solvents.

Applications

DNA purification: The ability to purify DNA from a sample is important for many modern biotechnology processes. However, samples often contain nucleases that degrade the target DNA before it can be purified. It has been shown that DNA fragments will partition into the light phase of a polymer–salt separation system. If ligands known to bind and deactivate nucleases are incorporated into the polymer phase, the nucleases will then partition into the heavy phase and be deactivated. Thus, this polymer–salt system is a useful tool for purifying DNA from a sample while simultaneously protecting it from nucleases.

- Food industry: The PEG–NaCl system has been shown to be effective at partitioning small molecules, such as peptides and nucleic acids. These compounds are often flavorants or odorants. The system could then be used by the food industry to isolate or eliminate particular flavours.
- Analytical chemistry: Often there are chemical species present or necessary at one stage of sample processing that will interfere with the analysis. For example, some air monitoring is performed by drawing air through a small glass tube filled with sorbent particles that have been coated with a chemical to stabilize or derivatize the analyte of interest. The coating may be of such a concentration or characteristics that it would damage the instrumentation or interfere with the analysis. If the sample can be extracted from the sorbent using a non-polar solvent (such as toluene or carbon disulfide), and the coating is polar (such as HBr or phosphoric acid) the dissolved coating will partition into the aqueous phase. Clearly the reverse is true as well, using polar extraction solvent and a non-polar solvent to partition a non-polar interferent. A small aliquat of the organic phase (or in the latter case, (polar phase) can then be injected into the instrument for analysis.
- Purification of amines: Amines (analogously to ammonia) have a lone pair of electrons on the nitrogen atom that can for a relatively weak bond to a hydrogen atom. It is therefore the case that under acidic conditions amines are typically protonated, carrying a positive charge and under basic conditions they are typically deprotonated and neutral. Amines of sufficiently low molecular weight are rather polar and can form hydrogen bonds with water and therefore will readily dissolve in aqueous solutions. Deprotonated amines on the other hand, are neutral and have *greasy*, non-polar organic substituents, and therefore have a higher affinity for non-polar inorganic solvents. As such

purification steps can be carried out where an aqueous solution of an amine is neutralized with a base such as sodium hydroxide, then shaken in a separatory funnelwith a non-polar solvent that is immiscible with water. The organic phase is then drained off. Subsequent processing can recover the amine by techniques such as recrystallization, evaporation or distillation; subsequent extraction back to a polar phase can be performed by adding HCl and shaking again in a separatory funnel (at which point the ammonium ion could be recovered by adding an insoluble counterion), or in either phase, reactions could be performed as part of a chemical synthesis.

Kinetics of Extraction

It is important to investigate the rate at which the solute is transferred between the two phases, in some cases by an alteration of the contact time it is possible to alter the selectivity of the extraction. For instance, the extraction of palladium or nickel can be very slow because the rate of ligand exchange at these metal centres is much lower than the rates for iron or silver complexes.

Aqueous Complexing Agents

If a complexing agent is present in the aqueous phase then it can lower the distribution ratio. For instance, in the case of iodine being distributed between water and an inert organic solvent such as carbon tetrachloride then the presence of iodide in the aqueous phase can alter the extraction chemistry.

Instead of D_{I+2} being a constant it becomes $= k[[I_{2 \cdot Organic}]]/[I_{2 \cdot Aqueous}]$ $[[I^{-}{}_{\cdot Aqueous}]]$ This is because the iodine reacts with the iodide to form I_3^-. The I_3^- anion is an example of a polyhalide anion that is quite common.

Industrial Process Design

In a typical scenario, an industrial process will use an extraction step in which solutes are transferred from the aqueous phase to the organic phase; this is often followed by a scrubbing stage in which unwanted solutes are removed from the organic phase, then a stripping stage in which the wanted solutes are removed from the organic phase. The organic phase may then be treated to make it ready for use again.

After use, the organic phase may be subjected to a cleaning step to remove any degradation products; for instance, in PUREX plants, the used organic phase is washed withsodium carbonate solution to remove any dibutyl hydrogen phosphate or butyl dihydrogen phosphate that might be present.

11

PH and Buffers

WATER, PH, AND NON-COVALENT BONDING

WATER

Water is essential for life. It covers 2/3 of the earth's surface and every living thing is dependent upon it. The human body is comprised of over 70 per cent water, and it is a major component of many bodily fluids including blood, urine, and saliva. What accounts for the ubiquitous use of water in living systems? If we take a step back and consider the structure of water and compare it to another substance, methane, we can understand the unique properties water possesses that make it well suited for biological systems.

Water (H_2O) is made up of 2 hydrogen atoms and one oxygen atom, with a total atomic weight of 18 daltons. The structure of the electrons surrounding water is tetrahedral, resembling a pyramid. For comparison, Methane (CH_4) is made up of one carbon and 4 hydrogens. Note that methane is similar to water in that it weighs 16 daltons and also has a tetrahedral structure, yet has very different physical properties.

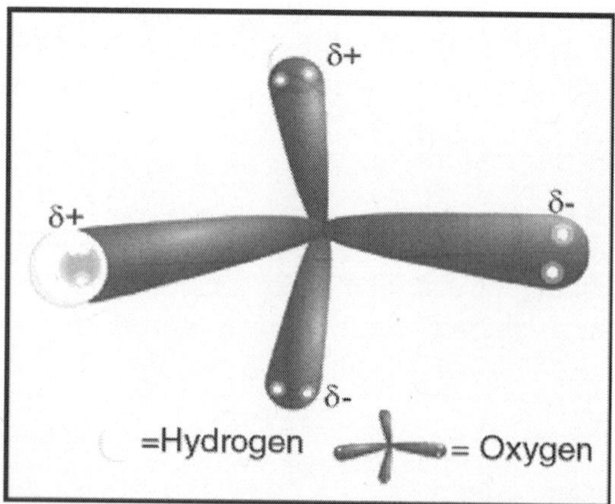

Fig. Molecular Geometry of Water: The Orbitals are Pyramidal, Yet the Atoms form a Bent Shape.

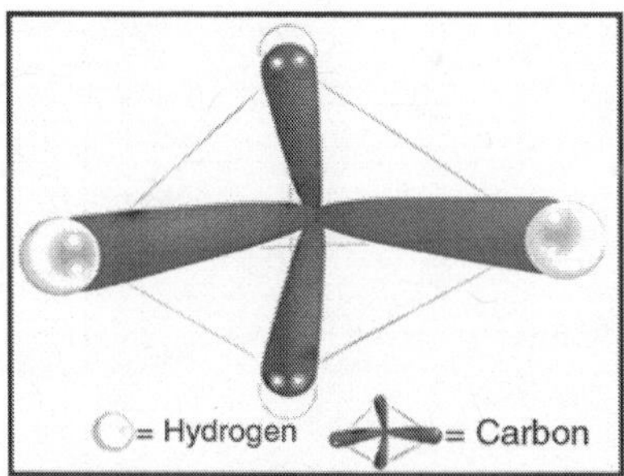

Fig. Molecular Geometry of Methane: Pyramidal

All the vertices on the methane that form the points of a pyramid are occupied by a hydrogen. However, only two vertices in water are occupied with hydrogens. The other two vertices are each occupied by a lone pair of electrons. This fact causes the water molecule to have a bent molecular shape. Oxygen is highly electrophilic (electron loving).

This means that even though the oxygen in water is bound to each of the hydrogens by a covalent bond (sharing a pair of electrons), the oxygen "pulls" the shared electrons closer to itself. This unequal sharing of the electrons in the O-H bond in water causes the hydrogens to have a partial positive charge (positive dipole), and the oxygen has a partial negative charge (negative dipole). Water is called a polar molecule because it has a positive side and a negative side, called a dipole moment. In contrast, the carbon in methane shares electrons equally with the 4 hydrogens. Methane's does not have a dipole moment, and in contrast with water is non-polar.

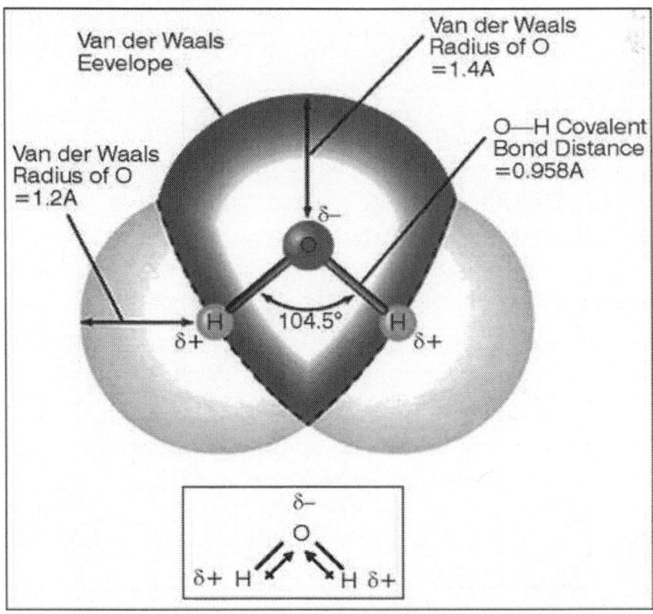

Fig. The "Magnet-like" Dipole Moment in Water

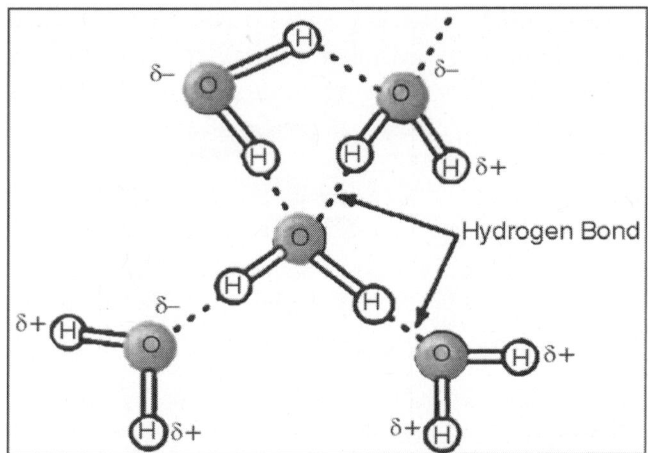

Water molecules, then, have both a partial positive and a partial negative charge. Since opposites attract, this means that water molecules are attracted to each other, like the positive and negative ends of a magnet. Because of these attractive forces, water molecules are in close proximity to one another, making water very dense. Methane molecules do not have a dipole attraction for one another, and therefore and are spaced farther apart.

Therefore, despite its similar size and mass to water, methane is much less dense than water. This is the reason that under room temperature situations, water exists as a liquid while methane is a gas. We know that water can be converted to its gaseous phase, steam, but only by applying a lot of energy in the form of heat to disrupt the large attraction of the water molecules for one another.

Hydrogen Bonding

Water molecules are bound together through hydrogen bonds. Hydrogen bonds arise when a hydrogen that is covalently bound to either an oxygen, sulfur, or nitrogen atom nears an electrophilic atom, such as oxygen. The electrophilic atom "pulls" the hydrogen closer to itself. The end result is that the hydrogen is now shared (unequally) between the atom to which it is covalently bound and the electrophilic atom to which it is attracted (O-H...O).

Since water is made up solely of hydrogens covalently bound to oxygens and electrophilic oxygens, it is easy to see why there are so many hydrogen bonds in water.

The molecules in water are in constant motion and the hydrogen bonds are constantly being broken and reformed. In fact, the average hydrogen bond lasts only a one-trillionth of a second (10^{-12} s)! Each hydrogen bond has an average energy of 20 kJ/mol. This is much less than an O-H covalent bond, which is 460 kJ/mol. Even though an individual hydrogen bond is relatively weak, the large number of hydrogen bonds that exist in water which pull the molecules together give water its special density and phase transition properties. This is much like the story of Gulliver's Travels, when Gulliver awoke on the island of Lilliput to find himself tied down with many tiny strings. Each string was individually weak, but together they were enough to bind Gulliver.

On average, a liquid water molecule will have 3 out of 4 possible hydrogen bonds. This is because the molecules in liquid are in constant motion. In ice, water will form all 4 possible hydrogen bonds because the molecules in a solid are essentially locked in place. Water molecules are further apart from one another in ice than they are in the liquid state. Ice is therefore not as dense as liquid water.

This is why icebergs float in the ocean and why ice floats in a beverage. This contrasts greatly from the normal relationship between a compound's solid and liquid forms, with the solid form usually being of a higher density. While the unusual phase density properties of water might not seem too important in biological reactions, it is very important for life on earth. If ice did not float, then in very cold bodies of water ice would sink, removing its insulating effects from the water's surface.

Lakes, rivers, and major portions of the oceans thus would freeze completely from the bottom up. The world would be a much different place, and much less suited for life.

Water is a Good Solvent

The polar nature of water, with its partial positive and partial negative dipole, allows it to dissolve charged molecules (ions) easily. Water is thus an excellent solvent for charged compounds. The positive side of water surrounds negatively charged molecules, and the negatively charged side of water surrounds positively charged molecules.

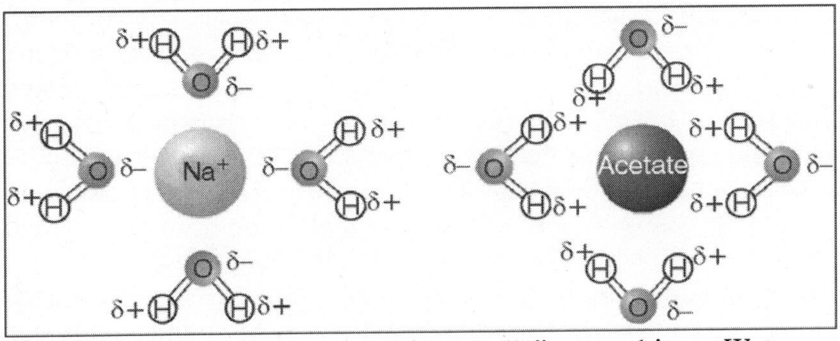

In this way, water makes "solvation shells" around ions. Water can also readily dissolve other polar molecules, even if they are not positively or negatively charged. The solvent properties of water allow for dissolved metals and buffering systems that are very important for the workhorses of life, enzymes. However, the saying "oil and water don't mix" is true—water cannot dissolve oil. This is because oily substances are non-polar. Non-polar substances (which lack dipoles) are also called *hydrophobic* (water fearing). Hydrophobic substances gather together to exclude water as best they can. This is why you see oil droplets in water. This is also important for the stability and structure of enzymes.

NON-COVALENT BONDING

Non-covalent bonds are not at strong as covalent bonds, but they are important in the stabilization of molecules. In contrast to covalent bonds, non-covalent bonds do not share electrons.

Noncovalent bonds include:

- Electrostatic interactions
- van der Waals forces
- Hydrophobic interactions
- Hydrogen bonds

Electrostatic interactions are formed between positive and negative ions. The bond is non-directional, meaning that the pull of the electrons does not favour one atom over another.

An example is NaCl, which is formed between the positively charged Na^+ ion and the negatively charged Cl^- ion. The average strength of an electrostatic interaction is 15 kilojoules/mol (kJ/mol). The bond strength lessens when the distance between the two ions increases.

Van der Waals forces are weak forces between temporary dipoles. These forces may be attractive or repulsive. They are also non-directional. The average bond energy for van der Waals forces is on the order of 10 kJ/mol.

Hydrophobic interactions result when non-polar molecules are in a polar solvent, *e.g.* H_2O. The non-polar molecules group together to exclude water (hydrophobic means water fearing). By doing so they minimize the surface area in contact with the polar solvent.

A hydrogen bond results when a hydrogen atom that is covalently bound to an electronegative atom (*e.g.* O, N, S) is shared with another electronegative atom. A hydrogen bond is directional towards the electronegative atom. An example of this is the hydrogen bonds formed in water. Hydrogen bonds are constantly being made and remade. The half-life of a hydrogen bond is about 10 seconds. A hydrogen bond has an energy of 21 kJ/mol.

As stated above, non-covalent bonds are not as strong as covalent bonds, but the additive effect of many non-covalent bonds can stabilize a molecule. Non-covalent bonds are very important in the structure of proteins.

Bond type	Energy (kJ/mol)
Covalent, *e.g.* C-C	350
Electrostatic	15
van der Waal's	10
Hydrogen	21

TEST AND PROBLEMS

pH Questions

Below are a few sample questions to elucidate the concepts discussed in the book.

- A simple pH problem
- pH of a weak acid
- A buffer problem
- A more involved buffer problem

A Simple pH Problem

What is the pH 0.1 M HCl?

Answer: The pH of a solution is the negative logarithm of the concentration of H^+ ions. In this case, HCl is a very strong acid. HCl completely dissociates to H^+ and Cl^-.

$$HCl \rightarrow H^+ + Cl^-$$

The concentration of HCl is 0.1 M, if all of it dissociates to H^+, then

$$[H^+] = 0.1$$

$$-\log(0.1) = 1.26$$

$$pH = 1.26$$

Remember that a pH of 7 is neutral, and that the pH scale is logarithmic, meaning each pH unit represents a factor of 10. This HCl solution has a difference of 5.74 pH units from water (7 - 1.26). Taking the inverse log of 5.74 means that this HCl solution is more than 500,000 times as acidic as water.

pH of a Weak Acid Solution

What is the pH of a 0.05 M solution of acetic acid? (The K_a of acetic acid $= 1.8 \times 10^{-5}$)

Answer: Determining the pH of a weak acid solution is more complicated than for a strong acid. This is because a weak acid does not completely dissociate.

$$CH_3COOH \rightleftharpoons H^+ + CH_3COO^-$$

To simplify matters, it is helpful to construct a grid:

Reaction	CH_3COOH	H^+	CH_3COO^-
Initial	0.05 M	0 M	0 M
Change	(–x) M	+ x M	+x M
Equilibrium	(0.05–x) M	x M	x M

To determine the $[H^+]$, we must employ the K_a of acetic acid, which is 1.8×10^{-5}.

$$Ka = \frac{[H+][CH_3COO-]}{[CH_3COOH]} = 1.8 \times 10-5$$

Substituting the equilibrium values in the table above, we get:

$$1.8 \times 10{-}5 = \frac{(x\ M)(x\ M)}{(0.05 - x)\ M}$$

Since acetic acid is a weak acid, we can assume that very little of it will dissociate. *This simplifies the above equation to:*

$$1.8 \times 10^{-5} = \frac{(x\ M)(x\ M)}{0.05}$$

This can be easily solved:
9×10^{-7} M = x^2 M
x = 9.5×10^{-4} M
x = 0.00095 M
Now that we have the [H+], we can solve for pH:
pH = –log[0.00095]
pH = 3

Buffer Problem

Question 1: What is the ratio of [lactate]/[lactic acid] at pH = 5? (The pK_a of lactic acid = 3.85)

Answer: To solve this problem, we need to utilize the Henderson-Hasselbalch equation:

$$pH = pKa + \log \frac{[H+][A-]}{[HA]}$$

Table in the book lists the pKa of lactic acid as 3.85. Now just substitute into the equation:

$$5 = 3.85 + \log \frac{[\text{lactate}]}{[\text{lactic acid}]}$$

$$1.15 = \log \frac{[\text{lactate}]}{[\text{lactic acid}]}$$

$$14 = \frac{[\text{lactate}]}{[\text{lactic acid}]}$$

There is 14 times as much lactate as lactic acid at pH 5.

Question 2: What happens if 2 ml of 100 mM HCl is added to 100 ml of a 10 mM solution of phosphoric acid at pH = 7?

Answer: This is a more involved problem. The first thing to Realise is that phosphoric acid is polyprotic, meaning it has more than one H+ to donate, and therefore more than one pKa.

$$\begin{array}{ccccccc} & pK_a1 & & pK_a2 & & pK_a3 & \\ H_3PO_4 & \rightleftharpoons & H_2PO_4^- + H^+ & \rightleftharpoons & HPO_4^{-2} + H^+ & \rightleftharpoons & PO_4^{-3} + H^+ \end{array}$$

Since this problem is at pH = 7, it is the pKa2 value that is important, for this is the pKa nearest the pH.

First we need to find the ratio of,

$$[HPO_4^{-2}]/[H_2PO_4-].$$

Again, use the Henderson-Hasselbalch equation:

$$7 = 7.2 + \log[HPO_4^{-2}]/[H_2PO_4^-]$$

$$0.63 = [HPO_4^{-2}]/[H_2PO^{4-}]$$

$$0.63[H_2PO^{4-}] = [HPO_4^{-2}]$$

We know we have a total of: (0.01 moles/L) × (0.1 L) = 0.001 moles phosphoric acid that at pH = 7 is made up of HPO_4^{-2} and H_2PO_{4-}. (Note 10 mM = 0.01 moles/L and 100 mL = 0.1 L)

$$0.001 \text{ moles} = 1 \text{ mmole} = HPO_4^{-2} + H_2PO^{4-}$$

We know from above that $[HPO_4^{-2}] = 0.63[H_2PO^{4-}]$

$$\text{so } 1 \text{ mmole} = 0.63[H_2PO_4-] + [H_2PO_4-] = 1.63[H_2PO_4-]$$

$$[H_2PO_4-] = 0.6 \text{ mmoles}$$

and so then $[HPO_4^{-2}] = 0.4$ mmoles

So far, this problem is much like the previous one. Now let's answer the question about the addition of 1 ml of 0.1 M HCl. From the high pK_a2 value, we know that phosphoric acid is a weak acid.

It will react completely with the HCl (which is the same as H+ in solution):

$$\begin{array}{c} pK_a3 \\ HPO_4^{2-} + H^+ \rightleftharpoons H_2PO_4- \end{array}$$

By adding 2 ml of 0.1 M HCl, we are adding:

$$(1 \times 10^{-3} \text{ L}) \times (100 \text{ mmoles/L}) = 0.2 \text{ mmoles } H^+$$

If we set up the grid as shown above,

Reaction	HPO4-2	H+	H2PO4-
Initial	0.6 mmoles	0.2 mmoles	0.4 M
Change	–0.2 mmoles	–0.2 mmoles	+ 0.2 M
Change after equilibrium	0.4 mmoles	0 mmoles	0.6 M

What is the pH of the phosphoric acid solution after the addition of HCl? Again, use the Henderson-Hasselbalch equation to solve,

$$pH = 7.2 + \log(0.6/0.4)$$

$$pH = 7.37$$

Note that this is not a big change. This is because of phosphoric acid is a buffer near its pKa and as such is able to resist changes in pH.

PH - POTENTIAL OF HYDROGEN DESCRIPTION

WHAT IS HYDROGEN?

Hydrogen (H) is the lightest and simplest element. A Hydrogen atom is made up of one proton and one electron. At room temperature, Hydrogen is gas - as it cools it becomes liquid. Hydrogen is one of 9 essential macronutrients and is a major component of organic molecules. Hydrogen supplies building blocks of all organisms whether plant, animal or human. Each water molecule contains two Hydrogen molecules and one Oxygen molecules (H20).

Ideal pH Values

When pH = 8: There are 6 billion H+ ions in solution.
When pH = 7: There are 60 billion H+ ions in solution.
When pH = 6: There is 600 billion H+ ions in solution.

Ideal pH Values Of Blood, Urine and Saliva

- Arterial Blood – 7.40 – 7.45
- Capillary – 7.35 – 7.45
- Venous – 7.30 – 7.35
- Saliva – 6.5
- Urine – 6.8

pH Values Of Body

Saliva – (6.0 – 7.4)	Skeletal Muscle - (6.9 - 7.2)	
Gastric Secretion - (1.0 - 3.5)	Heart - (7.0 - 7.4)	
Pancreatic Secretion - (8.0 - 8.3)	Liver - (7.2)	
Bile - (7.8)	Brain (7.1)	
Small Intestine Secretion - (7.5 - 8.0)		
Urine - (4.5 - 8.0)	Arterial Blood - (7.4 - 7.45)	
Capillary Blood - (7.35 - 7.4)	Venous Blood - (7.3 - 7.35)	Feces -
(4.6-8.4)		

CLINICAL ACIDOSIS AND ALKALOSIS

The blood pH is supposed to remain "practically constant" according to our best text books. The physician who starts to check his patients, however, cannot occur. Instead of the normal figure of pH 7.3 to 7.5 he finds it varying from 7.2 to 7.8 (1).

The measurement of pH is attended with difficulty because the blood loses carbon dioxide (CO_2) and thereby rises the pH rapidly after it has been withdrawn. In two minutes a rise of.2 may occur (1). pH metres are available where the electrodes are built into a syringe so the pH may be read directly before the blood is exposed to air. (National Technical Labouratories, Pasadena, California).

The importance of clinical pH testing was illustrated by the following report of a pharmacist. He had been supplying for a number of months a capsule formula of aspirin, caffeine, phenacetin and ammonium chloride to an arthritic patent where the remedy had been successful in controlling arthritic pains. The capsule dispensed was a standard formulas of a pharmaceutical house, with a specific catalog number. The patient returned two days after receiving a renewed supply of capsules, complaining that they were ineffective, and different from the previously used kind. On checking the records, it was found that the original formulas had been revised, and the supposedly unnecessary ammonium chloride had been omitted. On supplying capsules of ammonium chloride alone, the patient reported the usual successful results.

What was the biochemistry behind this: Simply that a high pH tends to throw calcium out of solution in the body fluids, and bursitis, arthritis, neuritis, lumbago, sciatica, and a host of other painful syndromes develop as a consequence. Cheilosis (lips redden, fissures at the angles) may occur and herpes simplex, and other virus types of diseases become active. The low calcium bicarbonate in the body fluids seems responsible, Vitamin F and Vitamin D as well as the Vitamin C complex are also factors, their deficiency aggravates the situation. Allergies become acute, the calcium deficiency aspect of allergic sensitivity is well known. Normal rates of wound healing are greatly reduced, ulcers then too become static.*

Ammonium chloride is a natural constituent of gastric juice, calcium chloride is the form of calcium found in arrowroot starch, so both of these materials are physiological therapeutic agents, factors common to body biochemistry and not new and foreign substances to the physiological economy (as are synthetics and antibiotics).

It is described that Ammonium Chloride is useful in the treatment of bronchial affections, hepatic congestion, pelvic cellulitis, muscular rheumatism, gout, sciatica, chronic glandular enlargement, hemicrania, senile gangrene, dysmenorrhea, leucorrhea.

Calcium Chloride is listed as useful in the treatment of hemorrhage, hemorrhagic endometritis, menorrhagia, erytherma nodosum, tetanus, spasmophillia, blackwater fever, hay fever, asthma, hemophilia, albuminuria, nephritis, typhoid, coryza, tuberculosis, osteomalacia, scrofula, rachitis, arthritis, spasm of glottis, infant convulsions, urticaria, pruritis, among others.

All of these effects are no doubt in the main simply accomplished by correcting the unbalanced state of the buffers and mineral salts in the tissue fluids. It will be noted that Vitamin C complex is a desirable synergist in most of the conditions listed. Gout is mentioned as one situation where ammonium chloride may help. Recently it has been discovered that the juice of cherries contains some active agent effective in gout, 4 to 6 ounces of the juice (preferably unsweetened, but may be canned or cooked form) promoting relief with in a few days with consistent improvement.

This is the first time that the possibility of gout being a deficiency disease has been suspected. We can say that the ingestion of sodium compounds particularly sodium phosphate - is definitely aggravative. Gout patients should see that their diet contains a preponderance of potassium. Sodium promotes the precipitation of gouty concretions of sodium urate in the tissues.

A very important aspect of alkalosis is the fact the habitual use of milk of magnesia promotes the condition. It also tends to cause epistaxis (nosebleed - can be secondary to nasal and sinus issues; scarlet fever or typhoid; arteriosclerosis; hypertension and anemias), very specifically. You will be surprised how many of the patients who ask for a remedy for nosebleed are using milk of magnesia as a laxative.

Citrus fruits being of a highly alkaline ash, with a high content of organic acid (citric), need special attention. The first and immediate effect of ingestion of a few ounces of grapefruit or lemon juice is to lower (acidity) the blood pH. The patient with alkalosis feels temporarily better. Later in the day, after the citric acid has been destroyed by oxidation as a fuel (it is classed as carbohydrate) the aggravation of the alkaline state become apparent. The temporary effect of the citric acid is to cause calcium to be picked up - no doubt from some bone reserves - and after the pH change this calcium is deposited elsewhere - in physiologically undesirable spots. The treatment for bursitis obviously becomes simple. No longer can we recommend X ray treatment with equanimity. Let the physiological treatment be the preferred method.

Guanidine, a fatigue and tissue poison, and end product of the breakdown of creatine (combated by Vitamin E complex which stops the loss of creatine from the tissues), is the most potent organic alkaline substance know. It specifically precipitates calcium (recall our comment that it is diffusion from tired heart muscle into coronary vessels is the cause of precipitation therein of calcium), is normally reconverted into creatine by the influence of the parathyroid hormones, and with the assistance of the thyroid (2). Blood guanidine levels rise eight fold after parathyroidectomy (3).*

We can see where the guanidine effect as a fatigue poison is contributory to the arthritis, sciatica, etc., the patient often tell us how fatigue aggravates his state of misfortune. Local inflammation also can release guanidine, its presence insure a spastic state of blood vessels, contributing to gangrene, of necessity.

There seems no doubt that the effect of oxygen metabolizing vitamins - the Vitamin C and Vitamin E complexes - is vital in combating these morbid reactions by preventing the degradation of tissue elements into guanidine. Where Vitamin E deficiency has caused the tissue demand for oxygen to rise up to 250 per cent, of normal (4), the augmented release of end products certainly includes guanidine, and we see here how the find of Dr. Shute of London, Ontario, that Vitamin E reduces capillary hemorrhage may be rationalized.*

A vicious cycle is set up, the more oxygen demand the less supplied by reason of the constrictive effect of guanidine on blood vessels. Degeneration, rupture of capillaries is inevitable. As Vitamin E promotes more oxygen supply, that is why it also is good in gangrene (especially the diabetic type) and in "capillary fragility". (The true antifragility vitamin is Vitamin P, which provides a special calcium to promote collagen formation.)

Vitamin F opposes the toxic effect of guanidine by providing more of the diffusible calcium (from the colloidal blood reserves) that is precipitated out of the body fluids by guanidine, thereby ameliorating the spastic, nervous and irritative (allergic) reactions.

The thyroid hormone physiologically promotes the resorption and dissolution of protein structures that have reached the end of their physiological cycle (the theory of the dynamic state of living tissue, replaced and rebuilt at specific time intervals). That is why in children, thyroid deficiency is the cause of delayed development. In the adult, a hyperactive thyroid may again cause toxicosis by promoting more tissue poisons than the eliminative system can tolerate. In old people, this may be a critical situation, thyroid sometimes becoming a violent poison.

Only because the normal synergists are not available to maintain the desirable complete cycle of activity. We can see that Vitamin F is vital to the prevention of this toxicosis. It tells why the thyroid puts its secretion into the blood after ingestion of Vitamin F to the extent of a doubling of the iodine content of the blood (5). The normal activity of the gland had been blocked by reason of Vitamin F deficiency. In this blocking, we have apparently the explanation for prostate hypertrophy. Once the thyroid is permitted to secrete, its hormone makes quick work of the fibrous tissue collected in the prostate. This also explains why we consider the Vitamin F complex, and Vitamin F2 the most friendly vitamins for the older person. The Vitamin F2 is a more highly developed compound of the basic F (a sensitized fatty acid), in which this fatty acid is combined into a phospholipid molecule that appears to have a specific function of catalyzing protective (Insulating) layers in nerve tissue and cell nuclear structure. Thereby protecting the basic controls of metabolic activity. Loss of appetite and wasting disease suggests the possible need for this vitamin. Appetite restoration is immediate where the deficiency exists (In adults and children alike.)

There seems to be as much clinical acidosis as alkalosis (1). The effects of acidosis may be aggravated by a diabetic state where organic acids cannot be oxidized normally, instead must be combined with reserve alkaline salts and excreted (5).

Where sodium bicarbonate is administered to relieve diabetic acidosis, the urine should be watched and dosage stopped when a neutral is reached, according to this authority. Normally, blood glutamine provides ammonia for this purpose. Beet root and leaf are best vegetable sources of glutamine, is very high if grown on high ammonia content soils. (Plant cells form glutamine to dispose of excess ammonia).

When abnormal blood pH is found, it must be considered only as a sign that something is wrong. The symptoms may be entirely different with the same abnormal pH figure in different patients, in view of the fact that there is an infinite number of combination of chemical situations that could alter pH - as many as there are different acids and alkalies. Just as a plump patient may be bloated with gas, water logged, fat, or hyper muscular. It is a preliminary classification, not a final.

The pH of the blood is a resultant of the forces acting upon it. The regulation of the blood CO_2 (carbon dioxide) by respiration control Centre of the brain is a major factor. The parathyroid by eliminating the alkaline guanidine, and the kidney, eliminating mineral salt (both acid and alkaline as may be required) no doubt exercises a supplementary control.

The clinical value of vinegar is more or less well known as an alternative of merit. Creatine is methylguanidine acetic acid, and is often craved by the patient, very logically it would seem.

Cider vinegar being the preferred form, no doubt the malic acid of the apple is a factor, calcium malate being a catalyzer of polyphenol detoxification.

ACIDOSIS AND ALKALOSIS INFORMATION

Regulatory Factors

The following factors tend to promote normalization of pH through physiological mechanisms. They do not supply mineral elements which directly influence body chemistry.

To directly influence pH values through acid or alkaline mineral administration, see: Alkalosis or Acidosis. Normalization of the function of the glands and kidneys which regulates pH balance should be the primary consideration. However, in most cases specific support of either acid or alkaline mineral therapy is necessary where immediate results are to be obtained, particularly in acute situations.

Primary

- *Renatrophin PMG*: Kidneys are thought to be the most important organ in acid-base balance regulation.
- *Drenatrophin PMG*: Adrenal support regulation of sodium-potassium-chloride balance.
- *Thytrophin PMG*: Thyroid supports regulation of calcium-magnesium-phosphorus balance.

Normally the endocrine glands, along with the kidneys regulate the pH of the blood, just as the pancreas with it insulin regulated the blood sugar levels.

The sex glands, along with the adrenals, seem to be the main endocrines involved. Particularly after menopause or male climacteric there may be serious changes in the body's economy whereby acid-base disorders develop, which may be aggravated, if not caused by dietary circumstance - just as diabetes is aggravated by a high sugar intake.

BUFFERS

As shown above, the pH of a solution is dependent on the concentration of H^+ ions. Addition or removal of H^+ ions, then, can greatly affect the pH of a solution.

In the body, the pH of cells and extracellular fluids can vary from pH 8 in pancreatic fluid to pH 1 in stomach acids. The average pH of blood is 7.4, and of cells is 7 – 7.3. Although there is great variation in pH between the fluids in the body, there is little variation within each system. For example, blood pH only varies between 7.35 – 7.45 in a healthy individual.

Large changes in pH can be life threatening. How does the body maintain a constant blood pH? The body uses a buffer system to withstand changes in pH.

Buffers are made up of a mixture of a weak acid with its conjugate base or a weak base with its conjugate acid. Remember that an acid donates a H^+. A weak acid does not donate its H^+ as easily. Similarly, a weak base will not accept a H^+ as well as a strong base.

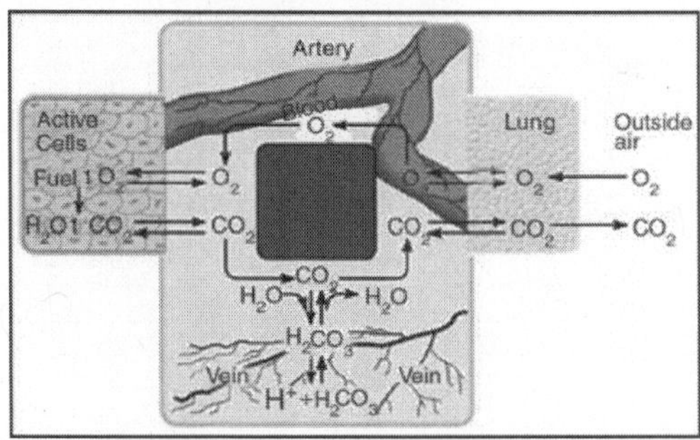

Fig. The Bicarbonate Blood Buffering System

Buffers maintain pH by binding H^+ or OH^- ions. This stabilizes changes in pH. The bicarbonate buffer system maintains blood pH near pH 7.4. The carbonic acid, H_2CO_3, in the blood is in equilibrium with the carbon dioxide (CO_2), in the air. Buffers are most effective in a pH range near its pK_a. This is where the titration curve is most shallow, and where the pH is least affected by added acid or base. For example, in the titration curve of phosphoric acid (another blood buffering system):

$$H_3PO_4 \underset{}{\overset{pK_a1}{\rightleftharpoons}} H_2PO_4^- + H^+ \overset{pK_a2}{\rightleftharpoons} HPO_4^{-2} + H^+ \overset{pK_a3}{\rightleftharpoons} PO_4^{-3} + H^+$$

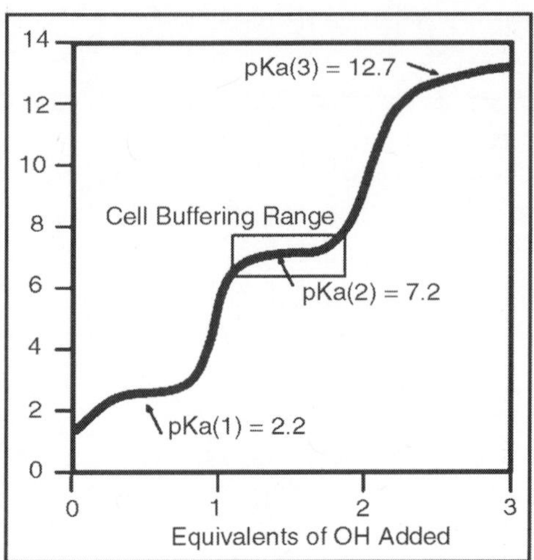

Fig. The Titration of a Polyprotic Weak Acid

The region in the rectangle is the buffering region of biological importance. Note that the slope of the curve is shallow within one pH unit of the pK_a value. This is where the buffering range is most effective, meaning that the buffer is able to resist changes in pH.

In chemistry, the degrees of acidity or alkalinity of a substance are expressed in pH values. The neutral point, where a solution would be neither acid nor alkaline is pH 7. Increasing acidity is expressed as a number less than 7, and increasing alkalinity is a number great than 7. Maximum acidity is pH 0 and maximum alkalinity is pH 14. Because each unit on the scale represents a logarithm, there is 10-fold difference between each unit (pH 5 is 10 times as acid as pH 6 and pH 4 is 100 times as acid as pH 6. Expressed mathematically, pH is logarithm of the hydrogen ion concentration divided into one - Log 1/H+

An acid is a substance that releases hydrogen into a solution. An alkaline substance is one that removes hydrogen from a solution. Most body functions occur at only certain levels of acidity or alkalinity, as well as, many enzymes and chemical reactions work best at a specific pH. A minute change in pH can have a profound effect on body function.*

Some of the pH values of a number of biological materials are given below:

Table A - pH Values of Representative Biological Materials

Material	pH value
Blood, normal limits	7.3 - 7.5
Blood, extreme limits	7.0 - 7.8
Enzymes, Activity Range of	
Amylopsin, optimum	7.0
Erepsin, optimum	7.8
Invertase, optimum	5.5
Lipase, optimum	7.0 - 8.0

Maltase, optimum	6.1 - 6.8
Pepsin, optimum	1.5 - 2.4
Trypsin, optimum	8.0 - 9.0
Fruit Juices	
Apple	3.8
Banana	4.6
Grapefruit	3.0 - 3.3
Orange	3.1 -4.1
Tomato	4.2
Gastric juice (adult)	0.9 - 1.6
Milk (cows)	6.2 - 7.3
Plants (Extracted Juice)	
Alfalfa tops	5.9
Carrot	5.2
Cucumber	5.2
Peas, field	6.8
Potato	6.1
Rhubarb, stalks	3.4
String beans	5.2
Sweat	4.5 - 7.1
Saliva	6.2 - 7.6
Urine (human)	4.2 - 8.0
Tears	7.2

STABILIZING pH OF AQUEOUS SOLUTIONS

In example, the pH of the buffer solution changed by less than 0.5 units when the NaOH was added. Yet adding the same amount of NaOH to pure water resulted in a 5 unit pH shift in example.

Clearly, the conjugate acid/base pair was able to protect the solution from a large change in pH.

A buffer is a solution that resists a change in pH when acids or bases are added to it. How is this possible? Buffers work by acting a little bit like a sponge, soaking up excess H3O+ or OH– ions when they are added to a solution. The dissociation reaction for acetic acid shows that a solution of acetic acid contains acetate ions:

$$CH_3COOH + H_2O \rightleftharpoons CH_3COO- + H_3O^+$$

How do acetic acid and the acetate ion work to buffer the solution? If a strong acid, such as hydrochloric acid, is added to this buffer, the H3O+ ions generated will react with the acetate ions, removing them from the solution:

$$CH_3COO^- + H_3O+ \rightleftharpoons CH_3COOH + H_2O$$

acetate ion	hydronium ion	acetic acid	water

Similarly, if a strong base, generating lots of OH^- ions, were added to the acetic acid buffer solution.

$$CH_3COOH + HO^- \rightleftharpoons CH_3COO - H_2O$$

acetate acid | hydroxide ion | acetate acid | water

The OH- ions are removed via a reaction with acetic acid. It is important to remember that buffers cannot maintain their pH indefinitely as more acid or base is added. Imagine slowly pouring a bucketful of water onto a small sponge on the kitchen floor. At first, as the water pours out of the bucket it is absorbed by the sponge, keeping the kitchen floor dry.

But once the sponge is soaked through, it can hold no more and the water spills onto the floor as fast as if the sponge weren't there. In the same way, buffers will protect the pH of the solution to some extent, but if they are inundated by large amounts of H_3O+ or OH– ions, all the available conjugate acid or base molecules will have been used up and pH changes will rapidly occur. Thus, buffers only work as long as the amount of conjugate acid and base ions are large compared to the H_3O+amount of or OH– ions to be removed.

Remember from the Henderson-Hasselbalch equation that the numbers of conjugate acid molecules is equal to the number of conjugate base molecules when the pH = pKa:

$$pH = pK_a + \log([Base]/[Acid])$$

$$\text{if } [Base] = [Acid], \text{ then}$$

$$pH = pK_a + \log 1$$

$$pH = pK_a$$

Because the conjugate acid and conjugate base molecules are available in equal amounts when the pH = pKa, the buffer solution has the strongest ability to protect against pH changes caused by incoming H_3O^+ or OH-ions when the pH of the solution is close to the pK_a of the conjugate acid.

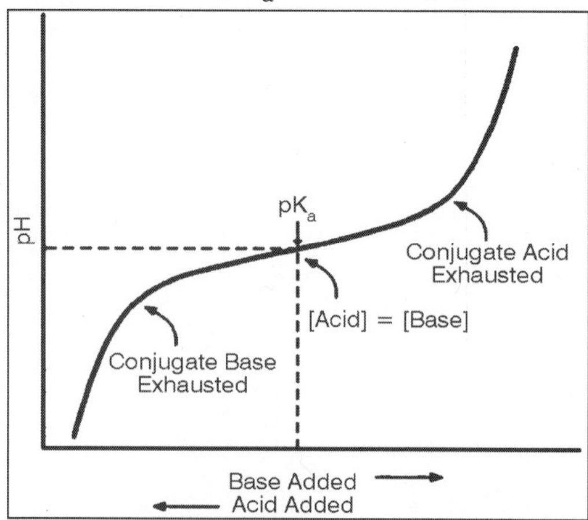

Fig. Buffer Titration: Within One Unit of the pKa, the Solution Resists Changes in pH.

If the pH of the solution strays too far from the pKa point, the buffer uses up all its available conjugate acid or base molecules, and the pH starts to fall dramatically. At this point, the buffering capacity of the buffer has been exceeded. this can be seen in the buffer titration graph below. Buffers protect the pH of a solution best within one pH unit of the pKa.

Buffers are important for biochemistry because the structures (and therefore the functions) of biological molecules are stable within a relatively narrow range of pH values.The evolution of life, which is believed to have begun in water, was likely due in part to the stable pH of seawater. Even today, the largest buffered systems in the world are the Earth's oceans. Carbon dioxide (CO2) from the Earth's atmosphere reacts with H2O to produce carbonic acid, H2CO3. Carbonic acid in turn reacts with water to form bicarbonate:

$$H_3CO_3 + H_2O \rightleftharpoons HCO_3 - H_3O^+$$

crabonic acid, water, bicarbonate ion, Hydronium ion

Furthermore, the bicarbonate ion can also react with water to form carbonate:

$$H_3CO_3^- + H_2O \rightleftharpoons HCO_3 - H_3O^+$$

bicarbonate ion, water, Carbonate ion, hydronium ion

The carbonate and bicarbonate ions act together as conjugate acid/base pair to keep the pH of the ocean at about 8.2. In the living organisms, proteins (the molecules which do most of the"work" of a cell) are very sensitive to acid concentration, because they only fold into their proper three-dimensional shapes within a small pH range. Some of the individual amino acid building blocks that make up a protein have ionizable side chains that change their ionization state, or charge, at different pH. Thus, if the pH changes, some of the attractive forces that hold the protein together will also change. The ionization states of the side chain of the amino acid lysine are shown below. The pKa of lysine's side chain amino group is approximately 10.8. Therefore, at a pH lower than the pKa, the side chain is in the protonated (conjugate acid) state; whereas at a pH higher than the pKa, the side chain is unprotonated (conjugate base).

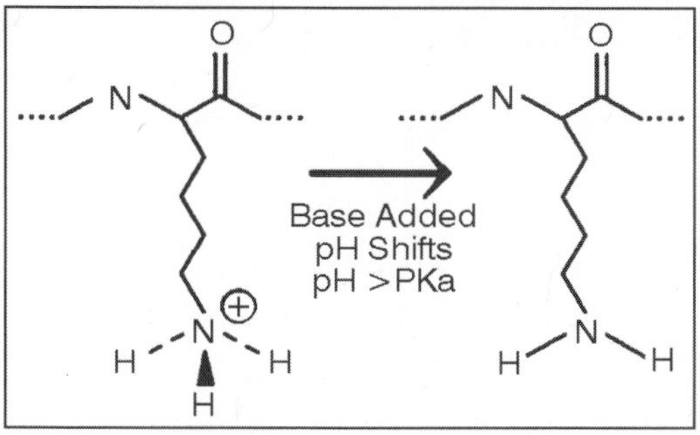

Fig. Amino Acid Charge changes.by Environment.

Proteins consist of long chains of amino acids, many of which include acidic or basic groups.

These groups are chaged at physiological pH, and form ionic interactions with each other that contribute to the stability of the folded protein. A change in pH of the protein's aqueous environment may alter the charge of some of these charged groups, disrupting the ionic interactions and the stability of the protein's structure.

Remember that the Henderson-Hasselbalch equation was used for calculating the pH of a solution containing a weak acid and its conjugate base. Since such a solution is a buffer, the Henderson-Hasselbalch equation is extremely useful for calculations involving buffers, allowing the following:

- Determination of the pH of buffered solutions.
- Determination of the ratio of conjugate base to conjugate acid at a given pH.

Example: A solution of lactic acid has twice as many conjugate acid molecules as conjugate base molecules. If the pKa of lactic acid is 3.85, what is the pH of such a solution?

***Answer*:**

We are given that the concentration of conjugate acid is twice the concentration of conjugate base.

Thus,

[acid] = 2 [base]

Substituting into the Henderson-Hasselbalch equation:

$pH = pK_a + \log([base]/[acid])$

$pH = 3.85 + \log([base]/2[base])$

$pH = 3.85 + \log(0.5)$

$pH = 3.85 + (-0.30)$

$pH = 3.55$

Question: If a buffer solution contains an acid as one component and a base as the other, can those two componenets be HCL and NaOH?

Answer: Your assumption is correct. A buffer is made of a weak acid and its conjugate base or a weak base and its conjugate acid. So neither HCl nor NaOH will be a component in a buffer; they are both strong. An example of a buffer is acetic acid (a weak acid) and sodium acetate (a conjugate base of acetic acid as a salt).

Incidentally, in the example you gave, HCl in water dissociates to hydronium ions and chloride ions. NaOH in water dissociates to hydroxide ions and sodium ions. When the two are mixed, the hydronium ions and the hydroxide ions combine to form water leaving behind the sodium and chloride ions. If you have the same amount and concentration of HCl as NaOH you will end up with salt water because sodium chloride is simply table salt.

BUFFERS AND IONIZATION OF AMINO ACIDS

- Buffers resist change. pH buffers resist change in pH when acid (H+) or base (OH–) is added.
- pH = –log [H+] –where bracket indicates molar concentration. If [H+] = 10^–7 M, then pH = 7. This is neutral pH since [H+] = [OH–] and it is middle of pH scale. Draw a pH scale by calculating the pH of 1 M HCl and 1 M NaOH. Use the equation Kw = 10^–14 = [H+] × [OH–], to find the pH of 1 M NaOH.
- pH of a buffered solution can be found by using the Henderson-Hasselbach equation:

$$pH = pK + \log([A-]/[HA])$$

All acids and bases have a conjugate acid (HA) and a conjugate base (A–). At [HA] = [A–], pH = pK (prove this with the above equation). At the pK, a buffer maximally resists change in pH and...the buffering zone is considered to be 1 pH unit above and below the pK.

CONCEPTS

Buffering zone = pK +/–1 pH unit:

- At pH = pK, [HA] = [A–].
- At pH below pK, [HA] > [A–].
- At pH above pK, [HA] <[A–].
- Amino acids (AAs) are more complex than simple buffers and...have at least two pK values because AAs have at least two ionizing groups:

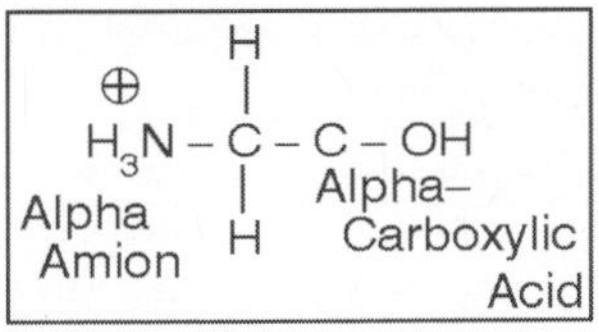

H
⊕ |
H_3N – C – C – OH
| Alpha–
Alpha H Carboxylic
Amion Acid

Glycine Structure

Example: Glycine has an alpha-amino group and alpha-carboxylic acid group. Each ionizable group has a pK (called pK–1 and pK–2). Glycine pK–1 = 2.3 (alpha-carboxylic acid); pK–2 = 9.6 (alpha amine).

Draw a titration curve for Glycine for practice. In this class, we treat all ionizing groups with in a molecule, as acting independently of other ionizing groups.

The pH on the titration curve where an AA has no net charge is called the pI or isoelectric point. The pI is calculated by averaging the two pK values on either side of the neutral form (ie form with no net charge).

All AAs must have at least 3 ionic forms:

- "AA+1","AA0", and"AA-1".
- Where"AA+1" = AA form with net charge of +1,

- "AA0" = neutral form with no net charge, and
- "AA-1" = AA form with net charge of –1.

While an AA may have other ionic forms, only these 3 forms count when finding pI:

- AA+1 goes to AA0 via pK–below.
- AA0 goes to AA-1 via pK–above.

Idea

pI = average of (pK-below) + (pK–above).
where pK-below is the pK between AA+1 and AA0.
and pK-above is the pK between AA0 and AA–1.

For more complex AAs, like Asp, Glu, Lys, Arg and His - with 3 ionizing groups, Other forms exist with charges other than +1, zero, and -1. For these AAs, the best approach is to start with the fully protonated form (ie. One with all the protons it can take on its ionizable groups).

Calculate the net charge on this group (it must be positively charged), then titrate the AA to remove the first proton and find net charge again. Continue doing this until all the protons have been removed. Keep track of this by making a simple model. Then, find the form with no net charge (ie. AA0) and use the pK values, which govern the transition from AA+1 to AA0 and AA0 to AA-1, as the ones to calculate the pI.

His has 3 ionizing groups, alpha-carboxylic acid (pK 1.8), side-chain amino (pK 6.0) and alpha-amino (pK 9.2):

- His+2 goes to His+1 via pK 1.8;
- His+1 goes to His0 via pK 6.0;
- His0 goes to His-1 via pK 9.2.

Therefore, pI = (6.0 + 9.2)/2 = 7.6:

- Some of you have trouble figuring out which group ionizes first. Here are some rules to apply to help you.
  - Carboxylic acids ionize at acidic pH; ie. Carboxylic acids give up their protons at acid pHs.
  - Amino groups ionize at basic pH; ie. Amines give up their protons at basic or alkaline pHs.
  - When groups with a similar chemical nature are present:
    (a) Carboxylic acids near an amino group have a more acidic pK than isolated carboxylic acids.
    (b) Amino groups near a carboxylic acid have a more acidic pK than isolated amines.
    (b) Aromatic amines (like in His side chain) have a pK near neutral.

Apply the above rules and concepts to calculate the pI for all the other complex AAs.

Asp, Glu, Lys, and Arg. Their pK values and structures are shown below:

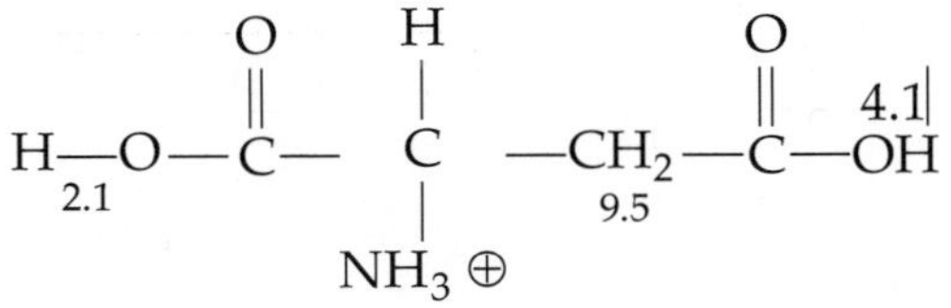

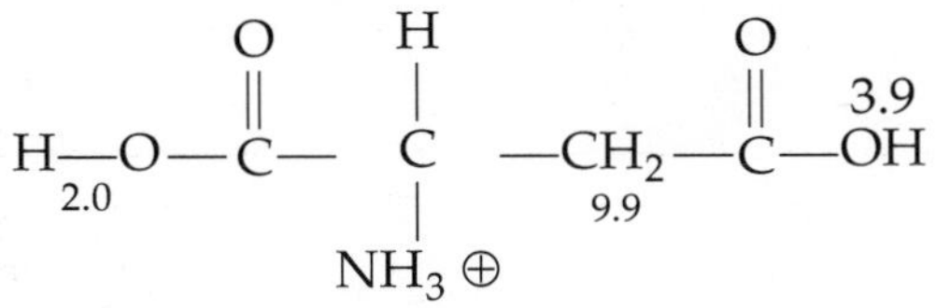

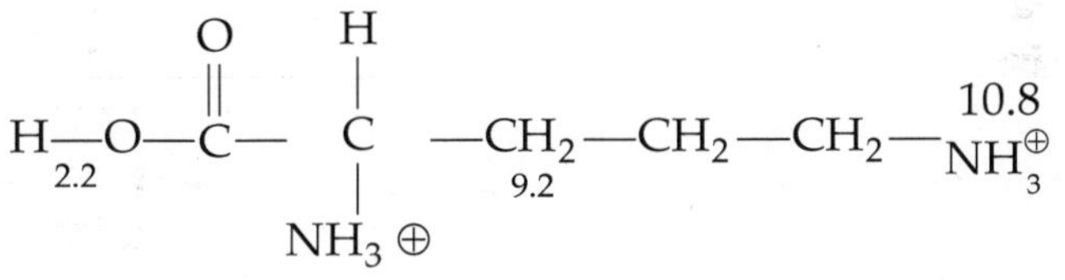

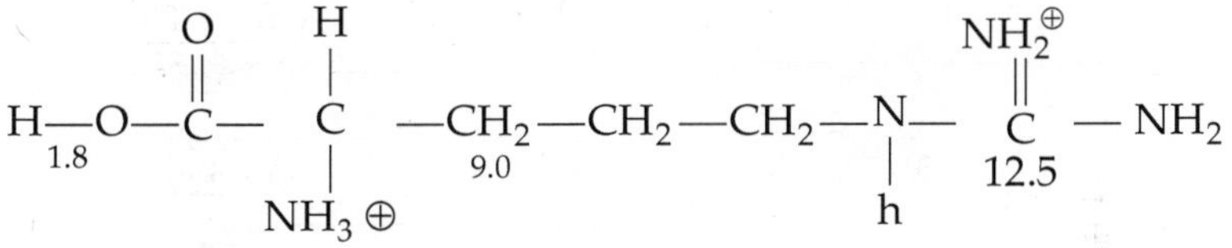

O H O
|| | || 4.1
H—O—C— C —CH_2—C—OH
2.1 | 9.5
NH_3 ⊕

O H O
|| | || 3.9
H—O—C— C —CH_2—C—OH
2.0 | 9.9
NH_3 ⊕

O H
|| | 10.8
H—O—C— C —CH_2—CH_2—CH_2—$NH_3^{\oplus}$
2.2 | 9.2
NH_3 ⊕

O H $NH_2^{\oplus}$
|| | ||
H—O—C— C —CH_2—CH_2—CH_2—N— C — NH_2
1.8 | 9.0 | 12.5
NH_3 ⊕ h

MOLECULAR STRUCTURE OF POLARITY WATER

Water has a simple molecular structure. It is composed of one oxygen atom and two hydrogen atoms. Each hydrogen atom is covalently bonded to the oxygen via a shared pair of electrons.

Oxygen also has two unshared pairs of electrons. Thus there are 4 pairs of electrons surrounding the oxygen atom, two pairs involved in covalent bonds with hydrogen, and two unshared pairs on the opposite side of the oxygen atom. Oxygen is an"electronegative" or electron"loving" atom compared with hydrogen.

Water is a"polar" molecule, meaning that there is an uneven distribution of electron density. Water has a partial negative charge (δ^-) near the oxygen atom due the unshared pairs of electrons, and partial positive charges (δ^+) near the hydrogen atoms.

An electrostatic attraction between the partial positive charge near the hydrogen atoms and the partial negative charge near the oxygen results in the formation of a hydrogen bond as shown in the illustration.

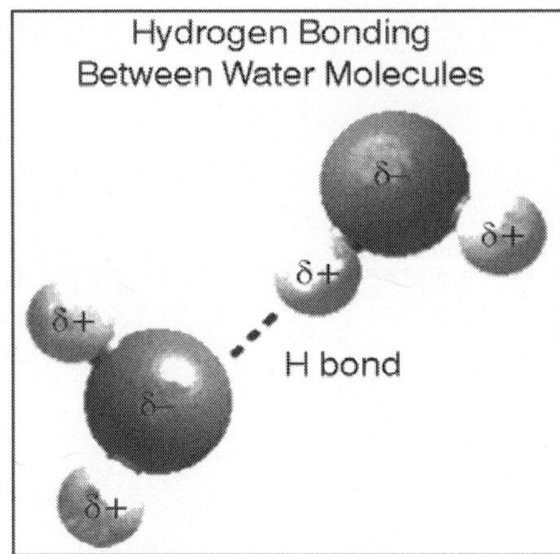

The ability of ions and other molecules to dissolve in water is due to polarity. For example, in the illustration below sodium chloride is shown in its crystalline form and dissolved in water.

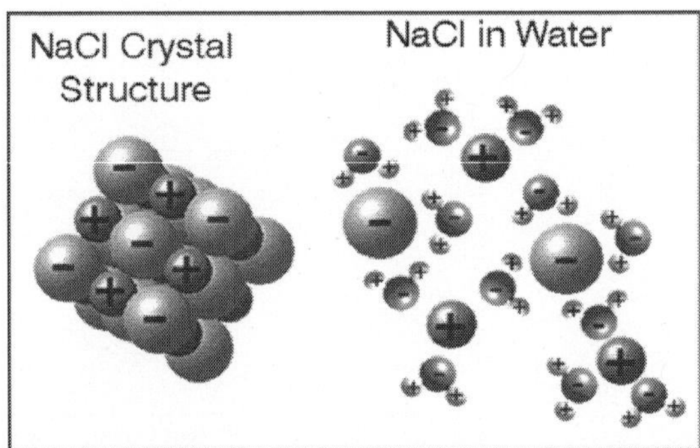

Many other unique properties of water are due to the hydrogen bonds. For example, ice floats because hydrogen bonds hold water molecules further apart in a solid than in a liquid, where there is one less hydrogen bond per molecule. The unique physical properties, including a high heat of Vapourization, strong surface tension, high specific heat, and nearly universal solvent properties of water are also due to hydrogen bonding.

The hydrophobic effect, or the exclusion of compounds containing carbon and hydrogen (nonpolar compounds) is another unique property of water caused by the hydrogen bonds. The hydrophobic effect is particularly important in the formation of cell membranes. The best description is to say that water"squeezes" nonpolar molecules together.

ACIDS AND BASES, IONIZATION OF WATER

$$H_2O \xrightarrow{\text{Limited}} H^+ + OH^-$$

- Acid release H^+
- Bases accept H^+

We define the pH of a solution as the negative logarithm of the hydrogen ion concentration.

- At pH 7.0, a solution is neutral.
- At lower pH (1-6), a solution is acidic.
- At higher pH (8-14), a solution is basic.

IONIZATION OF WATER

Sometimes the hydrogen of one water molecule will "jump" to another water molecule:

$$H_2O + H_2O = H_3O^+ + OH^-$$

This proton hopping is called the ionization of water (an ion is a positively or negatively charged atom or molecule). This ionization creates a H_3O^+ and a OH^- molecule. The H_3O^+ is often written as simply H^+. This is because a H_3O^+ is just a H^+ that jumps from one water molecule to another:

$$H_2O = H^+ + OH^-$$

So remember, $H_3O^+ = H^+$

Looking at either of the two chemical equations above, it is important to note that the reverse reaction is also occurring.

How much H^+ and OH^- exist in water? Very, very little! The ratio of either H^+ or OH^- to H_2O in neutral water is 1:1,000,000,000! Since this is such a small amount of either H^+ or OH^-, they rarely meet and neutralize each other.

The equilibrium constant, K_{eq} describes the ionization equilibrium of water:

$$K_{eq} = [H^+][OH^-]$$

Because of this relationship it is important to note that if the $[H^+]$ goes up then the $[OH^-]$ must go down, and vice-versa, for the value for the K_{eq} of water must remain constant. For neutral water, the K_{eq} is 1×10^{-14} M and the concentrations of $[H^+]$ and $[OH^-]$ are each 1×10^{-7} M.

Let's look at that last number without the exponent:

$$0.0000001 \text{ M}$$

This is obviously a very small number. A more manageable way to discuss small numbers such as this is to take the negative logarithm. For the concentration of $[H^+]$, this is called the pH. In this case:

$$-\log(0.0000001 \text{ M}) = 7$$

The pH of a solution is simply the negative logarithm of $[H^+]$. The pH of a solution describes the acidity of a solution. Acidic solutions are those with a pH of less than 7 and basic solutions have a pH greater than 7. A solution, like H_2O, with a pH = 7 is neutral. Similarly, the pH could be used to describe a solution in terms of its OH^- concentration. pOH is the negative logarithm of the OH^- concentration.

One useful thing to remember is:

pH + pOH = 14.

In the body, the pH of blood is 7.4. This corresponds to a [H^+] of about 40 nM. This value can only vary from 37 nM to 43 nM without serious metabolic consequences. pKa In living systems, much of the chemistry involves interactions between acids and bases. Acids are H^+ donors andbases are H^+ acceptors.

$$HA = H^+ + A^-$$

Acid and base reactions are made up of conjugate acid-base pairs. Strong acids are those that readily give up a H^+. Strong bases readily accept a H^+. Weak acids do not readily give up a H^+, but will under the right conditions. A weak base is one that does not easily take up a H^+.

The conjugate base for a strong acid is a weak base. In contrast, a strong base has a weak acid as its conjugate. Think of a strong acid as a person with an ugly wig. This person can't wait to get rid of this! So now we have a wig (H^+) and a person without the wig (the weak base). You can understand that the person without the wig doesn't really want to take it back!

However, what about strong base-weak acid conjugate pairs?

$$A^- + H^+ \rightleftharpoons HA$$

Think of it in similar terms: a person in dire need of a wig will grab just about anything. This is our strong base. Now on the right side of the equation we have someone with a wig who really doesn't want to give it up, just as a weak acid does not want to part with its H^+. We can think of a strong acid or base as having a "flip side", which is its weak base or weak acid conjugate, respectively. *How readily an acid gives up its* H^+ *is expressed by the acid dissociation constant, or* K_a:

$$K_a = \frac{[H+][A-]}{[]}$$

The pK_a is the negative logarithm of the K_a. Strong acids have small pK_as. Looking at the above equation, it can be seen that when [A^-] = [HA], then K_a = [H^+]. Then the pK_a = pH. This is the basis for the Henderson-Hasselbalch equation:

$$pH = pK_a + \log\frac{[H+][A-]}{[HA]}$$

This is Henderson-Hasselbalch Equation.

TITRATION CURVES

Acid with a Strong base: When acids or base are added to a solution, the pH changes. The controlled addition of an acid or base to a solution is called a titration. A titration curve can be made by plotting the pH changes against the volume of acid or base added to a solution. Many titration curves are made by plotting pH on the y axis and the volume of base added to the solution on the x axis.

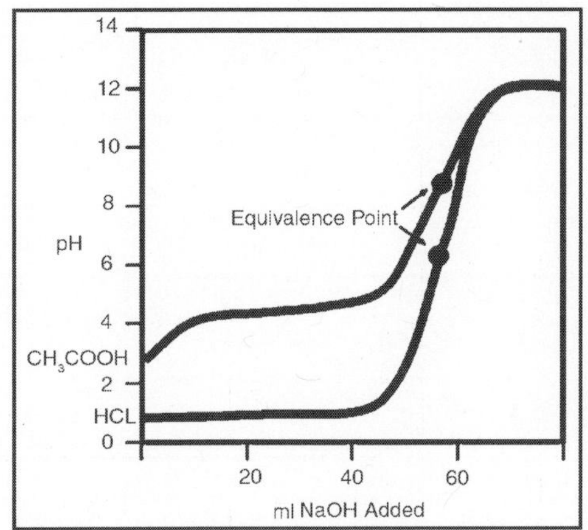

Fig. Titration Curves of Both a Strong and Weak.

For example, pH changes to acetic acid, CH_3COOH, and hydrochloric acid, HCl, with 0.1M sodium hydroxide, NaOH, is shown to the left. Notice the shape of the curves. The CH_3COOH curve starts out at a higher pH than the HCl curve. This is because CH_3COOH is a weaker acid than HCl.

As NaOH is added to the solution of CH_3COOH, the pH rises. However, the addition of NaOH to the HCl solution does not significantly change the pH. It takes less NaOH to change the pH of a weak acid solution than it does to change the pH of a strong acid solution. Eventually all of the acid in both solutions will have reacted with base, such that adding more base quickly increases the pH. This is called the equivalence point.

This is the point where all the acid has reacted with the base, such that further additions of base quickly raise the pH. The equivalence point for a strong acid-strong base titration is pH = 7. For weak acids-strong base titrations, the equivalence point is pH >7.

Bibliography

Y Anjaneyulu: *A Textbook of Analytical Chemistry*, K Chandrasekhar and Valli Manickam, PharmaMed Press, 2006.

Mahinder Singh: *A Textbook of Analytical Chemistry*, Instrumental Techniques, Dominant Publication Delhi, 2005.

S.A. Iqbal and M. Satake: *An Introduction to Analytical Chemistry*, Discovery Publication Delhi, 1999.

R.K. Soni, *Analytical Chemistry*, Shree Pub, 2007.

Dhruba Charan Dash: *Analytical Chemistry*, PHI Learning Delhi, 2011.

GARY D. Christian: *Analytical Chemistry*, Dominant Publication Delhi, 2005.

G Sharma: *Basic Analytical Chemistry*, Campus Books International, 2009.

Susan R. Mikkelsen and Eduardo Corton: *Bioanalytical Chemistry*, John Wiley and Sons, 2009.

Ulag Mahadevan: *Encyclopaedia of Analytical Chemistry*, Vols. I and II , Anmol Pub, 2010.

Satya Prakash Mohanty and Sushil Chauhan: *Experiments in Analytical Chemistry* , Campus Books Delhi , 2011.

Neelam SinglaNavneet KaurKanchan Kohli: *Practical Manual of Analytical Chemistry* , BS Publications , 2012.

Springer: *Principles of Analytical Chemistry: A Textbook*, Paperback Publication Delhi, 2006.

Springer: *Quality Assurance in Analytical Chemistry*: Training and Teaching (With CD) : Wenclawiak, Paperback Publication Delhi, 2006.

Harsh Malhotra: *Text Book of Analytical Chemistry*, Sonali Pub, 2011.

Y Anjaneyulu K Chandrasekhar and Valli Manickam: *Textbook of Analytical Chemistry*, Pharma Book Syndicate, 2006.

Milo Gibaldi: *Biopharmaceutics and Clinical Pharmacokinetics*, Pharma Book Syndicate, 2005.

Gayathri V. Patil and Harpal Singh: *Biopharmaceutics and Pharmacokinetics*, Shree Pub, 2009.

V. Venkateswarlu: *Biopharmaceutics and Pharmacokinetics*, PharmaMed Press, 2010.

K.K. Sahani: *Chemical Equilibrium and Chemical Kinetics*, Cyber Tech, 2009.

K Sarn: *Chemical Kinetics*, Rajat, 2005.

Springer: *Enzyme Kinetics and Mechanism : Taylor*, Paperback, 2006.

M. Nitya Devi and J. Jahir Hussain: *Handbook of Biosensors and Biosensor Kinetics*, SBS Pub., 2012.

Vidya Bhavani Suresh: *Kinetics of Kathakali and Kuchipudi : Demystifying Fine Arts, Vol. 29*, Skanda Pub, 2009.

Robert W. Balluffi, Samuel M. Allen and W. Craig Carter: *Kinetics of Materials*, Wiley, 2013.

G. Victor Rajamanickam and U. Subasini: *Pharmacokinetics and Biopharmaceutics*, New Academic Pub, 2009.

John G. Wagner: *Pharmacokinetics for the Pharmaceutical Scientist*, Technomic Publication, 2008.

Morris Sylvin: *Problems in Chemical Kinetics*, Sarup and Sons, 2001.

Syed Aftab Iqbal: *Problems in Chemical Kinetics and Solutions*, Discovery Pub, 2011.

S.A. Khan: *Text Book of Chemical Kinetics*, Sonali Pub, 2011.

Safaraz Niazi: *Textbook of Biopharmaceutics and Clinical Pharmacokinetics*, PharmaMed Press, 2011.

E. H. El-mossalamy: *Textbook Of Chemical Kinetics*, Discovery Publishing, 2011.

Pankaj Sethi: *Textbook Of Chemical Kinetics*, Campus Books, 2012.

Index